8-20-99

D0991469

ANALYTICAL MECHANICS

ANALYTICAL MECHANICS

Fifth Edition

Grant R. Fowles
University of Utah

George L. Cassiday
University of Utah

Saunders Golden Sunburst Series

Saunders College Publishing
Harcourt Brace College Publishers

Fort Worth   Philadelphia   San Diego   New York   Orlando   San Antonio
Austin   Toronto   Montreal   London   Sydney   Tokyo

Text Typeface: Times Roman and Caledonia
Compositor: G & S Typesetters, Inc.
Acquisitions Editor: John Vondeling
Associate Editor: Nanette Kauffman
Managing Editor: Carol Field
Project Editor: Laura Shur
Copy Editor: Sara Bernhardt-Black
Manager of Art and Design: Carol Bleistine
Art Director: Anne Muldrow
Associate Art Director: Caroline McGowan
Cover Designer: Louis Fuiano
Text Artwork: Vantage Art, Inc.
Director of EDP: Tim Frelick
Production Manager: Charlene Squibb
Marketing Manager: Marjorie Waldron

Cover Credit: John Lund/Tony Stone Images

Printed in the United States of America

Analytical Mechanics, 5/e.

ISBN: 0-03-096022-3

Library of Congress Catalog Card Number: 93-085193

789012   039   987654

PREFACE

This textbook is intended primarily for an undergraduate course in classical mechanics taken by students majoring in physics, physical science, or engineering. It is assumed that the student has taken a year of calculus-based general physics and a year of differential/integral calculus. It is highly recommended that a post calculus course that includes elementary differential equations and matrix algebra be taken prior to or concurrently with this course in mechanics.

This fifth edition is the same in general outline and content as the previous edition. New material has been introduced, including some history of the development of the subject, problems with solutions requiring the use of numerical techniques have been added, and many sections have been revised or expanded.

The text begins with a brief introduction to the essentials of vector algebra and vector differentiation. The latter leads naturally into the concepts of velocity and acceleration as the first and second derivatives of the position vector.

Newton's laws of motion are taken up in the second chapter with particular emphasis on the motion of a particle in one dimension.

Chapter 3 is devoted entirely to harmonic motion and includes resonance, the nonlinear oscillator, and the application of Fourier series to the oscillator driven by a nonsinusoidal force. The discussion of the physics of harmonic motion and resonance has been greatly expanded.

The general motion of a particle in three dimensions is discussed in Chapter 4. The concepts of conservative forces and potential energy are developed.

Chapter 5 deals with the appearance of fictitious forces and their effects on motion in noninertial frames of reference. Particular emphasis is placed on motion of particles in rotating frames of reference. A detailed numerical example of such motion is presented from the perspective of both a noninertial as well as an inertial observer. The example serves to illustrate the uniqueness of the result independent of frame of reference.

In Chapter 6, the student is introduced to Newton's theory of gravitation with emphasis on planetary and satellite motion. Central forces are discussed and criteria for orbital stability are developed. Also, an analysis of the Rutherford scattering problem is presented.

Systems of many articles are discussed in Chapter 7. Conservation laws are applied to collisions and the scattering of nuclei.

The study of rigid bodies is separated into two chapters. Chapter 8 deals with the study of rotation about a fixed axis and laminar motion of a rigid body. The concept of moments of inertia is introduced. The general motion of a rigid body in three dimensions is studied in Chapter 9. The equations of motion are developed using matrices and tensors. A detailed analysis of gyroscopic motion is presented.

Lagrangian mechanics is introduced in Chapter 10. Lagrange's equations are

derived from Hamilton's variational principle, initially via the simple example of mini-
mizing the time integral of the Lagrangian function for an object falling freely in a
gravitational field. Many examples are presented, and some are solved using both the
Lagrangian formalism as well as Newton's laws of motion.

Chapter 11 treats the subjects of normal coordinates for coupled oscillators, the
vibration of many coupled oscillators and, finally, traveling and standing waves along a
loaded string.

Metric units (either SI or CGS) are used for the most part in the examples and
problems. Occasionally, English units are reverted to since they are still in use in the
United States and students there are familiar with them.

Worked examples abound. They are usually found at the end of almost every section
in the book. Problems to be worked out by the student are found at the end of each
chapter. Also included are one or two problems that must be solved numerically, either
by computer or programmable calculator. A new appendix has been added to assist the
student in such endeavors. Answers to selected odd-numbered problems are given at
the end of the book. A list of units, physical constants, mathematical aids, formulae, and
discussions is also included in the appendices.

A solutions manual containing worked-out solutions is available to instructors upon
adoption of the text.

The authors wish to acknowledge with gratitude the efforts of those listed below.

Reviewers of the 5th Edition

David Benin
Arizona State University

Anthony Buffa
California Polytechnic State University

Graham Gutsche
United States Naval Academy

David Hertzog
University of Illinois, Urbana-Champaign

Peter B. Kahn
State University of New York, Stony Book

William Melton
University of North Carolina, Charlotte

Cindy Schwarz
Vassar College

Reviewers of the Fourth Edition

Jerry S. Faughn
Eastern Kentucky University

Don E. Harrison, Jr.
Naval Postgraduate School

Gerald R. Taylor
James Madison University

Charles D. Teague
Eastern Kentucky University

Gordon B. Thomson
Rutgers University

Jesse L. Weil
University of Kentucky

Raymond J. Winkel, Jr.
United States Naval Academy

G. L. Cassiday, G. R. Fowles

Salt Lake City
August 1993

CONTENTS OVERVIEW

CONTENTS

1

FUNDAMENTAL CONCEPTS. VECTORS

"Let no one unversed in geometry enter these portals."

(Plato's inscription over his Academy in Athens)

1.1 INTRODUCTION

The science of classical mechanics deals with the motion of objects through absolute *space* and *time* in the Newtonian sense. Although central to the development of classical mechanics, the concepts of space and time would remain arguable for more than two and one-half centuries following the publication of Sir Isaac Newton's *Philosophie naturalis principia mathematica* in 1687. As Newton put it in the first pages of the *Principia,* "absolute, true and mathematical time, of itself, and from its own nature, flows equably, without relation to anything external, and by another name is called duration. Absolute space, in its own nature, without relation to anything external, remains always similar and immovable."

The School of Athens—North Wind Picture Archives

1

Ernst Mach (1838–1916), who was to have immeasurable influence on Albert Einstein, questioned the validity of these two Newtonian concepts in his *The Science of Mechanics: A Critical and Historical Account of its Development* (1907), claiming that Newton had acted contrary to his expressed intention of "framing no hypotheses," that is, accepting as fundamental premises of a scientific theory nothing that could not be inferred directly from "observable phenomena" or induced from them by argument. Indeed, although Newton was on the verge of overtly expressing this intent in Book III of the *Principia* as the fifth and last rule of his *Regulae Philosophandi* (rules of reasoning in philosophy), it is significant that he actually refrained from doing so.

Throughout his scientific career he had exposed and rejected many hypotheses as false; he tolerated many as merely harmless; he put to use those that were verifiable. But he encountered a class of hypotheses that, neither "demonstrable from the phenomena nor following from them by argument based on induction," proved impossible to avoid. His concepts of space and time fell in this class. The acceptance of such hypotheses as fundamental was an embarrassing necessity; hence, he hesitated to adopt the frame-no-hypotheses rule. Newton certainly could be excused this sin of omission. After all, the adoption of these hypotheses and others of similar ilk (such as the "force" of gravitation) led to an elegant and comprehensive view of the world the likes of which had never been seen before.

It was not until the late 18th and early 19th century that experiments in electricity and magnetism would yield observable phenomena that could be understood only from the vantage point of a new space–time paradigm that arose from Albert Einstein's special relativity. This new paradigm was introduced by Hermann Minkowski in a semi-popular lecture in Cologne, Germany in 1908 with the words:

> *Gentlemen! The views of space and time which I wish to lay before you have sprung from the soil of experimental physics and therein lies their strength. They are radical. From now on, space by itself and time by itself are doomed to fade away into the shadows, and only a kind of union between the two will preserve an independent reality.*

Thus, even though his own concepts of space and time were superceded, Newton most certainly would have taken great delight in seeing the emergence of a new space–time concept based upon observed "phenomena," thus vindicating his unwritten frame-no-hypotheses rule.

1.2   MEASURE OF SPACE AND TIME. UNITS [1]

In this text we shall assume that space and time are described strictly in the Newtonian sense. Three-dimensional space is Euclidian, and positions of points in that space are specified by a set of three numbers (x, y, z) relative to the origin $(0, 0, 0)$ of a rectangular Cartesian coordinate system. Lengths are the spatial separation of two points relative to some standard length.

---

[1] A delightful account of the history of the standardization of units can be found in *The Science of Measurement—A Historical Survey*, by H. A. Klein, Dover Publ. (1988), ISBN 0-486-25839-4 (pbk).

Time is measured relative to the duration of reoccurrences of a given configuration of a cyclical system, say, a pendulum swinging to and fro, an earth rotating about its axis, or electromagnetic waves from a cesium atom vibrating inside a metallic cavity. The time of occurrence of any event is specified by a number t, which represents the number of reoccurrences of a given configuration of a chosen cyclical standard. For example, if 1 vibration of a standard physical pendulum is used to define 1 s, then to say that some event occurred at $t = 2.3$ s means that the standard pendulum executed 2.3 vibrations since its "start" at $t = 0$, when the event occurred.

All this sounds simple enough, but a substantial difficulty has been swept under the rug: Just what are the standard units? The choice of standards has usually been made more for political reasons than for scientific ones. For example, to say that a person is 6 feet tall is to say that the distance between the top of his head and the bottom of his foot is six times the length of something, which is taken to be the standard unit of 1 foot. In an earlier era that standard might have been the length of an actual foot or something that approximated that length as per the writing of Leonardo da Vinci on the views of the Roman architect–engineer Vitruvius Pollio (1st century B.C.):

> . . . *Vitruvius declares that Nature has thus arranged the measurements of a man: four fingers make 1 palm and 4 palms make 1 foot; six palms make 1 cubit; 4 cubits make once a man's height; 4 cubits make a pace, and 24 palms make a man's height* . . .

Clearly, the adoption of such a standard does not make for an accurately reproducible measure; homemakers might be excused their fits of anger upon being "short-footed" when purchasing bolts of cloth whose length was normalized to the foot of a current short-statured king.

The Unit of Length

Much that has been written about the French Revolution, which ended with the Napoleanic *coup d'etat* of 1799, has been distorted, but it gave birth to, among other things, an extremely significant plan for reform in measurement, namely the metric system, recently expanded into the Système International d'Unites (SI).

In 1791 toward the end of the first French National Assembly, Charles Maurice de Talleyrand-Perigord (1754–1838) proposed that a task of weight and measure reform be undertaken by a "blue ribbon" panel whose members were selected from the French Academy of Sciences. This problem was not trivial. Metrologically, as well as politically, France was still absurdly divided, confused, and complicated. A given unit of length recognized in Paris was about 4% longer than that in Bordeaux, 2% longer than that in Marseilles, and 2% shorter than that in Lille. The Academy of Sciences panel was to change all this. Invitations to Great Britain and the United States to take part in the process of unit standardization were rejected. Thus was born the beginnings of antipathy toward the metric system by English-speaking countries.

The panel chose 10 as the numerical base for all measure. The fundamental unit of length was taken to be one ten millionth of a quadrant, or a quarter of a full meridian. A surveying operation, extending from Dunkirk on the English Channel to a site near

Barcelona on the Mediterranean coast of Spain (a length equivalent to 10 degrees of latitude or one ninth of a quadrant), was carried out to determine this fundamental unit of length accurately. Ultimately, this monumental trek, which took from 1792 until 1799, changed the standard meter, estimated from previous, less ambitious surveys, by less than 0.3 mm or about 3 parts in 10,000. We now know that this result, too, was in error by a similar factor. The length of a standard quadrant of meridian is 10,002,288.3 m, a little over 2 parts in 10,000 greater than the quadrant length established by the Dunkirk–Barcelona expedition.

Interestingly enough, in 1799, the year in which the Dunkirk–Barcelona survey was completed, the national legislature of France ratified new standards, among them the meter. The standard meter was now taken to be the distance between two fine scratches made on a bar of a dense alloy of platinum and iridium shaped in an X-like cross section to minimize sagging and distortion. The United States has two copies of this bar, numbers 21 and 27, stored at the Bureau of Standards in Gaithersburg, MD, just outside Washington, DC. Measurements based on this standard are acccurate to about 1 part in 10^6. Thus, an object, a bar of platinum, rather than the concepts that led to it, was established as the standard meter. The earth might alter its circumference, if it so chose, but the standard meter would remain safe forever in a vault in Sevres, just outside Paris, France. This standard would persist until the 1960s.

The 11th General Conference of Weights and Measures, meeting in 1960, chose a reddish-orange radiation produced by atoms of krypton-86 as the next standard of length, with the meter defined in the following way:

> *The meter is the length equal to 1,650,763.73 wavelengths in vacuum of the radiation corresponding to the transition between the levels 2 p^{10} and 5 d^5 of the krypton-86 atom.*

Krypton is all around us; it comprises about 1 part per million of the earth's present atmosphere. Atmospheric krypton has an atomic weight of 83.8, being a mixture of six different isotopes, ranging in weight from 78 to 86. Krypton-86 comprises about 60% of these. Thus, the meter is defined in terms of the "majority kind" of krypton. Standard lamps contained no more than 1% of the other isotopes. Measurements based on this standard were accurate to about 1 part in 10^8.

Since 1983 the meter standard has been specified in terms of the velocity of light. A meter is the distance light travels in 1/299,792,458 s in a vacuum. In other words, the velocity of light is defined to be 299,792,458 m/s. Clearly, this makes the standard of length dependent on the standard of time.

The Unit of Time

Astronomical motions provide us with three great "natural" time units: the day, month, and year. The day is based on the earth's spin, the month on the moon's orbital motion about the earth, and the year on the earth's orbital motion about the sun. Why do we have ratios of 60:1 and 24:1 connecting the day, hour, minute, and second? These relationships were born about 6000 years ago on the flat, alluvial planes of Mesopotamia (now Iraq), where civilization and city-states first appeared on earth. The Mesopotamian number system was based on 60, not 10 like ours. It seems likely that the early Mesopotamians were more influenced by the 360 days in a year, the 30 days in a month, and the 12 months in a year than by the number of fingers on their hands. It was in such an environment that sky watching and measurement of stellar positions first became precise and continuous. Here, the movements of heavenly bodies across the sky were converted into clocks.

The second, the basic unit of time in SI, began as an arbitrary fraction (1/86,400) of a mean solar day ($24 \times 60 \times 60 = 86,400$). The trouble with astronomical clocks, though, is that they do not remain constant. The mean solar day is lengthening and the lunar month, or time between consecutive full phases, is shortening. In 1956, a new second was defined to be 1/31,556,926 of one particular and carefully measured mean solar year, that of 1900. That second would not last for long! In 1967, it would be redefined again, in terms of a specified number of oscillations of a cesium atomic clock.

A cesium atomic clock consists of a beam of cesium-133 atoms moving through an evacuated metal cavity and absorbing and emitting microwaves of a characteristic resonant frequency, 9,192,631,770 Hertz (Hz) or about 10^{10} cycles per second. This absorption and emission process occurs when a given cesium atom changes its atomic configuration and, in the process, either gains or loses a specific amount of energy in the form of microwave radiation. The two differing energy configurations correspond to situations where the spins of the cesium nucleus and that of its single outer shell electron are either opposed (lowest energy state) or aligned (highest energy state). This kind of a "spin-flip" atomic transition is called a *hyperfine transition*. The energy difference, and, hence, the resonant frequency, is precisely determined by the invariable structure of the cesium atom. It does not differ from one atom to another. A properly adjusted and maintained cesium clock can keep time with a stability of about 1 part in 10^{12}. Thus, in one year, its deviation from the right time should be no more than about 30 μs (30×10^{-6} s). When two different cesium clocks are compared, it will be found that they maintain agreement to about 1 part in 10^{10}.

It was inevitable then that in 1967, because of such stability and reproducibility, the 13th General Conference on Weights and Measures, would substitute the cesium-133 atom for any and all of the heavenly bodies as the primary basis for the unit of time with the historic words:

> *The second is the duration of 9,192,631,770 periods of the radiation corresponding to the transition between two hyperfine levels of the cesium-133 atom.*

So just as the meter is no longer bound to the surface of the earth, the second is no longer derived from the "ticking" of the heavens.

The Unit of Mass

The first sentence of this chapter stated that the science of mechanics dealt with the motion of objects. The final concept and accompanying unit needed to specify completely any physical quantity is mass.[2] Suffice it to say that the basic unit of mass is the *kilogram*. This primary standard is also stored in a vault in Sevres, France, with secondaries owned and kept by most major governments of the world. Note that the units of length and time are based on atomic standards. They are *universally* reproducible and virtually indestructible. Unfortunately, the unit of mass is not yet quite so robust.

A concept involving mass, which we shall have occasion to use throughout this text, is that of the *particle,* or point mass, an entity that possesses mass but no spatial extent. Clearly, the particle is a nonexistent idealization. Nonetheless, the concept serves as a useful approximation of physical objects in a certain context, namely, in a situation where the dimension of the object is small compared to the dimensions of its environment. Examples include a bug on a phonograph record, a baseball in flight, or the earth in orbit around the sun.

The above units (*kilogram, meter, second*) comprise a basis for the SI system.[3] In addition to the SI system, others are in common use, namely, the cgs (*gram, centimeter, second*) and the fps (pound, foot, second) systems. These latter systems may be regarded as secondary because they are defined relative to the SI standard. See Appendix A.

A physical quantity that is completely specified, in appropriate units, by a single number is called a *scalar.* Familiar examples of scalars are density, volume, and temperature. Mathematically, scalars are treated as ordinary real numbers. They obey all the regular rules of algebraic addition, subtraction, multiplication, division, and so on.

There are certain physical quantities that possess a directional characteristic, such as a displacement from one point in space to another. Such quantities require a direction *and* a magnitude for their complete specification. These quantities are called *vectors* if they combine with each other according to the parallelogram rule of addition as discussed in the next section.[4] Besides displacement in space, other familiar examples of vectors are velocity, acceleration, and force. The vector concept and the development of a whole mathematics of vector quantities have proved indispensible to the development of the science of mechanics. The remainder of this chapter will be devoted largely to a study of the mathematics of vectors.

1.3   NOTATION. FORMAL DEFINITIONS AND RULES OF VECTOR ALGEBRA

Vector quantities are denoted in print by boldface type, for example, **A,** whereas ordinary italic type represents scalar quantities. In written work it is customary to use a distinguishing mark, such as an arrow, **A** to designate a vector.

A given vector **A** is specified by stating its magnitude and its direction relative to some chosen reference system. A vector is represented diagrammatically by a directed

---

[2] The concept of mass will be treated in Chapter 2.

[3] Other basic and derived units are listed in Appendix A.

[4] An example of a directed quantity that does not obey the rule for addition is a finite rotation of an object about a given axis. The reader can readily verify that two successive rotations about *different* axes do not produce the same effect as a single rotation determined by the parallelogram rule. For the present we shall not be concerned with such non-vector-directed quantitites, however.

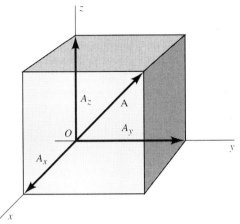

Figure 1.1 A vector **A** and its components in
Cartesian coordinates.

line segment, as shown in Figure 1.1. A vector can also be specified by listing its components or projections along the coordinate axes. The component symbol $[A_x, A_y, A_z]$ will be used as an alternate designation of a vector. The equation

$$\mathbf{A} = [A_x, A_y, A_z]$$

means that the vector **A** is expressed on the right in terms of its components in a particular coordinate system. For example, if the vector **A** represents a displacement from a point $P_1(x_1, y_1, z_1)$ to the point $P_2(x_2, y_2, z_2)$, then $A_x = x_2 - x_1$, $A_y = y_2 - y_1$, $A_z = z_2 - z_1$. If **A** represents a *force*, then A_x is the x component of the force, and so on. Clearly, the numerical values of the scalar components of a given vector depend on the choice of the coordinate axes.

If a particular discussion is limited to vectors in a plane, only two components are necessary. On the other hand, one can define a mathematical space of any number of dimensions. Thus, the symbol $[A_1, A_2, A_3, \ldots, A_n]$ denotes an n-dimensional vector. In this abstract sense a vector is an ordered set of numbers.

We begin the study of vector algebra with some formal statements concerning vectors.

I. *Equality of Vectors*

The equation

$$\mathbf{A} = \mathbf{B}$$

or

$$[A_x, A_y, A_z] = [B_x, B_y, B_z]$$

is equivalent to the three equations

$$A_x = B_x \qquad A_y = B_y \qquad A_z = B_z$$

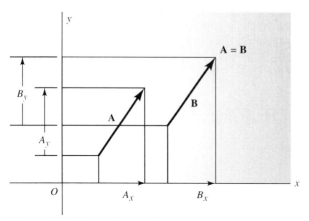

Figure 1.2   Illustrating equal vectors.

That is, two vectors are equal if, and only if, their respective components are equal. Geometrically, equal vectors are parallel and have the same length, but they do not necessarily have the same position. Equal vectors are shown in Figure 1.2, where only two components are drawn for clarity. Notice that the vectors form opposite sides of a parallelogram. (Equal vectors are not necessarily equivalent in all respects. Thus, two vectorially equal forces acting at *different* points on an object may produce different mechanical effects.)

II. *Vector Addition*

The addition of two vectors is defined by the equation

$$\mathbf{A} + \mathbf{B} = [A_x, A_y, A_z] + [B_x, B_y, B_z] = [A_x + B_x, A_y + B_y, A_z + B_z]$$

The sum of two vectors is a vector whose components are sums of the components of the given vectors. The geometric representation of the vector sum of two nonparallel vectors is the third side of a triangle, two sides of which are the given vectors. The vector sum is illustrated in Figure 1.3. The sum is also given by the

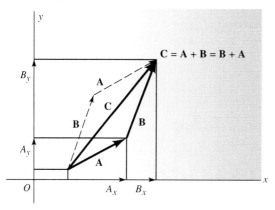

Figure 1.3   Addition of two vectors.

parallelogram rule, as shown in the figure. The vector sum is defined, however, according to the above equation even if the vectors do not have a common point.

III. *Multiplication by a Scalar*
If c is a scalar and **A** is a vector,

$$c\mathbf{A} = c[A_x, A_y, A_z] = [cA_x, cA_y, cA_z] - \mathbf{A}c$$

The product $c\mathbf{A}$ is a vector whose components are c times those of **A**. Geometrically, the vector $c\mathbf{A}$ is parallel to **A** and is c times the length of **A**. When $c = -1$, the vector $-\mathbf{A}$ is one whose direction is the reverse of that of **A**, as shown in Figure 1.4.

IV. *Vector Subtraction*
Subtraction is defined as follows:

$$\mathbf{A} - \mathbf{B} = \mathbf{A} + (-1)\mathbf{B} = [A_x - B_x, A_y - B_y, A_z - B_z]$$

That is, subtraction of a given vector **B** from the vector **A** is equivalent to adding $-\mathbf{B}$ to **A.**

V. *The Null Vector*
The vector $\mathbf{O} = [0, 0, 0]$ is called the *null* vector. The direction of the null vector is undefined. From (IV) it follows that $\mathbf{A} - \mathbf{A} = \mathbf{O}$. Since there can be no confusion when the null vector is denoted by a "zero," we shall hereafter use the notation $\mathbf{O} = 0$.

VI. *The Commutative Law of Addition*
This law holds for vectors; that is,

$$\mathbf{A} + \mathbf{B} = \mathbf{B} + \mathbf{A}$$

since $A_x + B_x = B_x + A_x$, and similarly for the y and z components.

VII. *The Associative Law*
The associative law is also true, because

$$\mathbf{A} + (\mathbf{B} + \mathbf{C}) = [A_x + (B_x + C_x), A_y + (B_y + C_y), A_z + (B_z + C_z)]$$
$$= [(A_x + B_x) + C_x, (A_y + B_y) + C_y, (A_z + B_z) + C_z]$$
$$= (\mathbf{A} + \mathbf{B}) + \mathbf{C}$$

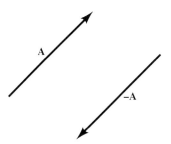

Figure 1.4 The negative of a vector.

VIII. *The Distributive Law*

Under multiplication by a scalar, the distributive law is valid, because, from (II) and (III),

$$c(\mathbf{A} + \mathbf{B}) = c[A_x + B_x, A_y + B_y, A_z + B_z]$$
$$= [c(A_x + B_x), c(A_y + B_y), c(A_z + B_z)]$$
$$= [cA_z + cB_x, cA_y + cB_y, cA_z + cB_z]$$
$$= c\mathbf{A} + c\mathbf{B}$$

Thus, vectors obey the rules of ordinary algebra as far as the above operations are concerned.

IX. *Magnitude of a Vector*

The magnitude of a vector **A**, denoted by $|\mathbf{A}|$ or by A, is defined as the square root of the sum of the squares of the components, namely,

$$A = |\mathbf{A}| = (A_x^2 + A_y^2 + A_z^2)^{1/2}$$

where the positive root is understood. Geometrically, the magnitude of a vector is its length, that is, the length of the diagonal of the rectangular parallelepiped whose sides are A_x, A_y, and A_z, expressed in appropriate units.

X. *Unit Coordinate Vectors*

A *unit vector* is a vector whose magnitude is unity. Unit vectors are often designated by the symbol **e** from the German word *einheit*. The three unit vectors

$$\mathbf{e}_x = [1, 0, 0] \qquad \mathbf{e}_y = [0, 1, 0] \qquad \mathbf{e}_z = [0, 0, 1]$$

are called *unit coordinate vectors* or *basis vectors*. In terms of basis vectors, any vector can be expressed as a vector sum of components as follows:

$$\mathbf{A} = [A_x, A_y, A_z] = [A_x, 0, 0] + [0, A_y, 0] + [0, 0, A_z]$$
$$= A_x[1, 0, 0] + A_y[0, 1, 0] + A_z[0, 0, 1]$$
$$= \mathbf{e}_x A_x + \mathbf{e}_y A_y + \mathbf{e}_z A_z$$

A widely used notation for Cartesian unit vectors are the letters **i, j,** and **k,** namely

$$\mathbf{i} = \mathbf{e}_x \qquad \mathbf{j} = \mathbf{e}_y \qquad \mathbf{k} = \mathbf{e}_z$$

We shall usually employ this notation hereafter.

The directions of the Cartesian Unit vectors are defined by the orthogonal coordinate axes as shown in Figure 1.5. They form a right-handed or a left-handed triad, depending on which type of coordinate system is used. It is customary to use right-handed coordinate systems. The system shown in Figure 1.5 is right-handed. (The handedness of coordinate systems will be defined in Section 1.5).

EXAMPLE 1.1

Find the sum and the magnitude of the sum of the two vectors $\mathbf{A} = [1, 0, 2]$ and $\mathbf{B} = [0, 1, 1]$.

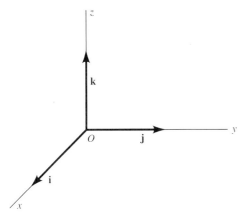

Figure 1.5   The unit vectors **ijk.**

Solution:
Adding components we have $\mathbf{A} + \mathbf{B} = [1, 0, 2] + [0, 1, 1] = [1, 1, 3]$

$$|\mathbf{A} + \mathbf{B}| = (1 + 1 + 9)^{1/2} = \sqrt{11}$$ ∎

EXAMPLE 1.2

For the above two vectors, express the difference in **ijk** form.

Solution:
Subtracting components, we have

$$\mathbf{A} - \mathbf{B} = [1, -1, 1] = \mathbf{i} - \mathbf{j} + \mathbf{k}$$ ∎

EXAMPLE 1.3

A helicopter flies 100 m vertically upward, then 500 m horizontally east, then 1000 m horizontally north. How far is it from a second helicopter that starts from the same point rising 200 m upward, 100 m west, and 500 m north?

Solution:
Choosing up, east, and north as basis directions, the final position of the first helicopter is expressed vectorially as $\mathbf{A} = [100, 500, 1000]$ and the second as $\mathbf{B} = [200, -100, 500]$, in meters. Hence, the distance between the final positions is given by the expression

$$\begin{aligned}
|\mathbf{A} - \mathbf{B}| &= |[(100 - 200), (500 + 100), (1000 - 500)] \text{ m} \\
&= (100^2 + 600^2 + 500^2)^{1/2} \text{ m} \\
&= 787.4 \text{ m}
\end{aligned}$$ ∎

$(\Lambda \ \text{NUMBER})$ $\left(\text{WORK} : \text{Force} / \text{DISP}\right.$

1.4   THE SCALAR PRODUCT

Given two vectors **A** and **B,** the scalar product or "dot" product, **A · B,** is the scalar defined by the equation

$$\mathbf{A} \cdot \mathbf{B} = A_x B_x + A_y B_y + A_z B_z \tag{1.1}$$

It follows from the above definition that scalar multiplication is *commutative*

$$\mathbf{A} \cdot \mathbf{B} = \mathbf{B} \cdot \mathbf{A} \tag{1.2}$$

since $A_x B_x = B_x A_x$, and so on. It also follows that it is *distributive*

$$\mathbf{A} \cdot (\mathbf{B} + \mathbf{C}) = \mathbf{A} \cdot \mathbf{B} + \mathbf{A} \cdot \mathbf{C} \tag{1.3}$$

because if we apply the definition [(1.1)] is detail

$$\mathbf{A} \cdot (\mathbf{B} + \mathbf{C}) = A_x(B_x + C_x) + A_y(B_y + C_y) + A_z(B_z + C_z)$$
$$= A_x B_x + A_y B_y + A_z B_z + A_x C_x + A_y C_y + A_z C_z$$
$$= \mathbf{A} \cdot \mathbf{B} + \mathbf{A} \cdot \mathbf{C}$$

The dot product **A · B** has a simple geometrical interpretation and can be used to calculate the angle θ between those two vectors. For example, shown in Figure 1.6 are the two vectors **A** and **B** separated by an angle θ, along with an x', y', z' coordinate system arbitrarily chosen as a basis for those vectors. However, since the quantity **A · B** is a scalar, its value is independent of choice of coordinates. With no loss of generality, we can rotate the x', y', z' system into an x, y, z coordinate system, such that the x-axis is aligned with the vector **A** and the z-axis is perpendicular to the plane defined by the two vectors. This coordinate system is also shown in Figure 1.6. The components of the

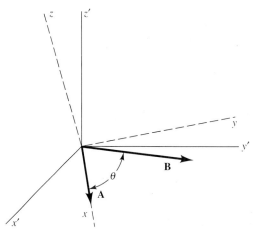

Figure 1.6   Evaluating a dot product between two vectors.

vectors, and, their dot product, are much simpler to evaluate in this system. The vector **A** is expressed as $[A, 0, 0]$ and the vector **B** as $[B_x, B_y, 0]$ or $[B \cos \theta, B \sin \theta, 0]$. Thus,

$$\mathbf{A} \cdot \mathbf{B} = A_x B_x = A(B \cos \theta) = |\mathbf{A}| \, |\mathbf{B}| \cos \theta \qquad (1.4)$$

Geometrically, it can easily be seen that $B \cos \theta$ is simply the projection of **B** onto **A.** If we had aligned the x-axis along **B,** we would have obtained the same result but with the geometrical interpretation that $\mathbf{A} \cdot \mathbf{B}$ is now the projection of **A** onto **B** times the length of **B.** Thus, $\mathbf{A} \cdot \mathbf{B}$ can be interpreted as either the projection of **A** onto **B** times the length of **B** or **B** onto **A** times the length of **A.** Either interpretation is correct. Perhaps more importantly, we can see that we have just proved that the cosine of the angle between two line segments is given by

$$\cos \theta = \frac{\mathbf{A} \cdot \mathbf{B}}{|\mathbf{A}| \, |\mathbf{B}|} \qquad (1.5)$$

This last equation may be regarded as an alternate definition of the dot product.

Note: If $\mathbf{A} \cdot \mathbf{B}$ is equal to zero and neither **A** nor **B** is null, then $\cos \theta$ is zero and **A** is perpendicular to **B.**

The square of the magnitude of a vector **A** is given by the dot product of **A** with itself,

$$A^2 = |\mathbf{A}|^2 = \mathbf{A} \cdot \mathbf{A} \qquad \Rightarrow |\vec{A}| = \sqrt{\vec{A} \cdot \vec{A}}$$

From the definitions of the unit coordinate vectors **i, j,** and **k,** it is clear that the following relations hold:

$$\mathbf{i} \cdot \mathbf{i} = \mathbf{j} \cdot \mathbf{j} = \mathbf{k} \cdot \mathbf{k} = 1 \qquad (1.6)$$
$$\mathbf{i} \cdot \mathbf{j} = \mathbf{i} \cdot \mathbf{k} = \mathbf{j} \cdot \mathbf{k} = 0$$

Expressing Any Vector as the Product of Its Magnitude by a Unit Vector. Projection

Consider the equation

$$\mathbf{A} = \mathbf{i}A_x + \mathbf{j}A_y + \mathbf{k}A_z$$

Multiply and divide on the right by the magnitude of **A**

$$\mathbf{A} = A\left(\mathbf{i}\frac{A_x}{A} + \mathbf{j}\frac{A_y}{A} + \mathbf{k}\frac{A_z}{A} \right)$$

Now $A_x/A = \cos \alpha$, $A_y/A = \cos \beta$, and $A_z/A = \cos \gamma$ are the *direction cosines* of the vector **A,** and α, β, and γ are the *direction angles.* Thus, we can write

$$\mathbf{A} = A(\mathbf{i} \cos \alpha + \mathbf{j} \cos \beta + \mathbf{k} \cos \gamma) = A[\cos \alpha, \cos \beta, \cos \gamma]$$

or

$$\mathbf{A} = A\mathbf{n} \qquad (1.7)$$

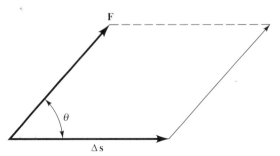

Figure 1.7 A force acting on a body undergoing a displacement.

where **n** is a unit vector whose components are cos α, cos β, and cos γ. Consider any other vector **B**. Clearly, the projection of **B** on **A** is just

$$B \cos \theta = \frac{\mathbf{B} \cdot \mathbf{A}}{A} = \mathbf{B} \cdot \mathbf{n} \qquad (1.8)$$

where θ is the angle between **A** and **B**.

EXAMPLE 1.4   *Component of a Vector. Work*
_____

As an example of the dot product, suppose that an object under the action of a constant force[5] undergoes a linear displacement $\Delta\mathbf{s}$, as shown in Fig. 1.7. By definition, the *work* ΔW done by the force is given by the product of the component of the force **F** in the direction of $\Delta\mathbf{s}$, multiplied by the magnitude Δs of the displacement, that is,

$$\Delta W = (F \cos \theta) \, \Delta s$$

where θ is the angle between **F** and $\Delta\mathbf{s}$. But the expression on the right is just the dot product of **F** and $\Delta\mathbf{s}$, that is,

$$\Delta W = \mathbf{F} \cdot \Delta\mathbf{s} \qquad \blacksquare$$

EXAMPLE 1.5   *Law of Cosines*
_____

Consider the triangle whose sides are **A, B,** and **C,** as shown in Figure 1.8. Then **C = A + B.** Take the dot product of **C** with itself

$$\begin{aligned} \mathbf{C} \cdot \mathbf{C} &= (\mathbf{A} + \mathbf{B}) \cdot (\mathbf{A} + \mathbf{B}) \\ &= \mathbf{A} \cdot \mathbf{A} + 2\mathbf{A} \cdot \mathbf{B} + \mathbf{B} \cdot \mathbf{B} \end{aligned}$$

_____

[5] The concept of force will be discussed in Chapter 2.

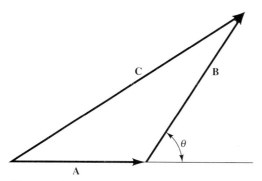

Figure 1.8   The law of cosines.

The second step follows from the application of the rules in Equations 1.2 and 1.3. Replace $\mathbf{A} \cdot \mathbf{B}$ by $AB \cos \theta$ to obtain

$$C^2 = A^2 + 2AB \cos \theta + B^2$$

which is the familiar law of cosines. This is just one example of the use of vector algebra to prove theorems in geometry.   ■

EXAMPLE 1.6

Find the cosine of the angle between a long diagonal and an adjacent face diagonal of a cube.

Solution:
We can represent the two diagonals in question by the vectors $\mathbf{A} = [1, 1, 1]$ and $\mathbf{B} = [1, 1, 0]$. Hence, from Equation 1.4

$$\cos \theta = \frac{\mathbf{A} \cdot \mathbf{B}}{AB} = \frac{1 + 1 + 0}{\sqrt{3}\sqrt{2}} = \sqrt{\frac{2}{3}} = 0.8165 \qquad ■$$

EXAMPLE 1.7

The vector $a\mathbf{i} + \mathbf{j} - \mathbf{k}$ is perpendicular to the vector $\mathbf{i} + 2\mathbf{j} - 3\mathbf{k}$. What is the value of a?

Solution:
If the vectors are perpendicular to each other, their dot product must vanish ($\cos 90° = 0$). Hence, we have

$$(a\mathbf{i} + \mathbf{j} - \mathbf{k}) \cdot (\mathbf{i} + 2\mathbf{j} - 3\mathbf{k}) = a + 2 + 3 = a + 5 = 0$$

Hence,

$$a = -5 \qquad ■$$

Torque $r \times F$

1.5   THE VECTOR PRODUCT

Given two vectors **A** and **B**, the vector product or cross product, $\mathbf{A} \times \mathbf{B}$, is defined as the vector whose components are given by the equation

$$\mathbf{A} \times \mathbf{B} = [A_y B_z - A_z B_y, \; A_z B_x - A_x B_z, \; A_x B_y - A_y B_x] \tag{1.9}$$

It can be shown that the following rules hold for cross multiplication:

$$\mathbf{A} \times \mathbf{B} = -\mathbf{B} \times \mathbf{A} \tag{1.10}$$
$$\mathbf{A} \times (\mathbf{B} + \mathbf{C}) = \mathbf{A} \times \mathbf{B} + \mathbf{A} \times \mathbf{C} \tag{1.11}$$
$$n(\mathbf{A} \times \mathbf{B}) = (n\mathbf{A}) \times \mathbf{B} = \mathbf{A} \times (n\mathbf{B}) \tag{1.12}$$

The proofs of these follow directly from the definition and are left as an exercise.

Note: The first equation states that the cross product is *anticommutative.*

According to the definitions of the unit coordinate vectors, Section 1.3, it readily follows that

$$
\begin{aligned}
\mathbf{i} \times \mathbf{i} &= \mathbf{j} \times \mathbf{j} = \mathbf{k} \times \mathbf{k} = 0 \\
\mathbf{j} \times \mathbf{k} &= \mathbf{i} = -\mathbf{k} \times \mathbf{j} \\
\mathbf{i} \times \mathbf{j} &= \mathbf{k} = -\mathbf{j} \times \mathbf{i} \\
\mathbf{k} \times \mathbf{i} &= \mathbf{j} = -\mathbf{i} \times \mathbf{k}
\end{aligned}
\tag{1.13}
$$

These latter three relations define a right-handed triad. For example,

$$\mathbf{i} \times \mathbf{j} = [0 - 0, \; 0 - 0, \; 1 - 0] = [0, 0, 1] = \mathbf{k}$$

The remaining equations are easily proved in a similar manner.

The cross product expressed in **ijk** form is

$$\mathbf{A} \times \mathbf{B} = \mathbf{i}(A_y B_z - A_z B_y) + \mathbf{j}(A_z B_x - A_x B_z) + \mathbf{k}(A_x B_y - A_y B_x)$$

Each term in parentheses is equal to a determinant

$$\mathbf{A} \times \mathbf{B} = \mathbf{i} \begin{vmatrix} A_y & A_z \\ B_y & B_z \end{vmatrix} + \mathbf{j} \begin{vmatrix} A_z & A_x \\ B_z & B_x \end{vmatrix} + \mathbf{k} \begin{vmatrix} A_x & A_y \\ B_x & B_y \end{vmatrix}$$

and finally

$$\mathbf{A} \times \mathbf{B} = \begin{vmatrix} \mathbf{i} & \mathbf{j} & \mathbf{k} \\ A_x & A_y & A_z \\ B_x & B_y & B_z \end{vmatrix} \tag{1.14}$$

which is readily verified by expansion. The determinant form is a convenient aid for remembering the definition of the cross product. From the properties of determinants, it can be seen at once that if **A** is parallel to **B**, that is, if $\mathbf{A} = c\mathbf{B}$, then the two lower rows of the determinant are proportional and so the determinant is null. Thus, the cross product of two parallel vectors is null.

Let us calculate the magnitude of the cross product. We have

$$|\mathbf{A} \times \mathbf{B}|^2 = (A_y B_z - A_z B_y)^2 + (A_z B_x - A_x B_z)^2 + (A_x B_y - A_y B_x)^2$$

With a little patience this can be reduced to

$$|\mathbf{A} \times \mathbf{B}|^2 = (A_x^2 + A_y^2 + A_z^2)(B_x^2 + B_y^2 + B_z^2) - (A_xB_x + A_yB_y + A_zB_z)^2$$

or, from the definition of the dot product, the above equation may be written in the form

$$|\mathbf{A} \times \mathbf{B}|^2 = A^2B^2 - (\mathbf{A} \cdot \mathbf{B})^2 \qquad (1.15)$$

Taking the square root of both sides of Equation 1.15 and using Equation 1.5, we can express the magnitude of the cross product as

$$|\mathbf{A} \times \mathbf{B}| = AB(1 - \cos^2 \theta)^{1/2} = AB \sin \theta \qquad (1.16)$$

where θ is the angle between **A** and **B**.

To interpret the cross product geometrically, we observe that the vector $\mathbf{C} = \mathbf{A} \times \mathbf{B}$ is perpendicular to both **A** and to **B** because

$$\begin{aligned}
\mathbf{A} \cdot \mathbf{C} &= A_xC_x + A_yC_y + A_zC_z \\
&= A_x(A_yB_z - A_zB_y) + A_y(A_zB_x - A_xB_z) + A_z(A_xB_y - A_yB_x) \\
&= 0
\end{aligned}$$

Similarly, $\mathbf{B} \cdot \mathbf{C} = 0$. Thus, the vector **C** is perpendicular to the plane containing the vectors **A** and **B**.

The sense of the vector $\mathbf{C} = \mathbf{A} \times \mathbf{B}$ is determined from the requirement that the three vectors **A**, **B**, and **C** form a right-handed triad, as shown in Figure 1.9. (This is consistent with the previously established result that in the right-handed triad **ijk** we have $\mathbf{i} \times \mathbf{j} = \mathbf{k}$.) Therefore, from Equation 1.16 we see that we can write

$$\mathbf{A} \times \mathbf{B} = (AB \sin \theta)\mathbf{n} \qquad (1.17)$$

$$= |\hat{A} \times \hat{B}| \, h$$

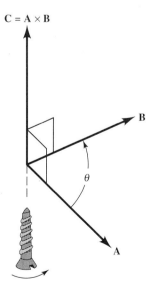

C = A × B

B

θ

A

Figure 1.9 The cross product of two vectors.

where **n** is a unit vector normal to the plane of the two vectors **A** and **B**. The sense of **n** is given by the *right-hand rule,* that is, the direction of advancement of a right-handed screw rotated from the positive direction of **A** to that of **B** through the smallest angle between them, as illustrated in Figure 1.9. Equation 1.17 may be regarded as an alternate definition of the cross product in a right-handed coordinate system.

EXAMPLE 1.8

Given the two vectors $\mathbf{A} = 2\mathbf{i} + \mathbf{j} - \mathbf{k}, \mathbf{B} = \mathbf{i} - \mathbf{j} + 2\mathbf{k}$, find $\mathbf{A} \times \mathbf{B}$.

Solution:
In this case it is convenient to use the determinant form

$$\mathbf{A} \times \mathbf{B} = \begin{vmatrix} \mathbf{i} & \mathbf{j} & \mathbf{k} \\ 2 & 1 & -1 \\ 1 & -1 & 2 \end{vmatrix} = \mathbf{i}(2 - 1) + \mathbf{j}(-1 - 4) + \mathbf{k}(-2 - 1)$$

$$= \mathbf{i} - 5\mathbf{j} - 3\mathbf{k} \qquad \blacksquare$$

EXAMPLE 1.9

Find a unit vector normal to the plane containing the two vectors **A** and **B** above.

Solution:

$$\mathbf{n} = \frac{\mathbf{A} \times \mathbf{B}}{|\mathbf{A} \times \mathbf{B}|} = \frac{\mathbf{i} - 5\mathbf{j} - 3\mathbf{k}}{[1^2 + 5^2 + 3^2]^{1/2}}$$

$$= \frac{\mathbf{i}}{\sqrt{35}} - \frac{5\mathbf{j}}{\sqrt{35}} - \frac{3\mathbf{k}}{\sqrt{35}} \qquad \blacksquare$$

EXAMPLE 1.10

Show by direct evaluation that $\mathbf{A} \times \mathbf{B}$ is a vector whose direction is perpendicular to **A** and **B** and whose magnitude is $AB \sin \theta$.

Solution:
Use the frame of reference discussed for Figure 1.6 in which the vectors **A** and **B** are defined to be in the x, y plane; **A** is given by $[A, 0, 0]$ and **B** is given by $[B \cos \theta, B \sin \theta, 0]$. Then

$$\mathbf{A} \times \mathbf{B} = \begin{vmatrix} \mathbf{i} & \mathbf{j} & \mathbf{k} \\ A & 0 & 0 \\ B \cos \theta & B \sin \theta & 0 \end{vmatrix} = \mathbf{k}AB \sin \theta \qquad \blacksquare$$

1.6 AN EXAMPLE OF THE CROSS PRODUCT: MOMENT OF A FORCE

A particularly useful application of the cross product is the representation of moments. Let a force **F** act at a point $P(x, y, z)$, as shown in Figure 1.10, and let the vector **OP** be designated by **r**, that is,

$$\mathbf{OP} = \mathbf{r} = \mathbf{i}x + \mathbf{j}y + \mathbf{k}z$$

The moment **N,** or the *torque vector,* about a given point O is defined as the cross product

$$\mathbf{N} = \mathbf{r} \times \mathbf{F} \qquad (1.18)$$

Thus, the moment of a force about a point is a vector quantity having a magnitude and a direction. If a single force is applied at a point P on a body that is initially at rest and is free to turn about a fixed point O as a pivot, then the body tends to rotate. The axis of this rotation is perpendicular to the force **F,** and it is also perpendicular to the line OP. Hence the direction of the torque vector **N** is along the axis of rotation.

The magnitude of the torque is given by

$$|\mathbf{N}| = |\mathbf{r} \times \mathbf{F}| = rF \sin \theta \qquad (1.19)$$

in which θ is the angle between **r** and **F**. Thus, $|\mathbf{N}|$ can be regarded as the product of the magnitude of the force and the quantity $r \sin \theta$, which is just the perpendicular distance from the line of action of the force to the point O.

When several forces are applied to a single body at different points, the moments

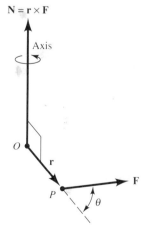

Figure 1.10 Illustrating the moment of a force about a point O.

add vectorially. This follows from the distributive law of vector multiplication. The condition for rotational equilibrium is that the vector sum of all the moments is zero:

$$\sum_i (r_i \times \mathbf{F}_i) = \sum_i \mathbf{N}_i = 0 \tag{1.20}$$

A more complete discussion of force moments will be given in Chapters 8 and 9 when we study the motion of rigid bodies.

1.7   TRIPLE PRODUCTS

The expression

$$\mathbf{A} \cdot (\mathbf{B} \times \mathbf{C})$$

is called the *triple scalar product* of **A, B,** and **C.** It is a scalar since it is the dot product of two vectors. Referring to the determinant expression for the cross product, Equation 1.14, we see that the triple scalar product may be written

$$\mathbf{A} \cdot (\mathbf{B} \times \mathbf{C}) = \begin{vmatrix} A_x A_y A_z \\ B_x B_y B_z \\ C_x C_y C_z \end{vmatrix} \tag{1.21}$$

From the well-known property of determinants that the exchange of the terms of two rows or of two columns changes the sign but does not change the absolute value of the determinant, we can easily derive the following useful equation:

$$\mathbf{A} \cdot (\mathbf{B} \times \mathbf{C}) = (\mathbf{A} \times \mathbf{B}) \cdot \mathbf{C} \tag{1.22}$$

Thus, the dot and the cross may be interchanged in the triple scalar product.

The expression

$$\mathbf{A} \times (\mathbf{B} \times \mathbf{C})$$

is called the *triple vector product.* It is left for the student to prove that the following equation holds for the triple vector product:

$$\mathbf{A} \times (\mathbf{B} \times \mathbf{C}) = \mathbf{B}(\mathbf{A} \cdot \mathbf{C}) - \mathbf{C}(\mathbf{A} \cdot \mathbf{B}) \tag{1.23}$$

This last result can be remembered simply as the "back minus cab" rule.

Triple products of vectors are particularly useful in the study of rotating coordinate systems and rotations of rigid bodies, which we shall take up in later chapters. A geometric application is given in Problem 1.9 at the end of this chapter.

EXAMPLE 1.11

Given the three vectors $\mathbf{A} = \mathbf{i}$, $\mathbf{B} = \mathbf{i} - \mathbf{j}$, and $\mathbf{C} = \mathbf{k}$, find $\mathbf{A} \cdot (\mathbf{B} \times \mathbf{C})$.

Solution:
Using the determinant expression, Equation 1.21, we have

$$\mathbf{A} \cdot (\mathbf{B} \times \mathbf{C}) = \begin{vmatrix} 1 & 0 & 0 \\ 1 & -1 & 0 \\ 0 & 0 & 1 \end{vmatrix} = 1(-1 + 0) = -1 \quad \blacksquare$$

EXAMPLE 1.12

Find $\mathbf{A} \times (\mathbf{B} \times \mathbf{C})$ above.

Solution:
From Equation 1.23 we have

$$\mathbf{A} \times (\mathbf{B} \times \mathbf{C}) = (\mathbf{A} \cdot \mathbf{C})\mathbf{B} - (\mathbf{A} \cdot \mathbf{B})\mathbf{C} = 0(\mathbf{i} - \mathbf{j}) - (1 - 0)\mathbf{k} = -\mathbf{k}$$

$$\blacksquare$$

EXAMPLE 1.13

Show that the vector triple product is nonassociative.

Solution:

$$(\mathbf{a} \times \mathbf{b}) \times \mathbf{c} = -\mathbf{c} \times (\mathbf{a} \times \mathbf{b}) = -\mathbf{a}(\mathbf{c} \cdot \mathbf{b}) + \mathbf{b}(\mathbf{c} \cdot \mathbf{a})$$
$$\mathbf{a} \times (\mathbf{b} \times \mathbf{c}) - (\mathbf{a} \times \mathbf{b}) \times \mathbf{c} = \mathbf{a}(\mathbf{c} \cdot \mathbf{b}) - \mathbf{c}(\mathbf{a} \cdot \mathbf{b})$$

which is not necessarily zero. $\qquad \blacksquare$

*1.8 CHANGE OF COORDINATE SYSTEM. THE TRANSFORMATION MATRIX

In this section we shall show how to represent a vector in different coordinate systems. Consider the vector $\mathbf{A}$ expressed relative to the triad **ijk**

$$\mathbf{A} = \mathbf{i}A_x + \mathbf{j}A_y + \mathbf{k}A_z$$

Relative to a new triad **i'j'k'** having a different orientation from that of **ijk,** the same vector $\mathbf{A}$ is expressed as

$$\mathbf{A} = \mathbf{i'}A_{x'} + \mathbf{j'}A_{y'} + \mathbf{k'}A_{z'}$$

Now the dot product $\mathbf{A} \cdot \mathbf{i'}$ is just $A_{x'}$, that is, the projection of $\mathbf{A}$ on the unit vector **i'.**

---

*This section may be omitted without loss of continuity.

Thus, we may write

$$A_{x'} = \mathbf{A} \cdot \mathbf{i'} = (\mathbf{i} \cdot \mathbf{i'})A_x + (\mathbf{j} \cdot \mathbf{i'})A_y + (\mathbf{k} \cdot \mathbf{i'})A_z$$
$$A_{y'} = \mathbf{A} \cdot \mathbf{j'} = (\mathbf{i} \cdot \mathbf{j'})A_x + (\mathbf{j} \cdot \mathbf{j'})A_y + (\mathbf{k} \cdot \mathbf{j'})A_z \qquad (1.24)$$
$$A_{z'} = \mathbf{A} \cdot \mathbf{k'} = (\mathbf{i} \cdot \mathbf{k'})A_x + (\mathbf{j} \cdot \mathbf{k'})A_y + (\mathbf{k} \cdot \mathbf{k'})A_z$$

The scalar products $(\mathbf{i} \cdot \mathbf{i'})$, $(\mathbf{i} \cdot \mathbf{j'})$, and so on, are called the *coefficients of transformation.* They are equal to the direction cosines of the axes of the primed coordinate system relative to the unprimed system. The unprimed components are similarly expressed as

$$A_x = \mathbf{A} \cdot \mathbf{i} = (\mathbf{i'} \cdot \mathbf{i})A_{x'} + (\mathbf{j'} \cdot \mathbf{i})A_{y'} + (\mathbf{k'} \cdot \mathbf{i})A_{z'}$$
$$A_y = \mathbf{A} \cdot \mathbf{j} = (\mathbf{i'} \cdot \mathbf{j})A_{x'} + (\mathbf{j'} \cdot \mathbf{j})A_{y'} + (\mathbf{k'} \cdot \mathbf{j})A_{z'} \qquad (1.25)$$
$$A_z = \mathbf{A} \cdot \mathbf{k} = (\mathbf{i'} \cdot \mathbf{k})A_{x'} + (\mathbf{j'} \cdot \mathbf{k})A_{y'} + (\mathbf{k'} \cdot \mathbf{k})A_{z'}$$

All the coefficients of transformation in Equation 1.25 also appear in Equation 1.24, because $\mathbf{i} \cdot \mathbf{i'} = \mathbf{i'} \cdot \mathbf{i}$ and so on, but those in the rows (equations) of Equation 1.25 appear in the columns of terms in Equation 1.24, and conversely. The transformation rules expressed in these two sets of equations are a general property of vectors. As a matter of fact, they constitute an alternative way of defining vectors.[6]

The equations of transformation are conveniently expressed in matrix notation.[7] Thus, Equation 1.24 is written

$$\begin{bmatrix} A_{x'} \\ A_{y'} \\ A_{z'} \end{bmatrix} = \begin{bmatrix} \mathbf{i} \cdot \mathbf{i'} & \mathbf{j} \cdot \mathbf{i'} & \mathbf{k} \cdot \mathbf{i'} \\ \mathbf{i} \cdot \mathbf{j'} & \mathbf{j} \cdot \mathbf{j'} & \mathbf{k} \cdot \mathbf{j'} \\ \mathbf{i} \cdot \mathbf{k'} & \mathbf{j} \cdot \mathbf{k'} & \mathbf{k} \cdot \mathbf{k'} \end{bmatrix} \begin{bmatrix} A_x \\ A_y \\ A_z \end{bmatrix} \qquad (1.26)$$

The 3 by 3 matrix in Equation 1.26 is called the *transformation matrix.* One advantage of the matrix notation is that successive transformations are readily handled by means of matrix multiplication.

The reader will observe that the application of a given transformation matrix to some vector **A** is also formally equivalent to rotating that vector within the unprimed (fixed) coordinate system, the components of the rotated vector being given by Equation 1.26. Thus, finite rotations can be represented by matrices. (Note that the sense of rotation of the vector in this context is opposite that of the rotation of the coordinate system in the previous context.)

EXAMPLE 1.14

Express the vector $\mathbf{A} = 3\mathbf{i} + 2\mathbf{j} + \mathbf{k}$ in terms of the triad $\mathbf{i'j'k'}$ where the $x'y'$-axes are rotated 45° around the z-axis, the z- and the z'-axes coinciding, as shown in

---

[6] See, for example, M. H. Hull, Jr., *The Calculus of Physics*, W. A. Benjamin, New York, NY, 1969.

[7] A brief review of matrices is given in Appendix H.

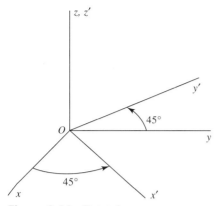

Figure 1.11 Rotated axes.

Figure 1.11. Referring to the figure, we have for the coefficients of transformation, $\mathbf{i} \cdot \mathbf{i}' = \cos 45°$ and so on; hence,

$$\mathbf{i} \cdot \mathbf{i}' = 1/\sqrt{2} \qquad \mathbf{j} \cdot \mathbf{i}' = 1/\sqrt{2} \qquad \mathbf{k} \cdot \mathbf{i}' = 0$$
$$\mathbf{i} \cdot \mathbf{j}' = -1/\sqrt{2} \qquad \mathbf{j} \cdot \mathbf{j}' = 1/\sqrt{2} \qquad \mathbf{k} \cdot \mathbf{j}' = 0$$
$$\mathbf{i} \cdot \mathbf{k}' = 0 \qquad \mathbf{j} \cdot \mathbf{k}' = 0 \qquad \mathbf{k} \cdot \mathbf{k}' = 1$$

These give

$$A_{x'} = \frac{3}{\sqrt{2}} + \frac{2}{\sqrt{2}} = \frac{5}{\sqrt{2}} \qquad A_{y'} = \frac{-3}{\sqrt{2}} + \frac{2}{\sqrt{2}} = \frac{-1}{\sqrt{2}} \qquad A_{z'} = 1$$

so that, in the primed system, the vector $\mathbf{A}$ is given by

$$\mathbf{A} = \frac{5}{\sqrt{2}}\mathbf{i}' - \frac{1}{\sqrt{2}}\mathbf{j}' + \mathbf{k}'$$

∎

EXAMPLE 1.15

Find the transformation matrix for a rotation of the primed coordinate system through an angle ϕ about the z-axis. (Example 1.14 is a special case of this.) We have

$$\mathbf{i} \cdot \mathbf{i}' = \mathbf{j} \cdot \mathbf{j}' = \cos \phi$$
$$\mathbf{j} \cdot \mathbf{i}' = -\mathbf{i} \cdot \mathbf{j}' = \sin \phi$$
$$\mathbf{k} \cdot \mathbf{k}' = 1$$

and all other dot products are zero. Hence, the transformation matrix is

$$\begin{bmatrix} \cos \phi & \sin \phi & 0 \\ -\sin \phi & \cos \phi & 0 \\ 0 & 0 & 1 \end{bmatrix}$$

∎

It is clear from Example 1.15 that the transformation matrix for a rotation about a different coordinate axis, say the y-axis through an angle θ, will be given by the matrix

$$\begin{bmatrix} \cos\theta & 0 & -\sin\theta \\ 0 & 1 & 0 \\ \sin\theta & 0 & \cos\theta \end{bmatrix}$$

Consequently, the matrix for the combination of two rotations, the first being about the z-axis (angle ϕ) and the second being about the new y'-axis (angle θ), is given by the matrix product

$$\begin{bmatrix} \cos\theta & 0 & -\sin\theta \\ 0 & 1 & 0 \\ \sin\theta & 0 & \cos\theta \end{bmatrix} \begin{bmatrix} \cos\phi & \sin\phi & 0 \\ -\sin\phi & \cos\phi & 0 \\ 0 & 0 & 1 \end{bmatrix} = $$

$$\begin{bmatrix} \cos\theta\cos\phi & \cos\theta\sin\phi & -\sin\theta \\ -\sin\phi & \cos\phi & 0 \\ \sin\theta\cos\phi & \sin\theta\sin\phi & \cos\theta \end{bmatrix}$$

Now matrix multiplication is, in general, noncommutative. Hence, we might expect that if the order of the rotations were reversed, and therefore the order of the matrix multiplication on the left, the final result would be different. This turns out to be the case, which the reader can verify. This is in keeping with a remark made earlier, namely that finite rotations do not obey the law of vector addition and, hence, are not vectors even though a single rotation has a direction (the axis) and a magnitude (the angle of rotation). However, we shall show later that infinitesimal rotations do obey the law of vector addition and can be represented by vectors.

1.9   DERIVATIVE OF A VECTOR

Up to this point we have been concerned mainly with vector algebra. We now begin the study of the calculus of vectors and its use in the description of the motion of particles.

Consider a vector $\mathbf{A}$, the components of which are functions of a single variable u. The vector may represent position, velocity, and so on. The parameter u is usually the time t, but it can be any quantity that determines the components of $\mathbf{A}$:

$$\mathbf{A}(u) = \mathbf{i}A_x(u) + \mathbf{j}A_y(u) + \mathbf{k}A_z(u)$$

The derivative of $\mathbf{A}$ with respect to u is defined, quite analogously to the ordinary derivative of a scalar function, by the limit

$$\frac{d\mathbf{A}}{du} = \lim_{\Delta u \to 0} \frac{\Delta\mathbf{A}}{\Delta u} = \lim_{\Delta u \to 0} \left(\mathbf{i}\frac{\Delta A_x}{\Delta u} + \mathbf{j}\frac{\Delta A_y}{\Delta u} + \mathbf{k}\frac{\Delta A_z}{\Delta u} \right)$$

where $\Delta A_x = A_x(u + \Delta u) - A_x(u)$ and so on. Hence,

$$\frac{d\mathbf{A}}{du} = \mathbf{i}\frac{dA_x}{du} + \mathbf{j}\frac{dA_y}{du} + \mathbf{k}\frac{dA_z}{du} \qquad (1.27)$$

The derivative of a vector, therefore, is a vector whose Cartesian components are ordinary derivatives.

It follows from Equation 1.27 that the derivative of the sum of two vectors is equal to the sum of the derivatives, namely,

$$\frac{d}{du}(\mathbf{A} + \mathbf{B}) = \frac{d\mathbf{A}}{du} + \frac{d\mathbf{B}}{du} \tag{1.28}$$

Rules for differentiating vector products will be treated later.

1.10 POSITION VECTOR OF A PARTICLE. VELOCITY AND ACCELERATION IN RECTANGULAR COORDINATES

In a given reference system the position of a particle can be specified by a single vector, namely, the displacement of the particle relative to the origin of the coordinate system. This vector is called the *position vector* of the particle. In rectangular coordinates, Figure 1.12, the position vector is simply

$$\mathbf{r} = \mathbf{i}x + \mathbf{j}y + \mathbf{k}z \qquad \vec{r}(t) = x(t)\,\hat{i} + y(t)\,\hat{j} + z(t)\,\hat{k}$$

The components of the position vector of a moving particle are functions of the time, namely,

$$x = x(t) \qquad y = y(t) \qquad z = z(t)$$

In Equation 1.27 we gave the formal definition of the derivative of any vector with respect to some parameter. In particular, if the vector is the position vector **r** of a moving particle and the parameter is the time *t*, the derivative of **r** with respect to *t* is called the *velocity*, which we shall denote by **v**:

$$\mathbf{v} = \frac{d\mathbf{r}}{dt} = \mathbf{i}\dot{x} + \mathbf{j}\dot{y} + \mathbf{k}\dot{z} \tag{1.29}$$

$$\underbrace{}_{V_x} \quad \underbrace{}_{V_y} \quad \underbrace{}_{V_z}$$

$$\dot{x} = \frac{dx}{dt} \qquad \dot{y} = \frac{dy}{dt} \qquad \dot{z} = \frac{dz}{dt}$$

$$\text{speed} = |\vec{v}|$$

$$= \sqrt{\dot{x}^2 + \dot{y}^2 + \dot{z}^2}$$

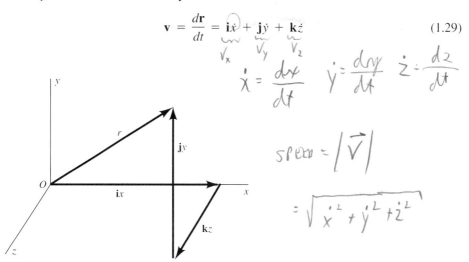

Figure 1.12 The position vector **r** and its components in a Cartesian coordinate system.

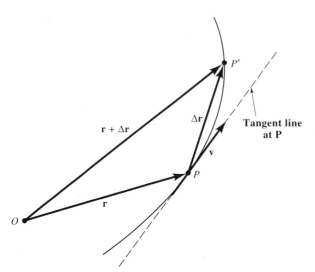

Figure 1.13   Velocity vector of a moving particle as the limit of the ratio $\Delta\mathbf{r}/\Delta t$.

where the dots indicate differentiation with respect to t. (This convention is standard and will be used throughout the book.) Let us examine the geometric significance of the velocity vector. Suppose a particle is at a certain position at time t. At a time Δt later, the particle will have moved from the position $\mathbf{r}(t)$ to the position $\mathbf{r}(t + \Delta t)$. The vector displacement during the time interval Δt is

$$\Delta\mathbf{r} = \mathbf{r}(t + \Delta t) - \mathbf{r}(t)$$

so the quotient $\Delta\mathbf{r}/\Delta t$ is a *vector* that is parallel to the displacement. As we consider smaller and smaller time intervals, the quotient $\Delta\mathbf{r}/\Delta t$ approaches a limit $d\mathbf{r}/dt$, which we call the *velocity*. The vector $d\mathbf{r}/dt$ expresses both the direction of motion and the rate. This is shown graphically in Figure 1.13. In the time interval Δt the particle moves along the path from P to P'. As Δt approaches zero, the point P' approaches P, and the direction of the vector $\Delta\mathbf{r}/\Delta t$ approaches the direction of the tangent to the path at P. The velocity vector, therefore, is always tangent to the path of motion.

The magnitude of the velocity is called the *speed*. In rectangular components the speed is just

$$v = |\mathbf{v}| = (\dot{x}^2 + \dot{y}^2 + \dot{z}^2)^{1/2} \tag{1.30}$$

If we denote the cumulative scalar distance along the path by s, then we can alternately express the speed as

$$v = \frac{ds}{dt} = \lim_{\Delta t \to 0} \frac{\Delta s}{\Delta t} = \lim_{\Delta t \to 0} \frac{[(\Delta x)^2 + (\Delta y)^2 + (\Delta z)^2]^{1/2}}{\Delta t}$$

which reduces to the expression on the right of Equation 1.30.

The time derivative of the velocity is called the *acceleration*. Denoting the acceleration by $\mathbf{a}$, we have

$$\mathbf{a} = \frac{d\mathbf{v}}{dt} = \frac{d^2\mathbf{r}}{dt^2} \;\;\dot{=}\; \overset{\rightarrow}{\mathbf{V}} \;=\; \overset{\rightarrow}{\mathbf{r}} \tag{1.31}$$

In rectangular components $= \dot{V}_x\,\hat{\lambda} + \dot{V}_y\,\hat{\imath} + \dot{V}_z\,\hat{k}$

$$\mathbf{a} = \mathbf{i}\ddot{x} + \mathbf{j}\ddot{y} + \mathbf{k}\ddot{z} \tag{1.32}$$

Thus, acceleration is a vector quantity whose components, in rectangular coordinates, are the second derivatives of the positional coordinates of a moving particle. The resolution of **a** into tangential and normal components will be discussed in Section 1.12.

EXAMPLE 1.16   *Projectile Motion*

Let us examine the motion represented by the equation

$$\mathbf{r}(t) = \mathbf{i}bt + \mathbf{j}\left(ct - \frac{gt^2}{2}\right) + \mathbf{k}0$$

$$\underset{V_x}{\underbrace{\qquad}} \quad \underset{V_y}{\underbrace{\qquad\qquad}}$$

This represents motion in the xy plane, since the z component is constant and equal to zero. The velocity **v** is obtained by differentiating with respect to t, namely,

$$\mathbf{v} = \frac{d\mathbf{r}}{dt} = \mathbf{i}b + \mathbf{j}(c - gt) \qquad @\; t=0 \;\; \overset{\rightarrow}{V_0} = b\,\hat{\imath} + c\,\hat{\jmath}$$

The acceleration, likewise, is given by

$$\mathbf{a} = \frac{d\mathbf{v}}{dt} = -\mathbf{j}g$$

Thus, **a** is in the negative y direction and has the constant magnitude g. The path of motion is a parabola, as shown in Fig. 1.14. (This equation actually represents the motion of a projectile.) The speed v varies with t according to the equation

$$v = [b^2 + (c - gt)^2]^{1/2} \qquad \blacksquare$$

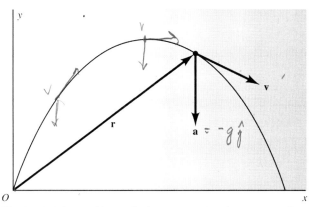

PARABOLIC
TRAJECTORY

$\mathbf{a} = -g\,\hat{\jmath}$

Figure 1.14   Position, velocity, and acceleration vectors of a particle (projectile) moving in a parabolic path.

EXAMPLE 1.17   *Circular Motion*

Suppose the position vector of a particle is given by

$$\mathbf{r} = \mathbf{i}b \sin \omega t + \mathbf{j}b \cos \omega t$$

where ω is a constant.

Let us analyze the motion. The distance from the origin remains constant

$$|\mathbf{r}| = r = (b^2 \sin^2 \omega t + b^2 \cos^2 \omega t)^{1/2} = b$$

So the path is a circle of radius b centered at the origin. Differentiating $\mathbf{r}$, we find the velocity vector

$$\mathbf{v} = \frac{d\mathbf{r}}{dt} = \mathbf{i}b\omega \cos \omega t - \mathbf{j}b\omega \sin \omega t$$

The particle traverses its path with constant speed:

$$v = |\mathbf{v}| = (b^2\omega^2 \cos^2 \omega t + b^2\omega^2 \sin^2 \omega t)^{1/2} = b\omega$$

The acceleration is

$$\mathbf{a} = \frac{d\mathbf{v}}{dt} = -\mathbf{i}b\omega^2 \sin \omega t - \mathbf{j}b\omega^2 \cos \omega t$$

In this case the acceleration is perpendicular to the velocity, since the dot product of $\mathbf{v}$ and $\mathbf{a}$ vanishes:

$$\mathbf{v} \cdot \mathbf{a} = (b\omega \cos \omega t)(-b\omega^2 \sin \omega t) + (-b\omega \sin \omega t)(-b\omega^2 \cos \omega t) = 0$$

Comparing the two expressions for $\mathbf{a}$ and $\mathbf{r}$, we see that we can write

$$\mathbf{a} = -\omega^2 \mathbf{r}$$

so $\mathbf{a}$ and $\mathbf{r}$ are oppositely directed, that is, $\mathbf{a}$ always points toward the center of the circular path, Figure 1.15.   ■

EXAMPLE 1.18   *Rolling Wheel*

Let us consider the following position vector of a particle *P:*

$$\mathbf{r} = \mathbf{r}_1 + \mathbf{r}_2$$

in which

$$\mathbf{r}_1 = \mathbf{i}b\omega t + \mathbf{j}b$$
$$\mathbf{r}_2 = \mathbf{i}b \sin \omega t + \mathbf{j}b \cos \omega t$$

Now $\mathbf{r}_1$ by itself represents a point moving along the line $y = b$ at constant velocity, provided ω is constant, namely,

$$\mathbf{v}_1 = \frac{d\mathbf{r}_1}{dt} = \mathbf{i}b\omega$$

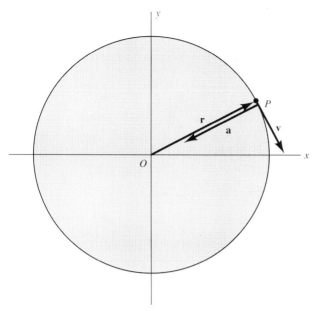

Figure 1.15 A particle moving in a circular path with constant speed.

The second part, $\mathbf{r}_2$, is just the position vector for circular motion, as discussed in Example 1.17. Hence, the vector sum $\mathbf{r}_1 + \mathbf{r}_2$ represents a point that describes a circle of radius b about a moving center. This is precisely what occurs for a particle on the rim of a rolling wheel, $\mathbf{r}_1$ being the position vector of the center of the wheel, and $\mathbf{r}_2$ is the position vector of the particle P *relative* to the moving center. The actual path is a *cycloid* as shown in Figure 1.16. The velocity of P is

$$\mathbf{v} = \mathbf{v}_1 + \mathbf{v}_2 = \mathbf{i}(b\omega + b\omega \cos \omega t) - \mathbf{j}b\omega \sin \omega t$$

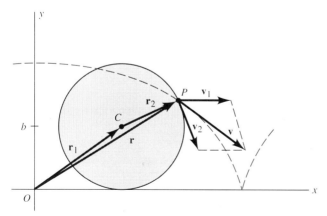

Figure 1.16 Cycloidal path of a particle on a rolling wheel.

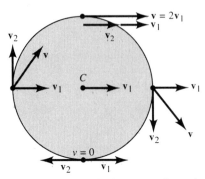

Figure 1.17   Velocity vectors for various points on a rolling wheel.

In particular, for $\omega t = 0, 2\pi, 4\pi, \ldots$, we find that $\mathbf{v} = \mathbf{i}2b\omega$, which is just twice the velocity of the center C. At these points the particle is at the uppermost part of its path. Further, for $\omega t = \pi, 3\pi, 5\pi, \ldots$, we obtain $\mathbf{v} = 0$. At these points the particle is at its lowest point and is instantaneously in contact with the ground. See Figure 1.17.   ∎

1.11   DERIVATIVES OF PRODUCTS OF VECTORS

It is often necessary to deal with derivatives of the products $n\mathbf{A}$, $\mathbf{A} \cdot \mathbf{B}$, and $\mathbf{A} \times \mathbf{B}$ where the scalar n and the vectors $\mathbf{A}$ and $\mathbf{B}$ are functions of a single parameter u, as in Section 1.9. From the general definition of the derivative, we have

$$\frac{d(n\mathbf{A})}{du} = \lim_{\Delta u \to 0} \frac{n(u + \Delta u)\mathbf{A}(u + \Delta u) - n(u)\mathbf{A}(u)}{\Delta u} \tag{1.33}$$

$$\frac{d(\mathbf{A} \cdot \mathbf{B})}{du} = \lim_{\Delta u \to 0} \frac{\mathbf{A}(u + \Delta u) \cdot \mathbf{B}(u + \Delta u) - \mathbf{A}(u) \cdot \mathbf{B}(u)}{\Delta u}$$

$$\frac{d(\mathbf{A} \times \mathbf{B})}{du} = \lim_{\Delta u \to 0} \frac{\mathbf{A}(u + \Delta u) \times \mathbf{B}(u + \Delta u) - \mathbf{A}(u) \times \mathbf{B}(u)}{\Delta u}$$

By adding and subtracting expressions like $n(u + \Delta u)\mathbf{A}(u)$ in the numerators, we obtain the following rules:

$n = f(t)$

$$\frac{d(n\mathbf{A})}{du} = \frac{dn}{du}\mathbf{A} + n\frac{d\mathbf{A}}{du} \tag{1.34}$$

$$\frac{d(\mathbf{A} \cdot \mathbf{B})}{du} = \frac{d\mathbf{A}}{du} \cdot \mathbf{B} + \mathbf{A} \cdot \frac{d\mathbf{B}}{du} \tag{1.35}$$

$ex \; r \times F$

$(Torque)$

$$\frac{d(\mathbf{A} \times \mathbf{B})}{du} = \frac{d\mathbf{A}}{du} \times \mathbf{B} + \mathbf{A} \times \frac{d\mathbf{B}}{du} \tag{1.36}$$

Notice that it is necessary to preserve the order of the terms in the derivative of the cross product. The steps are left as an exercise for the student.

1.12 TANGENTIAL AND NORMAL COMPONENTS OF ACCELERATION

In Section 1.4 it was shown that any vector can be expressed as the product of its magnitude and a unit vector giving its direction. Accordingly, the velocity vector of a moving particle can be written as the product of the particle's speed v and a unit vector $\boldsymbol{\tau}$ that gives the direction of the particle's motion. Thus,

$$\mathbf{v} = v\boldsymbol{\tau} \tag{1.37}$$

The vector $\boldsymbol{\tau}$ is called the *unit tangent vector*. As the particle moves, the speed v may change and the direction of $\boldsymbol{\tau}$ may change. Let us use the rule for differentiation of the product of a scalar and a vector to obtain the acceleration. The result is

$$\mathbf{a} = \frac{d\mathbf{v}}{dt} = \frac{d(v\boldsymbol{\tau})}{dt} = \dot{v}\boldsymbol{\tau} + v\frac{d\boldsymbol{\tau}}{dt} \tag{1.38}$$

The unit vector $\boldsymbol{\tau}$, being of constant magnitude, has a derivative $d\boldsymbol{\tau}/dt$ that must necessarily express the change in the direction of $\boldsymbol{\tau}$ with respect to time. This is illustrated in Figure 1.18a. The particle is initially at some point P on its path of motion. In a time interval Δt the particle moves to another point P' a certain distance Δs along the path. Let us denote the unit tangent vectors at P and P' by $\boldsymbol{\tau}$ and $\boldsymbol{\tau}'$, respectively, as shown. The directions of these two unit vectors differ by a certain angle $\Delta\psi$ as shown in Figure 1.18b.

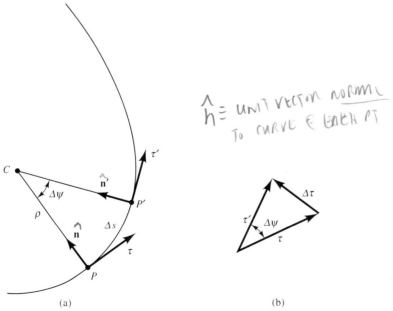

$\hat{h} \equiv$ UNIT VECTOR NORMAL TO CURVE E EACH PT

(a)   (b)

Figure 1.18 (a) The unit tangent and unit normal vectors. (b) Change of the unit tangent vector.

For small angles $\Delta\psi$ the ratio $|\Delta\tau/\tau|$ is equal to $\Delta\psi$. Since $|\tau| = 1$, $\Delta\tau$ approaches $\Delta\psi$ in magnitude as $\Delta\psi \to 0$. Also, in this limit, the direction of $\Delta\tau$ becomes perpendicular to the direction of τ. It follows that the derivative $d\tau/d\psi$ is of magnitude unity and is perpendicular to τ. We shall therefore call it the *unit normal vector* and denote it by $\mathbf{n}$:

$$\frac{d\tau}{d\psi} = \mathbf{n} \tag{1.39}$$

Next, in order to find the time derivative $d\tau/dt$, we use the chain rule as follows:

$$\frac{d\tau}{dt} = \frac{d\tau}{d\psi}\frac{d\psi}{dt} = \mathbf{n}\frac{d\psi}{ds}\frac{ds}{dt} = \mathbf{n}\frac{v}{\rho}$$

in which

$$\rho = \frac{ds}{d\psi}$$

is the radius of curvature of the path of the moving particle at the point P. The above value for $d\tau/dt$ is now inserted into Equation 1.38 to yield the final result

$$\mathbf{a} = \dot{v}\tau + \frac{v^2}{\rho}\mathbf{n} \tag{1.40}$$

Thus, the acceleration of a moving particle has a component

$$a_\tau = \dot{v} = \ddot{s}$$

in the direction of motion. This is the *tangential acceleration*. The other component

$$a_n = \frac{v^2}{\rho}$$

is the normal component. This component is always directed toward the center of curvature on the concave side of the path of motion. Hence, the normal component is also called the *centripetal acceleration*.

From the above considerations we see that the time derivative of the speed is only the tangential component of the acceleration. The magnitude of the total acceleration is given by

$$|\mathbf{a}| = \left|\frac{d\mathbf{v}}{dt}\right| = \left(\dot{v}^2 + \frac{v^4}{\rho^2}\right)^{1/2} \tag{1.41}$$

In particular if a particle moves on a circle with constant speed v, as in Example 1.17, the acceleration vector is of magnitude v^2/b where b is the radius of the circle. The acceleration vector always points to the center of the circle in this case. However, if the speed is not constant but increases as a certain rate $\dot{v}$, then the acceleration has a forward component of this amount and is slanted away from the center of the circle toward the forward direction as illustrated in Figure 1.19. If the particle is slowing down, then the acceleration vector is slanted in the opposite direction.

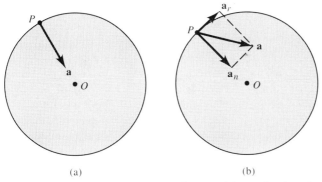

Figure 1.19 Acceleration vectors for a particle moving in a circular path with (a) constant speed and (b) increasing speed.

EXAMPLE 1.19

Let us consider, as in example 1.18, a wheel of radius b rolling without slipping along some surface, only this time assume that the center of the wheel is accelerating. Calculate the velocity and acceleration of a point on the rim of the wheel.

Solution:
Figure 1.17 showed the velocity vectors at four different points on the wheel when it is moving at constant velocity. The resultant velocity of any point on the rim is given by $\mathbf{v} = \mathbf{v}_1 + \mathbf{v}_2$ where $\mathbf{v}_1$ is the velocity of the center of the wheel and $\mathbf{v}_2$ is the velocity of the point relative to the center of the wheel. This relation is true even if $\mathbf{v}_1$ is changing. Examination of Figure 1.20a reveals that during a time interval δt the wheel moves a distance $\delta s = v_1 \delta t$ as it turns through an angle $\delta\theta = \delta s/b$. But $b \, \delta\theta/\delta t = v_2$. Hence, $v_1 = \delta s/\delta t = b \, \delta\theta/\delta t = v_2$. The magnitude of the vectors v_1 and v_2 are the same, even though v_1 is changing with time. Note that a point on the bottom of the wheel is instantaneously at rest but only when there is no slipping!

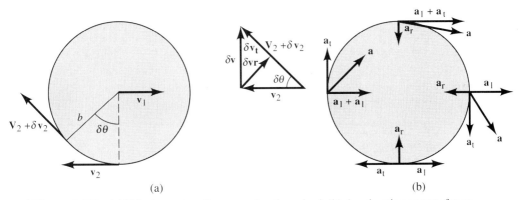

Figure 1.20 (a) Velocity vectors for an accelerating wheel. (b) Acceleration vectors for an accelerating wheel, rolling without slipping.

Now let us find the acceleration of a point on the rim as perceived by an observer at rest. In Figure 1.20a we see that during the time interval δt, the velocity of a point has changed from $\mathbf{v}_2$ to $\mathbf{v}_2 + \delta\mathbf{v}_2$. Also shown in that figure is an enlarged view of a vector triangle consisting of the sides $\mathbf{v}_2$, $\delta\mathbf{v}_2$, and $\mathbf{v}_2 + \delta\mathbf{v}_2$. Since $\mathbf{v}_2$ is always tangent to the wheel, the angle between $\mathbf{v}_2$ and $\mathbf{v}_2 + \delta\mathbf{v}_2$ is $\delta\theta$, the angle through which the wheel has turned. The change in $\mathbf{v}_2$ is given by $\delta\mathbf{v}_2 = \delta\mathbf{v}_t + \delta\mathbf{v}_r$. $\delta\mathbf{v}_t$ is a vector whose magnitude is the increase in the magnitude of $\mathbf{v}_2$ and whose direction is tangential to the rim. $\delta\mathbf{v}_r$ is a vector that represents the change in direction of $\mathbf{v}_2$ and is directed radially inward toward the center of the wheel. So, even though the radial velocity is zero, the change in velocity has a nonzero radial component.

Dividing the above equation for $\delta\mathbf{v}_2$ by δt and taking limits as $\delta t \to 0$, we get $\dot{\mathbf{v}}_2 = \dot{\mathbf{v}}_t + \dot{\mathbf{v}}_r$ or $\mathbf{a}_2 = \mathbf{a}_t + \mathbf{a}_r$ for the acceleration $\mathbf{a}_2$. The direction of the acceleration vectors $\mathbf{a}_t$ and $\mathbf{a}_r$ correspond to the directions of the above velocity vector increments $\delta\mathbf{v}_t$ and $\delta\mathbf{v}_r$. These directions are tangent to the wheel and radially inward toward the center of the wheel, respectively. $\mathbf{a}_t$ is nonzero only if the center of the wheel is accelerating. $\mathbf{a}_r$ is the familiar centripetal (or radial) acceleration that exists for any particle traveling along a curved path. (We emphasize again that even though the radial velocity of a point on the rim of the wheel is zero, its radial acceleration is not!) In Figure 1.20a we can see that, for small $\delta\theta$, $\delta v_r = v_2\,\delta\theta$. Again, dividing by δt and taking limits as $\delta t \to 0$, we get $a_r = \dot{v}_r = v_2\dot{\theta} = v_2^2/b$ for the particle's centripetal acceleration.

Since the wheel rolls without slipping, the increase in magnitude of the tangential velocity δv_t must equal any increase in magnitude of the velocity of the center of the wheel v_1. Therefore, $\delta\dot{v}_t = \delta\dot{v}_1$, and the acceleration of any point on the wheel is given by

$$\mathbf{a} = \mathbf{a}_1 + \mathbf{a}_t + \mathbf{a}_r$$

The centripetal acceleration has a magnitude v_2^2/b, or equivalently v_1^2/b, and arises because $\mathbf{v}_2$ is changing in direction. The acceleration of four points on the rim are shown in Figure 1.20b. These are the same four points as in Figure 1.17, but note that the acceleration of the center of the wheel at any instant of time is $\mathbf{a}_1$, not zero! The centripetal acceleration has the same magnitude at all four points on the rim. Muddy thinking could cause some problems here. For example, at the top of the wheel, the velocity of a point is $2v_1$, tangent to the wheel. Why is its centripetal acceleration not $(2v_1)^2/b$? Likewise, when the particle is at the bottom of the wheel, why is its centripetal acceleration not equal to zero, since its velocity is zero? We leave it to the student to figure this out. ∎

1.13   VELOCITY AND ACCELERATION IN PLANE POLAR COORDINATES

It is often convenient to employ polar coordinates r, θ to express the position of a particle moving in a plane. Vectorially, the position of the particle can be written as the product of the radial distance r by a unit radial vector $\mathbf{e}_r$:

$$\mathbf{r} = r\mathbf{e}_r \tag{1.42}$$

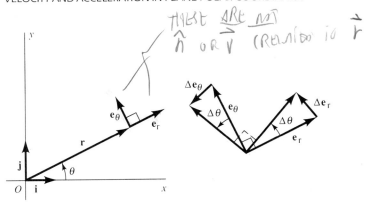

Figure 1.21   Unit vectors for plane polar coordinates.

As the particle moves, both r and $\mathbf{e}_r$ vary; thus, they are both functions of the time. Hence, if we differentiate with respect to t, we have

$$\mathbf{v} = \frac{d\mathbf{r}}{dt} = \dot{r}\mathbf{e}_r + r\frac{d\mathbf{e}_r}{dt} \tag{1.43}$$

In order to calculate the derivative $d\mathbf{e}_r/dt$, let us consider the vector diagram shown in Figure 1.21. A study of the figure shows that when the direction of $\mathbf{r}$ changes by an amount $\Delta\theta$, the corresponding change $\Delta\mathbf{e}_r$ of the unit radial vector is as follows: The magnitude $|\Delta\mathbf{e}_r|$ is approximately equal to $\Delta\theta$, and the direciton of $\Delta\mathbf{e}_r$ is very nearly perpendicular to $\mathbf{e}_r$. Let us introduce another unit vector $\mathbf{e}_\theta$ whose direction is perpendicular to $\mathbf{e}_r$. Then we have

$$\Delta\mathbf{e}_r \simeq \mathbf{e}_\theta\Delta\theta$$

If we divide by Δt and take the limit, we get

$$\frac{d\mathbf{e}_r}{dt} = \mathbf{e}_\theta\frac{d\theta}{dt} \tag{1.44}$$

for the time derivative of the unit radial vector. In a precisely similar way, we can argue that the change in the unit vector $\mathbf{e}_\theta$ is given by the approximation

$$\Delta\mathbf{e}_\theta \simeq -\mathbf{e}_r\Delta\theta$$

Here the minus sign is inserted to indicate that the direction of the change $\Delta\mathbf{e}_\theta$ is opposite to the direction of $\mathbf{e}_r$ as can be seen from Figure 1.21. Consequently, the time derivative is given by

$$\frac{d\mathbf{e}_\theta}{dt} = -\mathbf{e}_r\frac{d\theta}{dt} \tag{1.45}$$

By using Equation 1.44 for the derivative of the unit radial vector, we can finally write the equation for the velocity as

$$\mathbf{v} = \dot{r}\mathbf{e}_r + r\dot{\theta}\,\mathbf{e}_\theta \tag{1.46}$$

Thus, $\dot{r}$ is the radial component of the velocity vector, and $r\dot{\theta}$ is the transverse component.

In order to find the acceleration vector, we take the derivative of the velocity with respect to time. This gives

$$\mathbf{a} = \frac{d\mathbf{v}}{dt} = \ddot{r}\mathbf{e}_r + \dot{r}\frac{d\mathbf{e}_r}{dt} + (\dot{r}\dot{\theta} + r\ddot{\theta})\mathbf{e}_\theta + r\dot{\theta}\frac{d\mathbf{e}_\theta}{dt}$$

The values of $d\mathbf{e}_r/dt$ and $d\mathbf{e}_\theta/dt$ are given by Equations 1.44 and 1.45 and yield the following equation for the acceleration vector in plane polar coordinates:

$$\mathbf{a} = (\ddot{r} - r\dot{\theta}^2)\mathbf{e}_r + (r\ddot{\theta} + 2\dot{r}\dot{\theta})\mathbf{e}_\theta \qquad (1.47)$$

Thus, the radial component of the acceleration vector is

$$a_r = \ddot{r} - r\dot{\theta}^2 \qquad (1.48)$$

and the transverse component is

$$a_\theta = r\ddot{\theta} + 2\dot{r}\dot{\theta} = \frac{1}{r}\frac{d}{dt}(r^2\dot{\theta}) \qquad (1.49)$$

The above results show, for instance, that if a particle moves on a circle of constant radius b, so that $\dot{r} = 0$, then the radial component of the acceleration is of magnitude $b\dot{\theta}^2$ and is directed inward toward the center of the circular path. The transverse component in this case is $b\ddot{\theta}$. On the other hand, if the particle moves along a fixed radial line, that is, if θ is constant, then the radial component is just $\ddot{r}$ and the transverse component is zero. If r and θ both vary, then the general expression (1.47) gives the acceleration.

EXAMPLE 1.20

A honeybee homes in on its hive in a spiral path in such a way that the radial distance decreases at a constant rate: $r = b - ct$, while the angular speed increases at a constant rate: $\dot{\theta} = kt$. Find the speed as a function of time.

Solution:
We have $\dot{r} = -c$ and $\ddot{r} = 0$. Thus, from Equation 1.46

$$\mathbf{v} = -c\mathbf{e}_r + (b - ct)kt\mathbf{e}_\theta$$

so

$$v = [c^2 + (b - ct)^2k^2t^2]^{1/2}$$

which is valid for $t \leq b/c$. (Larger values of t make r become negative.) Note that $v = c$ both for $t = 0$, $r = b$ and for $t = b/c$, $r = 0$.

EXAMPLE 1.21

On a horizontal turntable that is rotating at constant angular speed, there is a bug crawling outward on a radial line such that the bug's distance from the center increases quadratically with time: $r = bt^2$, $\theta = \omega t$ where b and ω are constants. Find the acceleration of the bug. We have $\dot{r} = 2bt$, $\ddot{r} = 2b$, $\dot{\theta} = \omega$, $\ddot{\theta} = 0$.

Solution:
Substituting into Equation 1.47, we find

$$\mathbf{a} = \mathbf{e}_r(2b - bt^2\omega^2) + \mathbf{e}_\theta[0 + 2(2bt)\omega]$$
$$= b(2 - t^2\omega^2)\mathbf{e}_r + 4b\omega t\mathbf{e}_\theta$$

It is interesting to note that the radial component of the acceleration becomes negative for large t, in this example, although the radius is always increasing monotonically with time. ∎

1.14 VELOCITY AND ACCELERATION IN CYLINDRICAL AND SPHERICAL COORDINATES

Cylindrical Coordinates

In the case of three-dimensional motion, the position of a particle can be described in cylindrical coordinates R, ϕ, z. The position vector is then written as

$$\mathbf{r} = R\mathbf{e}_R + z\mathbf{e}_z \tag{1.50}$$

where $\mathbf{e}_R$ is a unit radial vector in the xy plane and $\mathbf{e}_z$ is the unit vector in the z direction. A third unit vector $\mathbf{e}_\phi$ is needed so that the three vectors $\mathbf{e}_R\mathbf{e}_\phi\mathbf{e}_z$ constitute a right-handed triad as illustrated in Figure 1.22. We note that $\mathbf{k} = \mathbf{e}_z$.

The velocity and acceleration vectors are found by differentiating, as before. This will again involve derivatives of the unit vectors. An argument similar to that used for

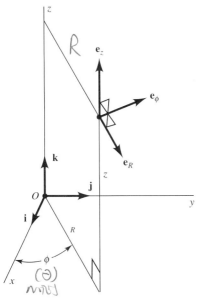

$$R = \sqrt{x^2 + y^2} \qquad x = R\cos\theta$$
$$\tan\phi\,(\theta) = \frac{y}{x} \qquad y = R\sin\theta$$
$$z = z \qquad z = z$$

Figure 1.22 Unit vectors for cylindrical coordinates.

the plane case shows that $de_R/dt = \mathbf{e}_\phi\dot\phi$ and $de_\phi/dt = -\mathbf{e}_R\dot\phi$. The unit vector $\mathbf{e}_z$ does not change in direction, so its time derivative is zero.

In view of these facts, the velocity and acceleration vectors are easily seen to be given by the following equations:

$$\mathbf{v} = \dot{R}\mathbf{e}_R + R\dot\phi\mathbf{e}_\phi + \dot{z}\mathbf{e}_z \tag{1.51}$$

$$\mathbf{a} = (\ddot{R} - R\dot\phi^2)\mathbf{e}_R + (2\dot{R}\dot\phi + R\ddot\phi)\mathbf{e}_\phi + \ddot{z}\mathbf{e}_z \tag{1.52}$$

These give the values of $\mathbf{v}$ and $\mathbf{a}$ in terms of their components in the *rotated* triad $\mathbf{e}_R\mathbf{e}_\phi\mathbf{e}_z$.

An alternative way of obtaining the derivatives of the unit vectors is to differentiate the following equations, which are the relationships between the fixed unit triad $\mathbf{ijk}$ and the rotated triad:

$$\mathbf{e}_R = \mathbf{i}\cos\phi + \mathbf{j}\sin\phi$$
$$\mathbf{e}_\phi = -\mathbf{i}\sin\phi + \mathbf{j}\cos\phi \tag{1.53}$$
$$\mathbf{e}_z = \mathbf{k}$$

The steps are left as an exercise. The result can also be found by use of the rotation matrix as given in Example 1.15.

Spherical Coordinates

When spherical coordinates r, θ, ϕ are employed to describe the position of a particle, the position vector is written as the product of the radial distance r and the unit radial vector $\mathbf{e}_r$, as with plane polar coordinates. Thus,

$$\mathbf{r} = r\mathbf{e}_r$$

The direction of $\mathbf{e}_r$ is now specified by the two angles ϕ and θ. We introduce two more unit vectors $\mathbf{e}_\phi$ and $\mathbf{e}_\theta$ as shown in Figure 1.23.

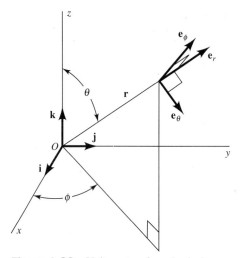

Figure 1.23   Unit vectors for spherical coordinates.

The velocity is

$$\mathbf{v} = \frac{d\mathbf{r}}{dt} = \dot{r}\mathbf{e}_r + r\frac{d\mathbf{e}_r}{dt} \tag{1.54}$$

Our next problem is how to express the derivative $d\mathbf{e}_r/dt$ in terms of the unit vectors in the rotated triad.

Referring to Figure 1.23, we can easily derive relationships between the $\mathbf{i}, \mathbf{j}, \mathbf{k}$ and $\mathbf{e}_r, \mathbf{e}_\theta, \mathbf{e}_\phi$ triads. For example,

$$\mathbf{e}_r = \mathbf{i}(\mathbf{e} \cdot \mathbf{i}) + \mathbf{j}(\mathbf{e} \cdot \mathbf{j}) + \mathbf{k}(\mathbf{e} \cdot \mathbf{k})$$

But, $\mathbf{e} \cdot \mathbf{j}$ is the projection of the unit vector $\mathbf{e}_r$ directly onto the unit vector $\mathbf{i}$. It is equal to $\cos \alpha$, the cosine of the angle between those two unit vectors. We need to express this dot product in terms of θ and ϕ, not α. We can obtain the desired relation by making two successive projections to get to the x-axis. First, project $\mathbf{e}_r$ onto the xy plane and then project from there onto the x-axis. The first projection gives us a factor of $\sin \theta$, while the second yields a factor of $\cos \phi$. The magnitude of the projection obtained this way is the desired dot product:

$$\mathbf{e}_r \cdot \mathbf{i} = \sin \theta \, \cos \phi$$

The remaining dot products can be evaluated in a similar way,

$$\mathbf{e}_r \cdot \mathbf{j} = \sin \theta \, \sin \phi \qquad \text{and} \qquad \mathbf{e}_r \cdot \mathbf{k} = \cos \theta$$

The relationships for $\mathbf{e}_\theta$ and $\mathbf{e}_\phi$ can be obtained as above yielding the desired relations

$$\mathbf{e}_r = \mathbf{i} \sin \theta \, \cos \phi + \mathbf{j} \sin \theta \, \sin \phi + \mathbf{k} \cos \theta$$

$$\mathbf{e}_\theta = \mathbf{i} \cos \theta \, \cos \phi + \mathbf{j} \cos \theta \, \sin \phi - \mathbf{k} \sin \theta \tag{1.55}$$

$$\mathbf{e}_\phi = -\mathbf{i} \sin \phi + \mathbf{j} \cos \phi$$

which express the unit vectors of the rotated triad in terms of the fixed trial $\mathbf{ijk}$. We note the similarity between this transformation and that of the second part of Example 1.15. The two are, in fact, identical if the correct identification of rotations is made. Let us differentiate the first equation with respect to time. The result is

$$\frac{d\mathbf{e}_r}{dt} = \mathbf{i}(\dot{\theta} \cos \theta \, \cos \phi - \dot{\phi} \sin \theta \, \sin \phi) + \mathbf{j}(\dot{\theta} \cos \theta \, \sin \phi$$

$$+ \dot{\phi} \sin \theta \, \cos \phi) - \mathbf{k}\dot{\theta} \sin \theta$$

Next, by using the expressions for $\mathbf{e}_\phi$ and $\mathbf{e}_\theta$ in Equation 1.55, we find that the above equation reduces to

$$\frac{d\mathbf{e}_r}{dt} = \dot{\phi}\mathbf{e}_\phi \sin \theta + \dot{\theta}\mathbf{e}_\theta \tag{1.56}$$

The other two derivatives are found by a similar procedure. The results are

$$\frac{d\mathbf{e}_\theta}{dt} = -\dot{\theta}\mathbf{e}_r + \dot{\phi}\mathbf{e}_\phi \cos \theta \tag{1.57}$$

$$\frac{d\mathbf{e}_\phi}{dt} = -\dot{\phi}\mathbf{e}_r \sin \theta - \dot{\phi}\mathbf{e}_\theta \cos \theta \tag{1.58}$$

The steps are left as an exercise. Returning now to the problem of finding **v,** we insert the expression for de_r/dt given by Equation 1.56 into Equation 1.54. The final result is

$$\mathbf{v} = \mathbf{e}_r\dot{r} + \mathbf{e}_\phi r\dot{\phi}\sin\theta + \mathbf{e}_\theta r\dot{\theta} \tag{1.59}$$

giving the velocity vector in terms of its components in the rotated triad.

To find the acceleration, we differentiate the above expression with respect to time. This gives

$$\mathbf{a} = \frac{d\mathbf{v}}{dt} = \mathbf{e}_r\ddot{r} + \dot{r}\frac{d\mathbf{e}_r}{dt} + \mathbf{e}_\phi\frac{d(r\dot{\phi}\sin\theta)}{dt} + r\dot{\phi}\sin\theta\frac{d\mathbf{e}_\phi}{dt} + \mathbf{e}_\theta\frac{d(r\dot{\theta})}{dt} + r\dot{\theta}\frac{d\mathbf{e}_\theta}{dt}$$

Upon using the previous formulas for the derivatives of the unit vectors, it is readily found that the above expression for the acceleration reduces to

$$\mathbf{a} = (\ddot{r} - r\dot{\phi}^2\sin^2\theta - r\dot{\theta}^2)\mathbf{e}_r + (r\ddot{\theta} + 2\dot{r}\dot{\theta} - r\dot{\phi}^2\sin\theta\cos\theta)\mathbf{e}_\theta$$
$$+ (r\ddot{\phi}\sin\theta + 2\dot{r}\dot{\phi}\sin\theta + 2r\dot{\theta}\dot{\phi}\cos\theta)\mathbf{e}_\phi \tag{1.60}$$

giving the acceleration vector in terms of its components in the triad $\mathbf{e}_r\mathbf{e}_\theta\mathbf{e}_\phi$.

EXAMPLE 1.22

A bead slides on a wire bent into the form of a helix, the motion of the bead being given in cylindrical coordinates by $R = b$, $\phi = \omega t$, $z = ct$. Find the velocity and acceleration vectors as functions of time.

Solution:
Differentiating, we find $\dot{R} = \ddot{R} = 0$, $\dot{\phi} = \omega$, $\ddot{\phi} = 0$, $\dot{z} = c$, $\ddot{z} = 0$. So, from Equations 1.51 and 1.52 we have

$$\mathbf{v} = b\omega\mathbf{e}_\phi + c\mathbf{e}_z$$

$$\mathbf{a} = -b\omega^2\mathbf{e}_R$$

Thus, in this case, both velocity and acceleration are constant in magnitude, but they vary in direction because both $\mathbf{e}_\phi$ and $\mathbf{e}_r$ change with time as the bead moves. ∎

EXAMPLE 1.23

A wheel of radius b is placed in a gimbal mount and is made to rotate as follows: The wheel spins with constant angular speed ω_1 about its own axis which, in turn, rotates with constant angular speed ω_2 about a vertical axis in such a way that the axis of the wheel stays in a horizontal plane and the center of the wheel is motionless. Use spherical coordinates to find the acceleration of any point on the rim of the wheel. In particular, find the acceleration of the highest point on the wheel. We can use the fact that spherical coordinates can be chosen such that $r = b$, $\theta = \omega_1 t$,

Figure 1.24   A rotating
wheel on a rotating
mount.

and $\phi = \omega_2 t$, Figure 1.24. Then we have $\dot{r} = \ddot{r} = 0$, $\dot{\theta} = \omega_1$, $\ddot{\theta} = 0$, $\dot{\phi} = \omega_2$, $\ddot{\phi} = 0$. Equation 1.60 gives directly

$$\mathbf{a} = (-b\omega_2^2 \sin^2 \theta - b\omega_1^2)\mathbf{e}_r - b\omega_2^2 \sin \theta \cos \theta \, \mathbf{e}_\theta + 2b\omega_1\omega_2 \cos \theta \, \mathbf{e}_\phi$$

The point at the top has coordinate $\theta = 0$, so at that point

$$\mathbf{a} = -b\omega_1^2 \mathbf{e}_r + 2b\omega_1\omega_2 \mathbf{e}_\phi$$

The first term on the right is the centripetal acceleration, and the last term is a transverse acceleration normal to the plane of the wheel.   ■

PROBLEMS

1.1 Given the two vectors $\mathbf{A} = \mathbf{i} + \mathbf{j}$ and $\mathbf{B} = \mathbf{j} + \mathbf{k}$ find the following:
(a) $\mathbf{A} + \mathbf{B}$ and $|\mathbf{A} + \mathbf{B}|$
(b) $3\mathbf{A} - 2\mathbf{B}$
(c) $\mathbf{A} \cdot \mathbf{B}$
(d) $\mathbf{A} \times \mathbf{B}$ and $|\mathbf{A} \times \mathbf{B}|$

1.2 Given the three vectors $\mathbf{A} = 2\mathbf{i} + \mathbf{j}$, $\mathbf{B} = \mathbf{i} + \mathbf{k}$, and $\mathbf{C} = 4\mathbf{j}$, find the following:
(a) $\mathbf{A} \cdot (\mathbf{B} + \mathbf{C})$ and $(\mathbf{A} + \mathbf{B}) \cdot \mathbf{C}$
(b) $\mathbf{A} \cdot (\mathbf{B} \times \mathbf{C})$ and $(\mathbf{A} \times \mathbf{B}) \cdot \mathbf{C}$
(c) $\mathbf{A} \times (\mathbf{B} \times \mathbf{C})$ and $(\mathbf{A} \times \mathbf{B}) \times \mathbf{C}$

1.3 Find the angle between the vectors $\mathbf{A} = a\mathbf{i} + 2a\mathbf{j}$ and $\mathbf{B} = a\mathbf{i} + 2a\mathbf{j} + 3a\mathbf{k}$. (*Note:* These two vectors define a face diagonal and a body diagonal of a rectangular block of sides a, $2a$, and $3a$.)

1.4 Given the time-varying vector

$$\mathbf{A} = \mathbf{i}\alpha t + \mathbf{j}\beta t^2 + \mathbf{k}\gamma t^3$$

where α, β, and γ are constants, find the first and second time derivatives $d\mathbf{A}/dt$ and $d^2\mathbf{A}/dt^2$.

1.5 For what value (or values) of q is the vector $\mathbf{A} = \mathbf{i}q + 3\mathbf{j} + \mathbf{k}$ perpendicular to the vector $\mathbf{B} = \mathbf{i}q - q\mathbf{j} + 2\mathbf{k}$?

1.6 Give an algebraic and a geometric proof of the following relations:

$$|\mathbf{A} + \mathbf{B}| \leq |\mathbf{A}| + |\mathbf{B}|$$
$$|\mathbf{A} \cdot \mathbf{B}| \leq |\mathbf{A}| \, |\mathbf{B}|$$

1.7 Prove the vector identity $\mathbf{A} \times (\mathbf{B} \times \mathbf{C}) = \mathbf{B}(\mathbf{A} \cdot \mathbf{C}) - \mathbf{C}(\mathbf{A} \cdot \mathbf{B})$.

1.8 Two vectors $\mathbf{A}$ and $\mathbf{B}$ represent concurrent sides of a parallelogram. Show that the area of the parallelogram is equal to $|\mathbf{A} \times \mathbf{B}|$.

1.9 Three vectors $\mathbf{A}$, $\mathbf{B}$, and $\mathbf{C}$ represent three concurrent edges of a parallelepiped. Show that the volume of the parallelepiped is equal to $|\mathbf{A} \cdot (\mathbf{B} \times \mathbf{C})|$.

1.10 Verify the transformation matrix for a rotation about the z-axis through an angle ϕ followed by a rotation abou the y'-axis through an angle θ, as given in Example 1.15.

1.11 Express the vector $2\mathbf{i} + 3\mathbf{j} - \mathbf{k}$ in the primed triad $\mathbf{i'j'k'}$ in which the $x'y'$-axes are rotated about the z axis (which coincides with the z'-axis) through an angle of 30°.

1.12 A racing car moves on a circle of constant radius b. If the speed of the car varies with time t according to the equation $v = ct$ where c is a positive constant, show that the angle between the velocity vector and the acceleration vector is 45° at time $t = \sqrt{b/c}$. (*Hint:* At this time the tangential and normal components of the acceleration are equal in magnitude.)

1.13 A small ball is fastened to a long rubber band and twirled around in such a way that the ball moves in an elliptical path given by the equation

$$\mathbf{r}(t) = \mathbf{i}b \cos \omega t + \mathbf{j}2b \sin \omega t$$

where b and ω are constants. Find the speed of the ball as a function of t. In particular, find v at $t = 0$ and at $t = \pi/2\omega$, at which times the ball is, respectively, at its minimum and maximum distance from the origin.

1.14 A buzzing fly moves in a helical path given by the equation

$$\mathbf{r}(t) = \mathbf{i}b \sin \omega t + \mathbf{j}b \cos \omega t + \mathbf{k}ct^2$$

Show that the magnitude of the acceleration of the fly is constant, provided b, ω, and c are constant.

1.15 A bee goes out from its hive in a spiral path given in plane polar coordinates by

$$r = be^{kt} \qquad \theta = ct$$

where b, k, and c are positive constants. Show that the angle between the velocity vector and the acceleration vector remains constant as the bee moves outward. (*Hint:* Find $\mathbf{v} \cdot \mathbf{a}/va$.)

1.16 Work Problem 1.14 using cylindrical coordinates where $R = b$, $\phi = \omega t$, and $z = ct^2$.

1.17 An ant crawls on the surface of a ball of radius b in a manner such that the ant's motion is given in spherical coordinates by the equations

$$r = b \qquad \phi = \omega t \qquad \theta = \frac{\pi}{2}\left[1 + \frac{1}{4}\cos(4\omega t)\right]$$

Find the speed of the ant as a function of the time t. What sort of path is represented by the above equations?

1.18 Prove that $\mathbf{v} \cdot \mathbf{a} = v\dot{v}$ and, hence, that for a moving particle $\mathbf{v}$ and $\mathbf{a}$ are perpendicular to each other if the speed v is constant. (*Hint:* Differentiate both sides of the equation $\mathbf{v} \cdot \mathbf{v} = v^2$ with respect to t. Remember that $\dot{v}$ is not the same as $|\mathbf{a}|$.)

1.19 Prove that

$$\frac{d}{dt}[\mathbf{r} \cdot (\mathbf{v} \times \mathbf{a})] = \mathbf{r} \cdot (\mathbf{v} \times \dot{\mathbf{a}})$$

1.20 Show that the tangential component of the acceleration is given by the expression.

$$a_\tau = \frac{\mathbf{v} \cdot \mathbf{a}}{v}$$

and the normal component is therefore

$$a_n = (a^2 - a_\tau^2)^{1/2} = \left[a^2 - \frac{(\mathbf{v} \cdot \mathbf{a})^2}{v^2}\right]^{1/2}$$

1.21 Use the above result to find the tangential and normal components of the acceleration as functions of time in Problems 1.14 and 1.15.

1.22 Prove that $|\mathbf{v} \times \mathbf{a}| = v^3/\rho$ where ρ is the radius of curvature of the path of a moving particle.

1.23 A wheel of radius b rolls along the ground with constant forward acceleration a_0. Show that, at any given instant, the magnitude of the acceleration of any point on the wheel is $(a_0^2 + v^4/b^2)^{1/2}$ relative to the center of the wheel and is also $a_0[2 + 2\cos\theta + v^4/a_0^2 b^2 - (2v^2/a_0 b)\sin\theta]^{1/2}$ relative to the ground. Here v is the instantaneous forward speed, and θ defines the location of the point on the wheel, measured forward from the highest point. Which point has the greatest acceleration relative to the ground?

COMPUTER/CALCULATOR APPLICATIONS

1.1 A particle is moving in uniform circular motion in the xy plane with an angular speed of 0.377 rad/s along the inside of a long, perfectly smooth cylinder whose axis is along the z direction. Suddenly, the particle is subjected to an acceleration given by $\ddot{z} = \ddot{z}_0 e^{-z}$ where $\ddot{z}_0 = 3$ m/s^2 and z is given in meters. Assuming $\dot{z}_0 = 0$, how far down the cylinder does the particle travel after completing one more revolution around the cylinder?

2

NEWTONIAN MECHANICS. RECTILINEAR MOTION OF A PARTICLE

"Salviati: *But if this is true, and if a large stone moves with a speed of, say, eight while a smaller one moves with a speed of four, then when they are united, the system will move with a speed less than eight; but the two stones when tied together make a stone larger than that which before moved with a speed of eight. Hence the heavier body moves with less speed than the lighter; an effect which is contrary to your supposition. Thus you see how, from your supposition that the heavier body moves more rapidly than the lighter one, I infer that the heavier body moves more slowly.*"

(Galileo—Dialogues Concerning Two New Sciences)

2.1 NEWTON'S LAW OF MOTION. HISTORICAL INTRODUCTION

In his *Principia* of 1687, Newton laid down three fundamental laws of motion, which would forever change mankind's perception of the world:

Galileo before The Tribunal—North Wind Picture Archives

I. *Every body continues in its state of rest, or of uniform motion in a straight line, unless it is compelled to change that state by forces impressed upon it.*
II. *The change of motion is proportional to the motive force impressed; and is made in the direction of the line in which that force is impressed.*
III. *To every action there is always imposed an equal reaction; or, the mutual actions of two bodies upon each other are always equal, and directed to contrary parts.*

These three laws of motion are now known collectively as Newton's laws of motion or more simply as Newton's laws. It is arguable whether or not these are indeed all his laws. However, no one before him stated them quite so precisely and certainly no one before him had such a clear understanding of the overall implication and power of these laws. The description of nature they portray seems to fly in the face of common experience. As any beginning student of physics soon discovers, they become "reasonable" only with the expenditure of great effort in attempting to understand thoroughly the apparent vagaries of physical systems.

Aristotle (384–322 B.C.) had frozen man's notion of the way the world works for almost 20 centuries by invoking powerfully logical arguments that led to a physics in which all moving, earthbound objects ultimately acquired a state of rest unless acted upon by some motive force. In his view a force was required to keep earthly things moving, even at constant speed, a law in distinct contradiction with Newton's first and second laws. On the other hand, heavenly bodies dwelt in a more perfect realm where perpetual circular motion was the norm and no forces were required to keep this celestial clockwork ticking.

Modern scientists heap untold amounts of scorn upon Aristotle for burdening us with such obviously flawed doctrine. He is particularly criticized for his failure to carry out even the most modest experiment that would have shown him the error of his ways. At that time, though, it was a commonly held belief that experiment was not a suitable enterprise for any self-respecting philosopher, and thus Aristotle, raised according to those mores, failed to acquire a true picture of nature. Such a viewpoint is a bit misleading. Even though he did no experiments in natural philosophy, Aristotle was a keen observer of nature, one of the first. If he was guilty of anything, it was not so much a failure to observe nature, but mostly a failure to follow through with a process of abstraction based upon observation. Indeed, bodies falling through air accelerate initially, but ultimately they attain a nearly constant velocity of fall. Heavy objects, in general, fall faster than lighter ones. It takes a sizable force to haul a ship through water, and the greater the force, the greater the ship's speed. A spear thrown vertically upward from a moving chariot will land behind the charioteer, not on top of him. And the motion of heavenly bodies does go on and on, apparently following a curved path forever without any visible motive means. Of course, nowadays we can understand these things if we pay close attention to all the variables that affect the motion of objects and then apply Newton's laws correctly.

That Aristotle failed to extract Newton's laws from such observations of the real world is a consequence only of the fact that he observed the world and interpreted its workings in a rather superficial way. He was basically unaware of the then subtle effects of air resistance, friction, and the like. It was only with the advent of our ability and motivation to carry out precise experiments followed by a process of abstraction that led us to the revolutionary point of view of nature represented by the Newtonian paradigm. Even today, the workings of that paradigm are most easily visualized in the artificial realm of our own minds, emptied of the real world's imperfections of friction and air resistance (look at any elementary physics book and see how often one encounters the phrase, "neglecting friction"). Aristotle's physics, much more than that of Newton, reflects the workings of a nature quite coincident with the common misconception of modern people in general (including the typical college student who chooses a curriculum curiously devoid of courses in physics).

There is no question that the first law, the so-called law of inertia, had already been set forth prior to the time of Newton. This law, commonly attributed to Galileo (1564–1642), was actually first formulated by René Descartes (1596–1650). According to Descartes, "inertia" made bodies persist in motion forever, not in perfect Aristotelian circles, but in a straight line. Descartes came to this conclusion, in fact, not by experiment but by pure thought. In contrast to belief in traditional authority (which at that time meant belief in the teachings of Aristotle), Descartes believed that only one's own thinking could be trusted. It was his intent to "explain effects by their causes, and not causes by their effects." For Descartes pure reasoning served as the sole basis of certainty. Such a paradigm would aid the transition from an Aristotelian worldview to that of a Newtonian one, but it contained within itself the seeds of its own destruction.

It was not too surprising that Descartes failed to grasp the implication of his law of inertia regarding planetary motion. Planets certainly did not move in straight lines. Descartes, more ruthless in his methods of thought than any of his predecessors, reasoned that some physical thing had to "drive" the planets along in their curved paths. Descartes rebelled in horror at the notion that the required physical force was some invisible entity reaching out across the void to grab the planets and hold them in their orbits. Moreover, having no knowledge of the second law, Descartes never realized that the required force was not a "driving" force but a force that had to be directed "inward" toward the sun. He, along with many others of that era, was certain that the planets had to be pushed along in their paths around the sun (or earth). Thus, he concocted the notion of an all-pervading, ether-like fluid made of untold numbers of unseen particles, rotating in vortices, within which the planets were driven round and round—an erroneous conclusion that arose from the fancies of a mind engaged only in pure thought, minimally constrained by experimental or observational data.

Galileo, on the other hand, mainly by clear argument based on actual experimental results, had gradually commandeered a fairly clear understanding of what would come to be the first of Newton's laws as well as the second. A necessary prelude to the final synthesis of a correct system of mechanics was his observation that a pendulum undergoing small oscillations was isochronous; that is, its period of oscillation was independent of its amplitude. This discovery led to the first clocks capable of making accurate measurements of small time intervals, a capability that Aristotle did not have. Galileo would soon exploit this capability in carrying out experiments of unprecedented precision with objects either freely falling or sliding down inclined planes. Generalizing brilliantly from the results of his experiments, Galileo came very close to formulating Newton's first two laws.

For example, concerning the first law, Galileo noted, as did Aristotle, that an object sliding along a level surface did, indeed, come to rest. But here Galileo made a wonderful mental leap that took him far past the dialectic of Aristotle. He imagined a second surface, more slippery than the first. An object given a push along the second surface would travel farther before stopping than it would if given a similar push along the first surface. Carrying this process of abstraction to its ultimate conclusion, Galileo reasoned that an object given a push along a surface of "infinite slipperiness" (i.e., "neglecting friction") would, in fact, go on forever, never coming to rest. Thus, contrary to Aristotle's physics, he reasoned that a force is not required merely to keep an object in motion. In fact, some force must be applied to stop it. This is very close to Newton's law of inertia but, astonishingly enough, Galileo did not argue that motion, in the absence of forces, would continue forever in a straight line!

The world, according to Galileo, was not an impersonal one ruled by mechanical laws. Instead, it was a cosmos that marched to the tune of an infinitely intelligent craftsman. Following the Aristotelian tradition, Galileo, too, saw a world ordered according to the perfect figure, the circle. Rectilinear motion implied disorder. Objects that found themselves in such a state of affairs would fly on in a straight line, not forever, but ultimately lapsing into their more natural state of perfect circular motion. The experiments necessary to discriminate between straight line motion forever and straight line motion ultimately evolving to pure circular motion could obviously not be performed in practice, but only within the confines of one's own mind, and only if that mind had been properly freed from the conditioning of centuries of ill-founded dogma. Galileo, brilliant though he was, but still doing battle with the ghosts of the past, had not yet reached that required state of mind.

Galileo's experiments with falling bodies led him to the brink of Newton's second law. Again, as Aristotle had known, Galileo saw that heavy objects, like stones, did fall faster than lighter ones, like feathers. However, by carefully timing similarly shaped objects, albeit of different weights, Galileo discovered that such objects accelerated as they fell and all more or less reached the ground at the same time! Indeed, very heavy objects, even though themselves differing greatly in weight, fell at almost identical rates, with a speed that increased about 10 m/s each second. (Incidentally, the famous experiment of dropping cannon balls from the Leaning Tower of Pisa was not carried out by Galileo but by one of his chief Aristotelian antagonists at Pisa, Giorgio Coressio, and not in hopes of refutation but in hopes of confirmation of the Aristotelian view that larger bodies must fall more quickly than small ones!) It was again, however, through a process of brilliant abstraction that Galileo realized that if the effects of air resistance could be eliminated, all objects would fall with the same acceleration, regardless of their weight or shape. Thus, even more of Aristotle's edifice was torn apart; a heavier weight does not fall faster than a light one, and a force causes objects to accelerate, not to move at constant speed.

Galileo's notions of mechanics on Earth were more closely on target with Newton's laws than were the conjectures of any of his predecessors. He sometimes applied them brilliantly in defense of the Copernican viewpoint, that is, a heliocentric model of the solar system. In particular, even though his notion of the law of inertia was somewhat flawed, he applied it correctly in arguing that terrestrial-based experiments could not be used to demonstrate that the earth could not be in motion around the sun. He pointed out that a stone dropped from the mast of a moving ship would not "be left behind" since the stone shares the ship's horizontal speed. By analogy, in contrast to Aristotelian argument, a stone dropped from a tall tower would not be left behind by an earth in motion. This powerful argument obviously implied that no such observation could be used to demonstrate whether the earth was rotating or not. The argument contained within itself the very seeds of relativity theory.

Unfortunately, as mentioned above, Galileo could not entirely break loose from the Aristotelian dogma of circular motion. In strict contradiction to the law of inertia, he postulated that a body left to itself will continue to move forever, not in a straight line, but in a circular orbit throughout eternity. His reasoning was as follows:

> . . . *straight motion being by nature infinite (because a straight line is infinite and indeterminate), it is impossible that anything should have by nature the principle of moving in a straight line; or, in other words, towards a place where it is impossible to arrive, there being no finite end. For that which cannot be done, nor endeavors to move whither it is impossible to arrive.*

This statement also contradicted his intimate knowledge of centrifugal forces, that is, the tendency of an object moving in a circle to fly off on a tangent in a straight line. He knew that earthbound objects could travel in circles only if this centrifugal force was either balanced or overwhelmed by some other offsetting force. Indeed, one of the Aristotelian arguments against a rotating earth was that objects on the earth's surface would be flung off it. Galileo argued that this conclusion was not valid because the earth's "gravity" overwhelmed this centrifugal tendency! Yet somehow, he failed to make the mental leap that some similar effect must keep the planets in circular orbit about the sun!

So ultimately it was to be Newton who would pull together all the fragmentary knowledge that had been accumulated about the motion of earthbound objects into the brilliant synthesis of the three laws and then demonstrate that the motion of heavenly objects obeyed those laws as well.

Newton's laws of motion can be thought of as a prescription for calculating or predicting the subsequent motion of a particle (or system of particles) given a knowledge of its position and velocity at some instant time. These laws, in and of themselves, say nothing about the reason why a given physical system behaves the way it does. Newton, himself, was quite explicit about that shortcoming. He refused to speculate (at least in print) why objects move the way they do. Whatever "mechanism" lay behind the workings of physical systems remained forever hidden from Newton's eyes. He simply stated that, for whatever reason, this is the way things work as demonstrated by the power of his calculational prescription to predict, with astonishing accuracy, the evolution of physical systems set in motion. Much has been learned since the time of Newton. But a basic fact of physical law persists. The laws of motion are simply prescriptions. They tell us how things work—not why.

Newton's First Law. Inertial Reference Systems

The first law describes a common property of matter, namely *inertia.* Loosely speaking, inertia is the name given to the characteristic that all matter resists having its motion changed. If a particle is at rest, it resists being moved; that is, it takes a force to do so. If the particle is in motion, it resists being brought to rest. Again, it takes a force to do so. It almost seems as though matter has been endowed with an innate abhorrence of acceleration. Be that as it may, for whatever reason, it takes a force to accelerate matter; in the absence of applied forces, matter simply persists in its current velocity state—forever.

A mathematical description of the motion of a particle requires the selection of a *frame of reference,* or a set of coordinates in configuration space that can be used to specify the position, velocity, and acceleration of the particle at any instant of time. A frame of reference in which Newton's first law of motion is valid is called an *inertial frame of reference.* This law rules out accelerated frames of reference as inertial since an object "really" at rest or moving at constant velocity would appear to be accelerated as seen from an accelerated frame of reference. Moreover, an object seen to be at rest in such a frame would be seen to be accelerated with respect to the inertial frame. So strong is our belief in the concept of inertia and the validity of Newton's laws of motion that we would be forced to invent "fictitious" forces in order to account for the apparent lack of acceleration of an object at rest in an accelerated frame of reference.

A simple example of a noninertial frame of reference should help clarify the situation. Consider an observer inside a railroad boxcar accelerating down the track with an acceleration **a.** Suppose a plumb bob was suspended from the ceiling of the boxcar. How would it appear to the observer? Take a look at Figure 2.1. (We will analyze this situation in complete detail in Chapter 5.) The point here is that the observer in the boxcar is in a noninertial frame of reference, who is at rest with respect to the boxcar and sees the plumb bob, also apparently at rest, hanging at an angle θ with respect to the vertical. He

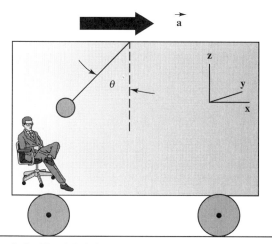

Figure 2.1   Plumb bob hangs at an angle θ in an accelerating frame of reference.

knows that, in the absence of any forces other than gravity and tension in the plumb line, such a device should align itself vertically. It does not, and he concludes that some unknown force must be pushing or pulling the plumb bob toward the back of the car. (Indeed, he too will experience such a force as anyone who has ever been in an accelerating vehicle knows from first-hand experience.)

The question naturally arises as to how it is possible to determine whether or not a given frame of reference constitutes an inertial frame. The answer is nontrivial! (For example, if the boxcar was sealed off from the outside world, how would the observer know that the apparent force causing the plumb bob to hang off-vertical was not due to the fact that the whole boxcar was "misaligned" with the direction of gravity; that is, the force due to gravity was actually in the direction indicated by the angle θ?) Observers would have to know that *all* external forces on a body had been eliminated before checking to see whether or not objects in their frame of reference obeyed Newton's first law. It would be necessary to isolate a body completely in order to eliminate all forces acting upon it. This is impossible, since there are always some gravitational forces acting unless the body were removed to an infinite distance from all other matter.

Is there a perfect inertial frame of reference? For most practical purposes, a coordinate system attached to the Earth's surface is approximately inertial. For example, a billiard ball seems to move in a straight line with constant speed as long as it does not collide with other balls or hit the cushion. If its motion were measured with very high precision, however, we would see that its path is slightly curved. This is due to the fact that the earth is rotating and its surface is therefore accelerating toward its axis. Hence, a coordinate system attached to the earth's surface is not inertial. A better system would be one that uses the center of the earth as coordinate origin with the sun and a star as reference points. But even this system would not be inertial because of the earth's orbital motion around the sun.

Suppose then we pick a system whose origin was the center of the sun with two distant stars as reference points. Strictly speaking, this would not be an inertial frame either since the sun and the two distant stars partake of the general rotational motion of the Milky Way galaxy. Again, we are stuck with a rotating frame of reference. We can continue our search for the perfect inertial frame in this fashion, but at each step it is not clear that we do not continue to encounter difficulties of this sort. For example, the Milky Way is part of a local group of galaxies whose other major member includes the Andromeda galaxy, and they are all rotating about their common center of gravity. This local group is part of a larger cluster of galaxies and groups of galaxies called the Virgo cluster whose center lies 20 Mpc away (about 60 million light years). This large cluster also undergoes rotational motion! Finally, we might choose a frame of reference based on the average background of all the matter in the universe. It is generally agreed that the choice of such a frame, in the sense of Newtonian mechanics, is about as good as we can do.

Mass and Force. Newton's Second and Third Laws

The quantitative measure of inertia is called *mass*. We are all familiar with the notion that the more massive an object is, the more resistive it is to acceleration. Go push a bike to get it rolling and then try the same thing with a car. Compare the effort. The car is much more massive and requires a much larger force to accelerate it than does the bike. A more quantitative definition may be constructed by considering two masses m_1 and m_2 attached by a spring and initially at rest in an inertial frame of reference. For example, we could imagine the two masses to be on a frictionless surface, almost achieved in practice by two carts on an air track, commonly seen in elementary physics class demonstrations. Now imagine someone pushing the two masses together, compressing the spring, and then suddenly releasing them such that they fly apart, attaining speeds v_1 and v_2. We *define* the ratio of the two masses to be

$$\frac{m_1}{m_2} = \left|\frac{\mathbf{v_2}}{\mathbf{v_1}}\right| \tag{2.1}$$

If we let m_2 be the standard of mass, then all other masses can be operationally defined in the above way relative to the standard. This operational definition of mass is consistent with Newton's second and third laws of motion as we shall soon see. Equation 2.1 is equivalent to

$$\Delta(m_1\mathbf{v_1}) = -\Delta(m_2\mathbf{v_2}) \tag{2.2}$$

since the initial velocities of each mass are zero and the final velocities $\mathbf{v_1}$ and $\mathbf{v_2}$ are in opposite directions. If we divide by Δt and take limits as $\Delta t \to 0$, we obtain

$$\frac{d}{dt}(m_1\mathbf{v_1}) = -\frac{d}{dt}(m_2\mathbf{v_2}) \tag{2.3}$$

The product of mass and velocity, $m\mathbf{v}$, is called *linear momentum*. The "change of motion" stated in the second law of motion was rigorously defined by Newton to be the time rate of change of the linear momentum of an object, and so the second law can be rephrased as *the time rate of change of an object's linear momentum is proportional to the impressed force,* **F.** Thus, the second law can be written as

$$\mathbf{F} = k\frac{d(m\mathbf{v})}{dt} \tag{2.4}$$

where k is a constant of proportionality. Considering the mass to be a constant, independent of velocity (which is not true of objects moving at "relativistic" speeds or speeds approaching the speed of light, 3×10^8 m/s, a situation that we will not consider in this text), we can write

$$\mathbf{F} = km\frac{d\mathbf{v}}{dt} = km\mathbf{a} \tag{2.5}$$

where **a** is the resultant acceleration of a mass m subjected to a force **F**. The constant of proportionality can be taken to be $k = 1$ by defining the unit of force in the SI system to be that which causes a 1 kg mass to be accelerated 1 m/s^2. This force unit is called 1 newton.

Thus, we finally express Newton's second law in the familiar form

$$\mathbf{F} = \frac{d(m\mathbf{v})}{dt} = m\mathbf{a} \tag{2.6}$$

The force $\mathbf{F}$ on the left side of the Equation 2.6 is the *net* force acting upon the mass m; that is, it is the vector sum of all of the individual forces acting upon m.

We note that Equation 2.3 is equivalent to

$$\mathbf{F}_1 = -\mathbf{F}_2$$

or Newton's third law, namely two interacting bodies exert equal and opposite forces upon one another. Thus, our prescription for defining mass is consistent with both Newton's second and third laws.

Linear Momentum

Linear momentum proves to be such a useful notion that it is given its own symbol

$$\mathbf{p} = m\mathbf{v} \tag{2.7}$$

Newton's second law may be written as

$$\mathbf{F} = \frac{d\mathbf{p}}{dt}$$

Thus, Equation 2.3, which describes the behavior of two mutually interacting masses, is equivalent to

$$\frac{d}{dt}(\mathbf{p}_1 + \mathbf{p}_2) = 0$$

or

$$\mathbf{p}_1 + \mathbf{p}_2 = \text{constant}$$

In other words, Newton's third law implies that the total momentum of two mutually interacting bodies is a constant. This constancy is a special case of the more general situation in which the total linear momentum of an isolated system (a system subject to no net externally applied forces) is a conserved quantity. The law of linear momentum conservation is one of the most fundamental laws of physics and is valid even in situations in which Newtonian mechanics fails.

Motion of a Particle

Equation 2.6 is the fundamental equation of motion for a particle subject to the influence of a *net* force, $\mathbf{F}$. We emphasize this point by writing $\mathbf{F}$ as $\mathbf{F}_{net}$, the vector sum of all the forces acting on the particle.

$$\mathbf{F}_{net} = \Sigma\mathbf{F}_i = m\frac{d^2\mathbf{r}}{d\mathbf{r}^2} = m\mathbf{a} \tag{2.8}$$

The usual problem of dynamics can be expressed in the following way: Given a knowledge of the forces acting on a particle (or system of particles), calculate the acceleration of the particle. Knowing the acceleration, calculate the velocity and position as a function of time. This process involves solving the second-order differential equation of motion represented by Equation 2.8. A complete solution will also require a knowledge of the "boundary" conditions of the problem, such as the initial conditions of the particle (the position and velocity at time $t = 0$). The initial conditions plus the dynamics dictated by the differential equation of motion of Newton's second law completely determine the subsequent motion of the particle. In some cases this procedure cannot be carried to completion in an analytic way. The solution of a complex problem will, in general, have to be carried out using numerical approximation techniques on a digital computer. Here we will mostly consider only those problems amenable to solution by fairly simple analytical techniques.

2.2 RECTILINEAR MOTION. UNIFORM ACCELERATION UNDER A CONSTANT FORCE

When a moving particle remains on a single straight line, the motion is said to be *rectilinear*. In this case, without loss of generality we can choose the *x*-axis as the line of motion. The general equation of motion is then

$$F_x(x, \dot{x}, t) = m\ddot{x}$$

Note: In the rest of this chapter, we shall usually use the single variable x to represent the position of a particle. To avoid excessive and unnecessary use of subscripts, we shall often use the symbols v and a for $\dot{x}$ and $\ddot{x}$, respectively, rather than v_x and a_x, and F rather than F_x.

The simplest situation is that in which the force is constant. In this case we have constant acceleration

$$\ddot{x} = \frac{dv}{dt} = \frac{F}{m} = \text{constant} = a \tag{2.9}$$

and the solution is readily obtained by direct integration with respect to time:

$$\dot{x} = v = at + v_0 \tag{2.9a}$$

$$x = \frac{1}{2}at^2 + v_0 t + x_0 \tag{2.9b}$$

where v_0 is the initial velocity and x_0 is the initial position, that is, the position at $t = 0$. By eliminating the time t between Equations 2.9a and 2.9b, we obtain

$$2a(x - x_0) = v^2 - v_0^2 \tag{2.9c}$$

The student will recall the above familiar equations of uniformly accelerated motion. There are a number of fundamental applications. For example, in the case of a body falling freely near the surface of the earth, neglecting air resistance, the acceleration is very nearly constant. We denote the acceleration of a freely falling body by **g**. (By measurement, $g = 9.8$ m/s^2 = 32 ft/sec^2.) The downward force of gravity (the *weight*) is, accordingly, equal to m**g**. The gravitational force is always present, regardless of the motion of the body, and is independent of any other forces that may be acting.[1] We shall henceforth call it *mg.*

EXAMPLE 2.1

Consider a particle that is sliding down a smooth plane inclined at an angle θ to the horizontal, as shown in Figure 2.2a.

Solution:
We choose the positive direction of the x-axis to be down the plane, as indicated. The component of the gravitational force in the x direction is equal to $mg \sin \theta$. This is a constant; hence, the motion is given by Equations 2.9a, 2.9b, and 2.9c where

$$\ddot{x} = \frac{F_x}{m} = g \sin \theta$$

Suppose that, instead of being smooth, the plane is rough; that is, it exerts a frictional force **f** on the particle. Then the net force in the x direction, as shown in Figure 2.2(b), is equal to $mg \sin \theta - f$. Now for sliding contact it is found that the magnitude of the frictional force is proportional to the magnitude of the normal force N, that is,

$$f = \mu_\kappa N$$

where the constant of proportionality μ is known as the *coefficient of sliding* or *kinetic friction.*[2] In the example under discussion the normal force, as shown in the figure, is equal to $mg \cos \theta$; hence,

$$f = \mu_\kappa mg \cos \theta$$

---

[1] Effects of the earth's rotation will be studied in Chapter 5.

[2] There is another coefficient of friction called the *static* coefficient μ_s, which, when multiplied by the normal force, gives the maximum frictional force under static contact, that is, the force required to barely start an object to move when it is initially at rest. In general $\mu_s > \mu_\kappa$.

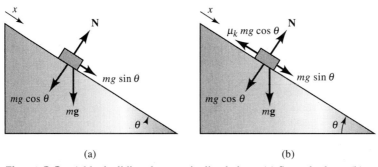

| (a) | (b) |

Figure 2.2   A block sliding down an inclined plane. (a) Smooth plane. (b) Rough plane.

Consequently, the net force in the x direction is equal to

$$mg \sin \theta - \mu_\kappa mg \cos \theta$$

Again the force is constant, and Equations 2.9a, 2.9b, and 2.9c apply where

$$\ddot{x} = \frac{F_x}{m} = g(\sin \theta - \mu_\kappa \cos \theta)$$

The speed of the particle will increase if the expression in parentheses is positive, that is, if $\theta > \tan^{-1} \mu_\kappa$. The angle, $\tan^{-1} \mu_\kappa$, usually denoted by ϵ, is called the *angle of kinetic friction*. If $\theta = \epsilon$, then $a = 0$, and the particle slides down the plane with constant speed. If $\theta < \epsilon$, a is negative, and so the particle will eventually come to rest. It should be noted that for motion *up* the plane the direction of the frictional force is reversed; that is, it is in the positive x direction. The acceleration (actually deceleration) is then $\ddot{x} = g(\sin \theta + \mu_\kappa \cos \theta)$. ∎

2.3 FORCES THAT DEPEND ON POSITION. THE CONCEPTS OF KINETIC AND POTENTIAL ENERGY

It is often true that the force a particle experiences depends on the particle's position with respect to other bodies. This is the case, for example, with electrostatic and gravitational forces. It also applies to forces of elastic tension or compression. If the force is independent of velocity or time, then the differential equation for rectilinear motion is simply

$$F(x) = m\ddot{x} \tag{2.10}$$

It is usually possible to solve this type of differential equation by one of several methods. One useful and significant method of solution is to use the chain rule and write the acceleration in the following way:

$$\ddot{x} = \frac{d\dot{x}}{dt} = \frac{dx}{dt}\frac{d\dot{x}}{dx} = v\frac{dv}{dx}$$

so the differential equation of motion may be written

$$F(x) = mv\frac{dv}{dx} = \frac{m}{2}\frac{d(v^2)}{dx} = \frac{dT}{dx} \tag{2.11}$$

The quantity $T = \frac{1}{2}mv^2$ is called the *kinetic energy* of the particle. We can now express Equation 2.11 in integral form

$$\int_{x_0}^{x} F(x)dx = T - T_0 \tag{2.11a}$$

Now the integral $\int F(x)dx$ is the *work* done on the particle by the impressed force $F(x)$. Thus, *the work is equal to the change in the kinetic energy of the particle.* Let us *define* a function $V(x)$ such that

$$-\frac{dV}{dx} = F(x) \tag{2.12}$$

The function $V(x)$ is called the *potential energy;* it is defined only to within an additive (arbitrary) constant. In terms of $V(x)$, the work integral is

$$\int_{x_0}^{x} F(x)dx = -\int_{x_0}^{x} dV = -V(x) + V(x_0) = T - T_0 \qquad (2.13)$$

Notice that Equation 2.13 remains unaltered if $V(x)$ is changed by adding *any* constant C because

$$-[V(x) + C] + [V(x_0) + C] = -V(x) + V(x_0)$$

We now transpose terms and write Equation 2.13 in the following form:

$$T + V(x) = T_0 + V(x_0) = \text{constant} = E \qquad (2.14)$$

or equivalently

$$\frac{1}{2}mv^2 + V(x) = E \qquad (2.14a)$$

This is known as the *energy equation.* We call the constant E the total energy. It is equal to the sum of the kinetic and the potential energies. In words: For one-dimensional motion, if the impressed force is a function of position only, then the sum of the kinetic and potential energies remains constant throughout the motion. The force in this case is said to be *conservative.*[3] Nonconservative forces, that is, those for which no potential function exists, are usually of a dissipational nature, such as friction.

The motion of the particle can be obtained by solving the energy equation (Equation 2.14a) for v

$$v = \frac{dx}{dt} = \pm\sqrt{\frac{2}{m}[E - V(x)]} \qquad (2.15)$$

which can be written in integral form

$$\int_{x_0}^{x} \frac{dx}{\pm\sqrt{\frac{2}{m}[E - V(x)]}} = t - t_0 \qquad (2.15a)$$

thus giving t as a function of x.

In view of Equation 2.15, we see that the expression for v is real only for those values of x such that $V(x)$ is less than or equal to the total energy E. Physically, this means that the particle is confined to the region, or regions, for which the condition $V(x) \le E$ is satisfied. Furthermore, v goes to zero when $V(x) = E$. This means that the particle must come to rest and reverse its motion at those points for which the equality

---

[3] A more complete discussion of conservative forces will be found in Chapter 4.

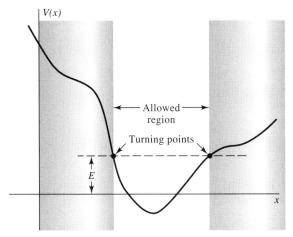

Figure 2.3   Graph of a one-dimensional potential energy function $V(x)$ showing allowed region of motion and the turning points for a given value of the total energy E.

holds. These points are called the *turning points* of the motion. The above facts are illustrated in Figure 2.3.

EXAMPLE 2.2 *Free Fall*

The motion of a freely falling body (discussed above under the case of constant acceleration) is an example of conservative motion. If we choose the x direction to be positive upward, then the gravitational force is equal to $-mg$. Therefore, $-dV/dx = -mg$, and $V = mgx + C$. The constant of integration C is arbitrary and merely depends on the choice of the reference level for measuring V. We can choose $C = 0$, which means that $V = 0$ when $x = 0$. The energy equation is then

$$\frac{1}{2}mv^2 + mgx = E$$

the energy constant E is determined from the initial conditions. For instance, let the body be projected upward with initial speed v_0 from the origin $x = 0$. These values give $E = mv_0^2/2 = mv^2/2 + mgx$, so

$$v^2 = v_0^2 - 2gx$$

The turning point of the motion, which is in this case the maximum height, is given by setting $v = 0$. This gives $0 = v_0^2 - 2gx_{max}$ or

$$h = x_{max} = \frac{v_0^2}{2g}$$

■

EXAMPLE 2.3   *Variation of Gravity with Height*

In Example 2.2 it was assumed that g was constant. Actually, the force of gravity between two particles is inversely proportional to the square of the distance between them (Newton's law of gravity).[4] Thus, the gravitational force that the earth exerts on a body of mass m is given by

$$F_r = -\frac{GMm}{r^2}$$

in which G is Newton's constant of gravitation, M is the mass of the earth, and r is the distance from the center of the earth to the body. By definition, this force is equal to the quantity $-mg$ when the body is at the surface of the earth, so $mg = GMm/r_e^2$. Thus, $g = GM/r_e^2$ is the acceleration of gravity at the earth's surface. Here r_e is the radius of the earth (assumed to be spherical). Let x be the distance above the surface so $r = r_e + x$. Then, neglecting any other forces such as air resistance, we can write

$$F(x) = -mg\frac{r_e^2}{(r_e + x)^2} = m\ddot{x}$$

for the differential equation of motion of a vertically falling (or rising) body with the variation of gravity taken into account. To integrate we set $\ddot{x} = v\,dv/dx$. Then

$$-mgr_e^2 \int_{x_0}^{x} \frac{dx}{(r_e + x)^2} = \int_{v_0}^{v} mv\,dv$$

$$mgr_e^2 \left(\frac{1}{r_e + x} - \frac{1}{r_e + x_0} \right) = \frac{1}{2}mv^2 - \frac{1}{2}mv_0^2.$$

This is, in fact, just the *energy equation* in the form of Equation 2.13. The potential energy is $V(x) = -mg[r_e^2/(r_e + x)]$ rather than mgx.

Maximum height. Escape speed

Suppose a body is projected upward with initial speed v_0 at the surface of the earth, $x_0 = 0$. The energy equation then yields, upon solving for v^2, the following result:

$$v^2 = v_0^2 - 2gx\left(1 + \frac{x}{r_e}\right)^{-1}$$

This reduces to the result for a uniform gravitational field of Example 2.2 if x is very small compared to r_e so that the term x/r_e can be neglected. The turning point (maximum height) is found by setting $v = 0$ and solving for x. The result is

$$x_{max} = h = \frac{v_0^2}{2g}\left(1 - \frac{v_0^2}{2gr_e}\right)^{-1} \tag{2.16}$$

Again we get the formula of Example 2.2 if the second term in the parentheses can be ignored, that is, if v_0^2 is much smaller than $2gr_e$.

---

[4] We shall study Newton's law of gravity in more detail in Chapter 6.

Finally, let us apply the exact formula (2.16) to find the value of v_0 that gives an infinite value of h. This is called the *escape speed,* and it is clearly found by setting the quantity in parentheses equal to zero. The result is

$$v_e = (2gr_e)^{1/2}$$

This gives, for $g = 9.8$ m/s^2 and $r_e = 6.4 \times 10^6$ m,

$$v_e \simeq 11 \text{ km/s} \simeq 7 \text{ mi/s}$$

for the numerical value of the escape speed from the surface of the earth.

In the earth's atmosphere, the average speed of air molecules (O_2 and N_2) is about 0.5 km/s, which is considerably less than the escape speed, so the earth retains its atmosphere. The moon, on the other hand, has no atmosphere; because the escape speed at the moon's surface, owing to the moon's small mass, is considerably smaller than that at the earth's surface, any oxygen or nitrogen would eventually disappear. The earth's atmosphere, however, contains no significant amount of hydrogen, even though hydrogen is the most abundant element in the universe as a whole. A hydrogen atmosphere would have escaped from the earth long ago, because the molecular speed of hydrogen is large enough (owing to the small mass of the hydrogen molecule) that at any instant a significant number of hydrogen molecules would have speeds exceeding the escape speed. ■

EXAMPLE 2.4

The *Morse function* $V(x)$ approximates the potential energy of a vibrating diatomic molecule as a function of x, the distance of separation of its constituent atoms, and is given by

$$V(x) = V_0(1 - e^{-(x-x_0)/\delta})^2 - V_0$$

where V_0, x_0, and δ are parameters chosen to describe the observed behavior of a particular pair of atoms. The force that each atom exerts on the other is given by the derivative of this function with respect to x. Show that x_0 is the separation of the two atoms when the potential energy function is a minimum and that its value for that distance of separation is $V(x_0) = -V_0$. (When the molecule is in this configuration, it is said to be in equilibrium.)

Solution:

The potential energy of the diatomic molecule is a minimum when its derivative with respect to r, the distance of separation, is zero. Thus,

$$F(x) = -\frac{dV(x)}{dx} = 0 =$$

$$2\frac{V_0}{\delta}(1 - e^{-(x-x_0)/\delta})(e^{-(x-x_0)/\delta}) = 0$$

$$1 - e^{-(x-x_0)/\delta} = 0$$

$$\ln(1) = -(x - x_0)/\delta = 0$$

$$\therefore x = x_0$$

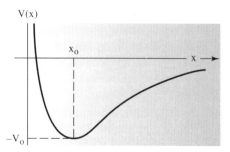

Figure 2.4   Potential energy function for a diatomic molecule.

The value of the potential energy at the minimum can be found by setting $x = x_0$ in the expression for $V(x)$. This gives $V(x_0) = -V_0$.   ■

EXAMPLE 2.5

Shown in Figure 2.4 is the potential energy function for a diatomic molecule. Show that, for separation distances x close to x_0, the potential energy function is parabolic and the resultant force on each atom of the pair is linear, always directed toward the equilibrium position.

Solution:
All we need do here is expand the potential energy function near the equilibrium position.

$$V(x) \approx V_0\left(1 - \left(1 - \left(\frac{x - x_0}{\delta}\right)\right)\right)^2 - V_0$$

$$\approx \frac{V_0}{\delta^2}(x - x_0)^2 - V_0$$

$$F(x) = -\frac{dV(x)}{dx} = -\frac{2V_0}{\delta^2}(x - x_0)$$

Note: The force is linear and is directed in such a way as to restore the diatomic molecule to its equilibrium position.   ■

EXAMPLE 2.6

The binding energy $(-V_0)$ of the diatomic hydrogen molecule H_2 is -4.52 eV (1 eV $= 1.6 \times 10^{-19}$ joules; 1 joule $= 1$ N $\cdot$ m). The values of the constants x_0 and δ are .074 and .036 nm, respectively (1 nm $= 10^{-9}$ m). Assume that at room temperature the total energy of the hydrogen molecule is about $\delta E = 1/40$ eV higher than its binding energy. Calculate the maximum separation of the two atoms in the diatomic hydrogen molecule.

Solution:

Since the molecule has a little more energy than its minimum possible value, the two atoms will vibrate between two values of x, where their kinetic energy is zero. At these turning points, all the energy is potential; hence,

$$V(x) = -V_0 + \delta E \approx \frac{V_0}{\delta^2}(x - x_0)^2 - V_0$$

$$x = x_0 \pm \delta \sqrt{\frac{\delta E}{V_0}}$$

Putting in numbers, we see that the hydrogen molecule vibrates at room temperature a distance of about $\pm 4\%$ of its equilibrium separation.

For this situation where the oscillation is small, the two atoms undergo a symmetrical displacement about their equilibrium position. This arises from approximating the potential function as a parabola near equilibrium. Note from Figure 2.4 that, farther away from the equilibrium position, the potential energy function is not symmetrical, being steeper at smaller distances of separation. Thus, as the diatomic molecule is "heated up," on the average it will spend an increasingly greater fraction of its time separated by a distance greater than their separation at equilibrium. This is why most substances tend to expand when heated. ∎

2.4 VELOCITY-DEPENDENT FORCES. FLUID RESISTANCE AND TERMINAL VELOCITY

It often happens that the force that acts on a body is a function of the velocity of the body. This is true, for example, in the case of viscous resistance exerted on a body moving through a fluid. If the force can be expressed as a function of v only, the differential equation of motion may be written in either of the two forms

$$F_0 + F(v) = m\frac{dv}{dt} \tag{2.17}$$

$$F_0 + F(v) = mv\frac{dv}{dx} \tag{2.18}$$

Here F_0 is any constant force that does not depend on v. Upon separating variables, integration yields either t or x as functions of v. A second integration can then yield a functional relationship between x and t.

For normal fluid resistance, including air resistance, $F(v)$ is not a simple function and generally must be found through experimental measurements. However, a fair approximation for many cases is given by the equation

$$F(v) = -c_1 v - c_2 v|v| = -v(c_1 + c_2|v|) \tag{2.19}$$

in which c_1 and c_2 are constants whose values depend on the size and shape of the body. (The absolute-value sign is necessary on the last term because the force of fluid resistance is always opposite to the direction of v.) If the above form for $F(v)$ is used to find the motion by solving Equations 2.17 or 2.18, the resulting integrals are somewhat messy. But for the limiting cases of either small v or large v, respectively, the linear or the quadratic term in $F(v)$ dominates, and the differential equations become somewhat more manageable.

For spheres in air, approximate values for the constants in the equation for $F(v)$ are, in SI units,

$$c_1 = 1.55 \times 10^{-4}D$$

$$c_2 = 0.22\ D^2$$

where D is the diameter of the sphere in meters. The ratio of the quadratic term $c_2 v|v|$ to the linear term $c_1 v$ is, thus,

$$\frac{0.22v|v|D^2}{1.55 \times 10^{-4}vD} = 1.4 \times 10^2|v|D.$$

This means that, for instance, with objects of baseball size ($D \sim 0.07$ m), the quadratic term dominates for speeds in excess of 0.01 m/s(1 cm/s), and the linear term dominates for speeds less than this value. For speeds around this value, *both* terms must be taken into account. (See Problem 2.13.)

EXAMPLE 2.7   *Horizontal Motion with Linear Resistance*

Suppose a block is projected with initial velocity v_0 on a smooth horizontal surface and that there is air resistance such that the linear term dominates. Then, in the direction of the motion, $F_0 = 0$ in Equations 2.17 and 2.18 and $F(v) = -c_1 v$. The differential equation of motion is then

$$-c_1 v = m\frac{dv}{dt}$$

which gives, upon integrating,

$$t = \int_{v_0}^{v} -\frac{m\ dv}{c_1 v} = -\frac{m}{c_1}\ln\left(\frac{v}{v_0}\right)$$

Solution:
We can easily solve for v as a function of t by multiplying by $-c_1/m$ and taking the exponential of both sides. The result is

$$v = v_0 e^{-c_1 t/m}$$

Thus, the velocity decreases exponentially with time. A second integration gives

$$x = \int_0^t v_0 e^{-c_1 t/m} dt$$

$$= \frac{mv_0}{c_1}(1 - e^{-c_1 t/m})$$

showing that the block approaches a limiting position given by $x_{lim} = mv_0/c_1$.   ∎

EXAMPLE 2.8 *Horizontal Motion with Quadratic Resistance*

If the parameters are such that the quadratic term dominates, then for positive v, we can write

$$-c_2 v^2 = m \frac{dv}{dt}$$

which gives

$$t = \int_{v_0}^{v} \frac{-m\,dv}{c_2 v^2} = \frac{m}{c_2}\left(\frac{1}{v} - \frac{1}{v_0}\right)$$

Solution:
Solving for v, we get

$$v = \frac{v_0}{1 + kt}$$

where $k = c_2 v_0/m$. A second integration gives us the position as a function of time

$$x(t) = \int_{0}^{t} \frac{v_0\,dt}{1 + kt} = \frac{v_0}{k} \ln(1 + kt)$$

Thus, as $t \rightarrow \infty$, v decreases as $1/t$, but the position does not approach a limit as was obtained in the case of a linear retarding force. Why might this be? You might guess that a quadratic retardation should be more effective in stopping the block than is a linear one. This is certainly true at large velocities, but as the velocity approaches zero, the quadratic retarding force goes to zero much faster than the linear one, enough to allow the block to continue on its merry way, albeit at a very slow speed. ∎

Vertical Fall Through a Fluid. Terminal Velocity

(a) *Linear case.* For an object falling vertically in a resisting fluid, the force F_0 in Equations 2.17 and 2.18 is the weight of the object, namely $-mg$ for the x-axis positive in the upward direction. For the linear case of fluid resistance, we then have for the differential equation of motion

$$-mg - c_1 v = m \frac{dv}{dt} \tag{2.20}$$

Separating variables and integrating, we find

$$t = \int_{v_0}^{v} \frac{m\,dv}{-mg - c_1 v} = -\frac{m}{c_1} \ln \frac{mg + c_1 v}{mg + c_1 v_0} \tag{2.21}$$

in which v_0 is the initial velocity at $t = 0$. Upon multiplying by $-c_1/m$ and taking the exponential, we can solve for v:

$$v = -\frac{mg}{c_1} + \left(\frac{mg}{c_1} + v_0\right) e^{-c_1 t/m} \tag{2.22}$$

The exponential term drops to a negligible value after a sufficient time ($t \gg m/c_1$), and the velocity approaches the limiting value $-mg/c_1$. The limiting velocity of a falling body is called the *terminal velocity;* it is that velocity at which the force of resistance is just equal and opposite to the weight of the body so that the total force is zero, and so the acceleration is zero. The magnitude of the terminal velocity is the *terminal speed.*

Let us designate the terminal speed mg/c_1 by v_t, and let us write τ (which we may call the *characteristic time*) for m/c_1. Equation 2.22 may then be written in the more significant form

$$v = -v_t(1 - e^{-t/\tau}) + v_0 e^{-t/\tau} \tag{2.23}$$

These two terms represent two velocities: the terminal velocity v_t, which exponentially "fades in," and the initial velocity v_0, which exponentially "fades out" due to the action of the viscous drag force.

In particular, for an object dropped from rest at time $t = 0$, $v_0 = 0$, we find

$$v = -v_t(1 - e^{-t/\tau}) \tag{2.24}$$

Thus, after one characteristic time the speed is $1 - e^{-1}$ times the terminal speed, in two characteristic times it is the factor $1 - e^{-2}$ of v_t, and so on. After an interval of 5τ, the speed is within 1% of the terminal value, namely $(1 - e^{-5})v_t = 0.993\, v_t$.

(b) *Quadratic case.* In this case, the magnitude of $F(v)$ is proportional to v^2. In order to ensure that the force remains resistive, we must remember that the sign preceding the $F(v)$ term depends on whether or not the motion of the object is upward or downward. This is the case for any resistive force proportional to an *even* power of velocity. A general solution usually involves treating the upward and downward motions separately. Here, we will simplify things somewhat by considering only the situation in which the body is either dropped from rest or projected downward with an initial velocity v_0. It will be left as an exercise for the student to treat the upward-going case. We will take the downward direction to be the positive y direction. The differential equation of motion is

$$
\begin{aligned}
m\frac{dv}{dt} &= mg - c_2 v^2 = mg\left(1 - \frac{c_2}{mg}v^2\right) \\
&= mg\left(1 - \frac{v^2}{v_t^2}\right) \\
\frac{dv}{dt} &= g\left(1 - \frac{v^2}{v_t^2}\right)
\end{aligned}
\tag{2.25}
$$

where

$$v_t = \sqrt{\frac{mg}{c_2}} \qquad \text{(\textit{the terminal speed})} \tag{2.26}$$

Integrating Equation 2.25 gives t as a function of v

$$t - t_0 = \int_{v_0}^{v} \frac{dv}{g\left(1 - \dfrac{v^2}{v_t^2}\right)} = \tau \left(\tanh^{-1} \frac{v}{v_t} - \tanh^{-1} \frac{v_0}{v_t} \right)$$

where

$$\tau = \frac{v_t}{g} = \sqrt{\frac{m}{c_2 g}} \qquad (\textit{the characteristic time})$$

Solving for v, we obtain

$$v = v_t \tanh\left(\frac{t - t_0}{\tau} - \tanh^{-1} \frac{v_0}{v_t} \right) \tag{2.27}$$

If the body is released from rest at time $t = 0$

$$v = v_t \tanh \frac{t}{\tau} = v_t \left(\frac{e^{2t/\tau} - 1}{e^{2t/\tau} + 1} \right) \tag{2.28}$$

The terminal speed is attained after the lapse of a few characteristic times, for example, at $t = 5\tau$, the speed is $0.99991\, v_t$. Graphs of speed versus time of fall for the linear and quadratic cases are shown in Figure 2.5.

In many instances we would like to know the speed attained upon falling a given distance. We could find this out by integrating Equation 2.27, obtaining y as a function of time and then eliminating the time parameter to find speed versus

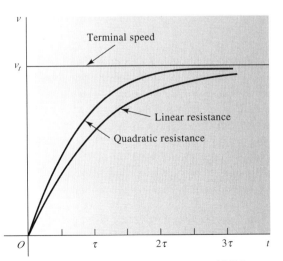

Figure 2.5   Graphs of speed versus time of fall for a falling body.

distance. A more direct solution can be obtained by direct modification of the fundamental differential equation of motion to one in which the independent variable is distance instead of time. For example, since

$$\frac{dv}{dt} = \frac{dv}{dy}\frac{dy}{dt} = \frac{1}{2}\frac{dv^2}{dy}$$

Equation 2.28 can be rewritten with y as the independent variable

$$\frac{dv^2}{dy} = 2g\left(1 - \frac{v^2}{v_t^2}\right) \tag{2.29}$$

We solve this equation as follows

$$u = 1 - \frac{v^2}{v_t^2} \qquad \text{so} \qquad \frac{du}{dy} = -\frac{1}{v_t^2}\frac{dv^2}{dy} = -\left(\frac{2g}{v_t^2}\right)u \tag{2.30}$$

$$u = u(y = 0)e^{-2gy/v_t^2} \qquad \text{but} \qquad u(y = 0) = 1 - \frac{v_0^2}{v_t^2}$$

$$u = \left(1 - \frac{v_0^2}{v_t^2}\right)e^{-2gy/v_t^2} = 1 - \frac{v^2}{v_t^2}$$

$$\therefore v^2 = v_t^2(1 - e^{-2gy/v_t^2}) + v_0^2 e^{-2gy/v_t^2}$$

Thus, we see that the squares of the initial velocity and terminal velocity exponentially fade in and out within a characteristic length of $v_t^2/2g$.

EXAMPLE 2.9 *Falling Raindrops and Basketballs*

Calculate the terminal speed in air and characteristic time for (a) a very tiny spherical raindrop of diameter 0.1 mm $= 10^{-4}$ m and (b) a basketball of diameter 0.25 m and mass 0.6 kg.

Solution:
To decide which type of force law to use, quadratic or linear, we recall the expression stated earlier giving the ratio of the quadratic to the linear force for air resistance, namely $1.4 \times 10^3|v|D$. For the raindrop this is $0.14v$, and for the basketball it is $350v$, numerically, where v is in meters per second. Thus, for the raindrop v must exceed $1/0.14 = 7.1$ m/s for the quadratic force to dominate. In the case of the basketball, v must exceed only $1/350 = 0.0029$ m/s for the quadratic force to dominate. We conclude that the linear case should hold for the falling raindrop, while the quadratic case should be correct for the basketball. (See also Problem 2.11.)

The volume of the raindrop is $\pi D^3/6 = 0.52 \times 10^{-12}$ m³, so, multiplying by the density of water, 10^3 kg/m³, gives the mass $m = 0.52 \times 10^{-9}$ kg. For the drag coefficient we get $c_1 = 1.55 \times 10^{-4}D = 1.55 \times 10^{-8}$ N · s/m. This gives a terminal speed

$$v_t = \frac{mg}{c_1} = \frac{0.52 \times 10^{-9} \times 9.8}{1.55 \times 10^{-8}} \text{ m/s} = 0.33 \text{ m/s}$$

The characteristic time is

$$\tau = \frac{v_t}{g} = \frac{0.33 \text{ m/s}}{9.8 \text{ m/s}^2} = 0.034 \text{ s}$$

For the basketball the drag constant is $c_2 = 0.22D^2 = 0.22 \times (0.25)^2 = 0.0138$ N $\cdot$ s^2/m^3, and so the terminal speed is

$$v_t = \left(\frac{mg}{c_2}\right)^{1/2} = \left(\frac{0.6 \times 9.8}{0.0138}\right)^{1/2} \text{ m/s} = 20.6 \text{ m/s}$$

the characteristic time is

$$\tau = \frac{v_t}{g} = \frac{20.6 \text{ m/s}}{9.8 \text{ m/s}^2} = 2.1 \text{ s}$$

Thus, the raindrop practically attains its terminal speed in less than 1 s when starting from rest, whereas it takes several seconds for the basketball to come to within 1% of the terminal value.

[For more information on aerodynamic drag, the reader is referred to an article by C. Frohlich in *Am. J. Phys.*, **52**, 325 (1984) and the extensive list of references cited therein.] ■

PROBLEMS

2.1 Find the velocity $\dot{x}$ and the position x as functions of the time t for a particle of mass m, which starts from rest at $x = 0$ and $t = 0$ subject to the following force functions:
(a) $F_x = F_0 + ct$
(b) $F_x = F_0 \sin ct$
(c) $F_x = F_0 e^{ct}$
where F_0 and c are positive constants.

2.2 Find the velocity $\dot{x}$ as a function of the displacement x for a particle of mass m, which starts from rest at $x = 0$ subject to the following force functions:
(a) $F_x = F_0 + cx$
(b) $F_x = F_0 e^{-cx}$
(c) $F_x = F_0 \cos cx$
where F_0 and c are positive constants.

2.3 Find the potential energy function $V(x)$ for each of the forces in Problem 2.2.

2.4 Given that the velocity of a particle in rectilinear motion varies with the displacement x according to the equation

$$\dot{x} = bx^{-3}$$

where b is a positive constant, find the force acting on the particle as a function of x. (*Hint:* $F = m\ddot{x} = m\dot{x}\, d\dot{x}/dx$.)

2.5 A baseball (radius = .0366 m, mass = .145 kg) is dropped from rest from the top of the Empire State Building (height = 1250 ft). Calculate (a) the initial potential energy of the baseball, (b) its final kinetic energy, and (c) the total energy dissipated by the falling baseball by computing the line integral of the force of air resistance along the baseball's total distance of fall. Compare this last result to the difference between the baseball's initial potential energy and its final kinetic energy. (*Hint:* In part (c) make approximations when evaluating the hyperbolic functions obtained in carrying out the line integral.)

2.6 A block of wood is projected up an inclined plane with initial speed v_0. If the inclination of the plane is $30°$ and the coefficient of sliding friction $\mu_K = 0.1$, find the total time for the block to return to the point of projection.

2.7 A metal block of mass m slides on a horizontal surface that has been lubricated with a heavy oil such that the block suffers a viscous resistance that varies as the three-halves power of the speed:

$$F(v) = -cv^{3/2}$$

If the initial speed of the block is v_0 at $x = 0$, show that the block cannot travel farther than $2mv_0^{1/2}/c$.

2.8 A gun is fired straight up. Assuming that the air drag on the bullet varies quadratically with speed, show that the speed varies with height according to the equations

$$v^2 = Ae^{-2kx} - \frac{g}{k} \quad (\textit{upward motion})$$

$$v^2 = \frac{g}{k} - Be^{2kx} \quad (\textit{downward motion})$$

in which A and B are constants of integration, g is the acceleration of gravity, and $k = c_2/m$ where c_2 is the drag constant and m is the mass of the bullet. (*Note:* x is measured positive upward, and the gravitational force is assumed to be constant.)

2.9 Use the above result to show that, when the bullet hits the ground on its return, the speed will be equal to the expression

$$\frac{v_0 v_t}{(v_0^2 + v_t^2)^{1/2}}$$

in which v_0 is the initial upward speed and

$$v_t = (mg/c_2)^{1/2} = \text{terminal speed} = (g/k)^{1/2}$$

(This result allows one to find the fraction of the initial kinetic energy lost through air friction.)

2.10 A particle of mass m is released from rest a distance b from a fixed origin of force that attracts the particle according to the inverse square law

$$F(x) = -kx^{-2}$$

Show that the time required for the particle to reach the origin is

$$\pi \left(\frac{mb^3}{8k} \right)^{1/2}$$

2.11 Show that the terminal speed of a falling spherical object is given by

$$v_t = [(mg/c_2) + (c_1/2c_2)^2]^{1/2} - (c_1/2c_2)$$

when *both* the linear and the quadratic terms in the drag force are taken into account.

2.12 Use the above result to calculate the terminal speed of a soap bubble of mass 10^{-7} kg and diameter 10^{-2} m. Compare with the value obtained by using Equation 2.26.

2.13 Given: The force acting on a particle is the product of a function of the distance and a function of the velocity: $F(x, v) = f(x)g(v)$. Show that the differential equation of motion can be solved by integration. If the force is a product of a function of distance and a function of time, can the equation of motion be solved by simple integration? Can it be solved if the force is a product of a function of time and a function of velocity?

2.14 The force acting on a particle of mass m is given by

$$F = kvx$$

in which k is a positive constant. The particle passes through the origin with speed v_0 at time $t = 0$. Find x as a function of t.

COMPUTER/CALCULATOR APPLICATIONS

2.1 A parachutist of mass 70 kg jumps from a plane at an altitude of 32 km above the surface of the earth. Unfortunately, the parachute fails to open. (In the parts below, neglect horizontal motion and assume that the initial velocity is zero.)

(a) Calculate the time of fall (accurate to 1 s) until ground impact, given no air resistance and a constant value of g.

(b) Calculate the time of fall (accurate to 1 s) until ground impact given constant g and a force of air resistance given by

$$F(v) = -c_2 v|v|$$

where c_2 is 0.5 in SI units for a falling man and is constant.

(c) Calculate the time of fall (accurate to 1 s) until ground impact given c_2 scales with atmospheric density as

$$c_2 = 0.5e^{-y/H}$$

where $H = 8$ km is the scale height of the atmosphere and y is the height above ground. Furthermore, assume that g is no longer constant but is given by

$$g = \frac{9.8}{\left(1 + \dfrac{y}{R_e}\right)^2} \text{ ms}^{-2}$$

where R_e is the radius of the earth and is 6370 km.

(d) For case (c), plot the acceleration, velocity, and altitude of the parachutist as a function of time. Explain why the acceleration becomes positive as the parachutist falls.

3

OSCILLATIONS

"These are the Phenomena of Springs and springy bodies, which as they have not hitherto been by any that I know reduced to Rules—It is very evident that the Rule or Law of Nature in every springing body is, that the force or power thereof to restore itself to its natural position is always proportionate to the distance or space it is removed therefrom—"

(Robert Hooke—De Potentia Restitutiva, 1678)

"If I have seen further, it is by standing upon ye shoulders of Giants."

(Sir Isaac Newton—letter to Robert Hooke, 1676)

Atomic Clock at The Naval Observatory UPI/BETTMANN

3.1   INTRODUCTION

The solar system was the most fascinating and intensively studied mechanical system known to early man. It is a marvelous example of periodic motion. It is not clear how long people would have toiled in mechanical ignorance were it not for this periodicity or had our planet been the singular observable member of the solar system. Everywhere around us we see systems engaged in a periodic dance: the small oscillations of a pendulum clock, a child playing on a swing, the rise and fall of the tides, the swaying of a tree in the wind, the vibrations of the strings on a fiddle or violin. Even things that we cannot see march to the tune of a periodic beat: the vibrations of the air molecules in the woodwind instruments of a symphony, the hum of the electrons in the wires of our modern civilization, the vibrations of the atoms and molecules that make up our bodies. It is ironic that we cannot even say the word *vibration* properly without the tip of the tongue oscillating.

The essential feature that all these phenomena have in common is *periodicity,* a pattern of movement or displacement that repeats itself over and over again. The pattern may be simple or it may be complex. For example, Figure 3.1a shows a record of the horizontal displacement of a supine human body, resting on a nearly frictionless surface,

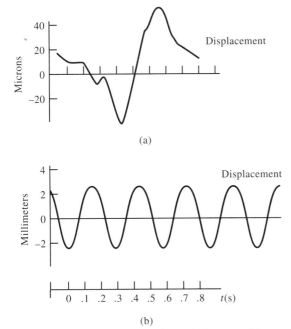

(a)

(b)

Figure 3.1   (a) Recoil vibrations of a human subject resting on a frictionless surface (in response to the pumping action of the heart).
(b) Horizontal displacement of a simple pendulum about equilibrium.

such as a thin layer of air. The body oscillates horizontally back and forth due to the mechanical action of the heart, pumping blood through and around the aortic arch. Such a recording is called a *ballistocardiogram.*[1] Figure 3.1b shows the almost perfect sine curve representing the horizontal displacement of a simple pendulum executing small oscillations about its equilibrium position. In both cases, the horizontal axis represents the steady advance of time. The period of the motion is readily identified as the time for one complete cycle of the motion to occur.

It is with the hope of being able to describe all the complicated forms of periodic motion Mother Nature exhibits, such as that shown in Figure 3.1a, that we undertake an analysis of her simplest—*simple harmonic motion* (exemplified in Figure 3.1b).

Simple harmonic motion exhibits two essential characteristics: (1) it is described by a second-order, linear differential equation with constant coefficients. Thus, the *superposition principle* holds; that is, if two particular solutions are found, their sum is also a solution. We will see evidence of this in the examples to come. (2) The period of the motion, or the time required for a particular configuration (not only position, but velocity as well) to repeat itself, is independent of the maximum displacement from equilibrium. We have already remarked that it was Galileo who first exploited this essential feature of the pendulum by using it as a clock. These features hold true only if the displacements from equilibrium are "small." "Large" displacements result in the appearance of nonlinear terms in the differential equations of motion, and the resulting oscillatory solutions no longer obey the principle of superposition or exhibit amplitude-independent periods. We will only briefly consider this situation toward the end of this chapter.

3.2   LINEAR RESTORING FORCE. HARMONIC MOTION

One of the simplest models of a system executing simple harmonic motion is a mass on a frictionless surface attached to a wall by means of a spring. Such a system is shown in Figure 3.2. If X_e is the unstretched length of the spring, the mass will sit at that position,

---

[1] *PHYSICS—With Illustrative Examples from Medicine and Biology,* George B. Benedek and Felix M. H. Villars, Addison-Wesley Publ. (1974).

The Harmonic Oscillator

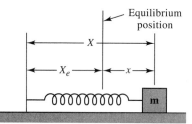

Figure 3.2   A model of the simple harmonic oscillator.

undisturbed, if initially placed there at rest. This position represents the equilibrium configuration of the mass, that is, the one in which its potential energy is a minimum or, equivalently, where the net force on it vanishes. If the mass is pushed, or pulled, away from this position, the spring will either be compressed or stretched. It will then exert a force on the mass which will always attempt to restore it to its equilibrium configuration.

We need an expression for this restoring force if we are to calculate the motion of the mass. We can make an estimate of the mathematical form of this force by appealing to arguments based on the presumed nature of the potential energy of this system. Recall from Example 2.4 that the Morse potential—the potential energy function of the di-atomic hydrogen molecule, a bound system of two particles—has the shape of a well or a cup. Mathematically, it was given by the following expression:

$$V(x) = V_0(1 - \exp(-x/\delta))^2 - V_0$$

We showed that this function exhibited quadratic behavior near its minimum and that the resulting force between the two atoms was linear, always acting to restore them to their equilibrium configuration. In general, any potential energy function can be de-scribed approximately by a polynomial function of the displacement x for displacements not too far from equilibrium

$$V(x) = a_0 + a_1x + a_2x^2 + a_3x^3 + \cdots$$

Furthermore, since only *differences* in potential energies are relevant for the behavior of physical systems, the constant term in each of the above expressions may be taken to be zero; this amounts to a simple reassignment of the value of the potential energy at some reference point. We also argue that the linear term in the above expression must be identically zero. This condition follows from the fact that the first derivative of any function must vanish at its minimum, presuming that the function and its derivatives are continuous, as they must be if the function is to describe the behavior of a real, physical system. Thus, the approximating polynomial takes the form

$$V(x) = a_2x^2 + a_3x^3 + \cdots$$

For example, Figure 3.3a is a plot of the Morse potential along with an approximating eighth-order polynomial "best fit." The width δ of the potential and its depth (the V_0 coefficient) were both set equal to 1.0 (the bare constant V_0 was set equal to 0). The fit was made over the rather sizable range $\Delta x = [-1, 4] = 5\delta$. The result is

$$V(x) = \sum_{i=0}^{8} a_ix^i$$

$a_0 = 1.015 \cdot 10^{-4}$	$a_1 = +0.007$	$a_2 = 0.995$
$a_3 = -1.025$	$a_4 = 0.611$	$a_5 = -0.243$
$a_6 = 0.061$	$a_7 = -0.009$	$a_8 = 5.249 \cdot 10^{-4}$

It can be seen that the polynomial function fits the Morse potential quite well throughout the quoted displacement range. Note that, if one examines closely the coefficients of the eighth-order fit, one will see that the first two terms are essentially zero, as we have argued they should be. Therefore, we also show plotted only the quadratic term $V(x) \approx a_2 \cdot x^2$. It seems as though this term does not agree very well with the Morse potential.

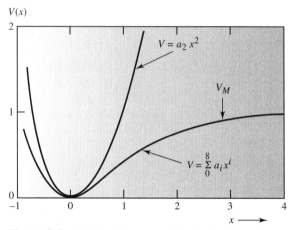

Figure 3.3   (a) The Morse potential, its 8th order approximating polynomial and the quadratic term only.

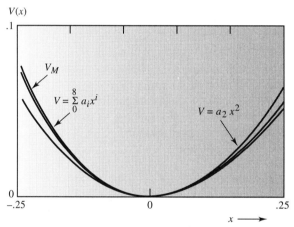

Figure 3.3   (b) Same as (a) but magnified in scale around $x = 0$.

However, if we "explode" the plot around $x = 0$ (see Figure 3.3b), we see that for small displacements, say, $-0.1\delta \leq x \leq +0.1\delta$, there is virtually no difference among the purely quadratic term, the eighth-order polynomial fit, or the actual Morse potential. For such small displacements, the potential function is, indeed, purely quadratic. One might argue that this example was contrived; however, it is fairly representative of many physical systems.

The potential energy function for the system of spring and mass must exhibit similar behavior near the equilibrium position at X_e, dominated by a purely quadratic term. The spring's restoring force is thus given by the familiar Hooke's law:

$$F(x) = -\frac{dV(x)}{dx} = -(2a_2)x = -kx \tag{3.1}$$

where $k = 2a_2$ is the *spring constant*. In fact, this is how we define small displacements from equilibrium, that is, those for which Hooke's law is valid or the restoring force is linear. That the derived force must be a restoring one is a consequence of the fact that the derivative of the potential energy function must be negative for positive displacements from equilibrium and vice versa for negative ones. Newton's second law of motion for the mass can now be written as

$$m\ddot{x} + kx = 0 \tag{3.2}$$

$$\ddot{x} + \frac{k}{m}x = 0 \tag{3.3}$$

Equation 3.3 can be solved in a wide variety of ways. It is a second-order, linear differential equation with constant coefficients. As previously stated, the principle of superposition holds for its solutions. Before solving the equation here, we point out those characteristics we expect the solution to exhibit. First, the motion is both periodic and bounded. The mass vibrates back and forth between two limiting positions. Suppose we pull the mass out to some position x_{m1} and then release it from rest. The restoring force, initially equal to $-kx_{m1}$, will pull the mass toward the left in Figure 3.2, where it will vanish at $x = 0$, the equilibrium position. The mass now finds itself moving to the left with some velocity v, and so it will pass on through equilibrium. But then the restoring force will begin to build up strength as the spring compresses, but now directed toward the right. It will slow the mass down until it stops, just for an instant, at some position, $-x_{m2}$. The spring, now fully compressed, will start to shove the mass back toward the right. But again, momentum will carry it through the equilibrium position until the now stretching spring finally manages to stop it, we might guess, at x_{m1}, the initial configuration of the system. This completes one cycle of the motion—a cycle that repeats itself, apparently forever! Clearly, the resultant functional dependence of x upon t must be represented by a periodic and bounded function. Sine and/or cosine functions come to mind, since they exhibit the sort of behavior we are describing here. In fact, sines and cosines are the real solutions of Equation 3.3. Later on, we will show that other functions, imaginary exponentials, are actually equivalent to sines and cosines and are easier to use in describing the more complicated systems soon to be discussed.

As can be verified by direct substitution into Equation 3.3, a solution is given by

$$x = A \sin(\omega_0 t + \phi_0) \tag{3.4}$$

where

$$\omega_0 = \sqrt{\frac{k}{m}} \tag{3.5}$$

is the *angular frequency* of the system. The motion represented by Equation 3.4 is a sinusoidal oscillation about equilibrium. A graph of the displacement x versus $\omega_0 t$ is shown in Figure 3.4. The motion exhibits the following features: (1) It is characterized by a single angular frequency ω_0. The motion repeats itself after the angular argument of the sine function ($\omega_0 t + \phi_0$) advances by 2π or after one cycle has occurred (hence,

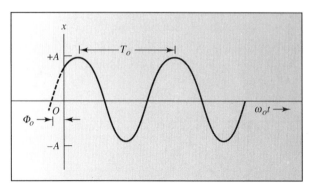

Figure 3.4   Displacement versus $\omega_0 t$ for the simple harmonic oscillator.

the name *angular frequency* given to ω_0). The time required for a phase advance of 2π is given by

$$\omega_0(t + T_0) + \phi_0 = \omega_0 t + \phi_0 + 2\pi \tag{3.6}$$

$$\therefore T_0 = \frac{2\pi}{\omega_0}$$

T_0 is called the period of the motion. (2) The motion is bounded; that is, it is confined within the limits $-A \le x \le +A$. A, the maximum displacement from equilibrium, is called the amplitude of the motion. It is independent of the angular frequency ω_0. (3) The phase angle ϕ_0 is the initial value of the angular argument of the sine function. It determines the value of the displacement x at time $t = 0$. For example, at $t = 0$ we have

$$x(t = 0) = A \sin (\phi_0) \tag{3.7}$$

The maximum displacement from equilibrium occurs at a time t_m given by the condition that the angular argument of the sine function is equal to $\pi/2$ or

$$\omega_0 t_m = \frac{\pi}{2} - \phi_0 \tag{3.8}$$

One commonly uses the term *frequency* to refer to the reciprocal of the period of the oscillation or

$$f_0 = \frac{1}{T_0} \tag{3.9}$$

where f_0 is the number of cycles of vibration per unit time. It is related to the angular frequency ω_0 by

$$2\pi f_0 = \omega_0 \tag{3.10}$$

$$f_0 = \frac{1}{T_0} = \frac{1}{2\pi} \sqrt{\frac{k}{m}} \tag{3.11}$$

The unit of frequency (cycles per second or s^{-1}) is called the *hertz* (or hz) in honor of Heinrich Hertz, who is given credit as the discoverer of radio waves. Note that 1 hz = 1 s^{-1}. The word *frequency* is used sloppily sometimes to mean either cycles per second or radians per second (angular frequency). The meaning is usually clear from the context.

Constants of the Motion and Initial Conditions

Equation 3.4, the solution for simple harmonic motion, contains two arbitrary constants A and ϕ_0. The value of each constant can be determined from knowledge of the initial conditions of the specific problem at hand. As an example of the simplest and most commonly described initial condition, consider a mass initially displaced from equilibrium to a position x_m, where it is then released from rest. The displacement at $t = 0$ is a maximum. Therefore, $A = x_m$ and $\phi_0 = \pi/2$.

As an example of another simple situation, suppose the oscillator is at rest at $x = 0$ and at time $t = 0$ it is struck sharply, imparting to it an initial velocity v_0 in the positive x direction. In such a case the initial phase is given by $\phi_0 = 0$. This automatically ensures that the solution yields $x = 0$ at $t = 0$. The amplitude can be found by differentiating x to get the velocity of the oscillator as a function of time and then demanding that the velocity equal v_0 at $t = 0$. Thus,

$$v(t) = \dot{x}(t) = \omega_0 A \cos(\omega_0 t + \phi_0)$$
$$v(0) = v_0 = \omega_0 A$$

$$\therefore A = \frac{v_0}{\omega_0}$$

For a more general scenario, consider a mass initially displaced to some position x_0 and given an initial velocity v_0. The constants can then be determined as follows:

$$x(0) = A \sin \phi_0 = x_0$$
$$\dot{x}(0) = \omega_0 A \cos \phi_0 = v_0 \qquad (3.12)$$

$$\therefore \tan \phi_0 = \frac{\omega_0 x_0}{v_0}$$

$$A^2 = x_0^2 + \frac{v_0^2}{\omega_0^2}$$

This more general solution reduces to either of those described above as can easily be seen by setting v_0 or x_0 equal to zero.

Simple Harmonic Motion as the Projection of a Rotating Vector

Imagine a vector **A** rotating at a constant angular velocity ω_0. Let this vector denote the position of a point P in uniform circular motion. The projection of the vector onto a line (which we will call the x-axis) in the same plane as the circle traces out simple harmonic motion. Suppose the vector **A** makes an angle θ with the x-axis at some time t as shown in Figure 3.5. Since $\dot{\theta} = \omega_0$, the angle θ increases with time according to

$$\theta = \omega_0 t + \theta_0$$

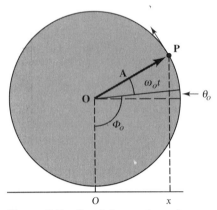

Figure 3.5   Simple harmonic motion as a
projection of uniform circular motion.

where θ_0 is the value of θ at $t = 0$. The projection of P onto the x-axis is given by

$$x = A \cos \theta = A \cos (\omega_0 t + \theta_0)$$

This point oscillates in simple harmonic motion as P goes around the circle in uniform
angular motion.

Our picture describes x as a cosine function of t. We can show the equivalence of
this expression to the sine function given by Equation 3.4, by measuring angles to the
vector **A** from the y-axis, instead of the x-axis as shown in Figure 3.5. If we do this, the
projection of **A** onto the x-axis is given by

$$x = A \sin(\omega_0 t + \phi_0)$$

We can see this equivalence in another way. We set the phase difference between ϕ_0 and
θ_0 to $\pi/2$ and then substitute into the above equation, obtaining

$$\phi_0 - \theta_0 = \frac{\pi}{2}$$

$$\cos(\omega_0 t + \theta_0) = \cos\left(\omega_0 t + \phi_0 - \frac{\pi}{2}\right)$$

$$= \sin(\omega_0 t + \phi_0)$$

We now see that simple harmonic motion can be described equally well by a sine function or a cosine function. The one we choose is largely a matter of taste; it depends upon
our choice of initial phase angle to within an arbitrary constant.

You might guess from the above commentary that we could use a sum of sine and
cosine functions to represent the general solution for harmonic motion. For example, we
can convert the sine solution of Equation 3.4 directly into such a form using the trigonometric identity for the sine of a sum of angles

$$x(t) = A \sin(\omega_0 t + \phi_0) = A \sin \phi_0 \cos \omega_0 t + A \cos \phi_0 \sin \omega_0 t$$
$$= C \cos \omega_0 t + D \sin \omega_0 t$$

Neither A nor ϕ_0 appears explicitly in the solution. They are there implicitly, that is,

$$\tan \phi_0 = \frac{C}{D} \qquad A^2 = C^2 + D^2 \qquad\qquad (3.12a)$$

There are occasions when this form may be the preferred one.

Effect of a Constant External Force on a Harmonic Oscillator

Suppose the same spring shown in Figure 3.2 is held in a vertical position supporting the same mass m, Figure 3.6. The total force acting is now given by adding the weight mg to the restoring force

$$F = -k(X - X_e) + mg \qquad\qquad (3.13)$$

where the positive direction is down. This equation could be written $F = -kx + mg$ by defining x to be $X - X_e$, as previously. However, it is more convenient to define the variable x in a different way, namely, as the displacement from the *new* equilibrium position X'_e obtained by setting $F = 0$ in Equation 3.13: $0 = -k(X'_e - X_e) + mg$ which gives $X'_e = X_e + mg/k$. We now define the displacement as

$$x = X - X'_e = X - X_e - mg/k$$

Putting this into Equation 3.13 gives, after a very little algebra,

$$F = -kx$$

so the differential equation of motion is again

$$m\ddot{x} + kx = 0$$

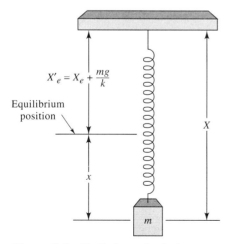

Figure 3.6 Vertical case for the harmonic oscillator.

and our solution in terms of our newly defined x is identical to that of the horizontal case. It should now be evident that *any* constant external force applied to a harmonic oscillator merely shifts the equilibrium position. The equation of motion remains unchanged if we measure the displacement x from the new equilibrium position.

EXAMPLE 3.1

When a light spring supports a block of mass m in a vertical position, the spring is found to stretch by an amount D_1 over its unstretched length. If the block is further pulled downward a distance D_2 from the equilibrium position and released, say at time $t = 0$, find (a) the resulting motion, (b) the velocity of the block when it passes back upward through the equilibrium position, and (c) the acceleration of the block at the top of its oscillatory motion.

Solution:
First, for the equilibrium position we have

$$F_x = 0 = -kD_1 + mg$$

where x is chosen positive downward. This gives us the value of the stiffness constant:

$$k = \frac{mg}{D_1}$$

From this we can find the angular frequency of oscillation:

$$\omega_0 = \sqrt{\frac{k}{m}} = \sqrt{\frac{g}{D_1}}$$

We shall express the motion in the form $x(t) = A \cos \omega_0 t + B \sin \omega_0 t$. Then $\dot{x} = -A\omega_0 \sin \omega_0 t + B\omega_0 \cos \omega_0 t$. From the initial conditions we find

$$x_0 = D_2 = A \qquad \dot{x}_0 = 0 = B\omega_0 \qquad B = 0$$

The motion is therefore given by

(a) $$x(t) = D_2 \cos\left(\sqrt{\frac{g}{D_1}}t\right)$$

in terms of the given quantities. Note that the mass m does not appear in the final expression. The velocity is then

$$\dot{x}(t) = -D_2 \sqrt{\frac{g}{D_1}} \sin\left(\sqrt{\frac{g}{D_1}}t\right)$$

and the acceleration

$$\ddot{x}(t) = -D_2 \frac{g}{D_1} \cos\left(\sqrt{\frac{g}{D_1}} t\right)$$

As the block passes upward through the equilibrium position, the argument of the sine term is $\pi/2$ (one-quarter period), so

(b) $\dot{x} = -D_2 \sqrt{\dfrac{g}{D_1}}$ (*center*)

At the top of the swing the argument of the cosine term is π (one-half period), which gives

(c) $\ddot{x} = D_2 \dfrac{g}{D_1}$ (*top*)

It is interesting to note that in the case $D_1 = D_2$ the downward acceleration at the top of the swing is just g. This means that the block, at that particular instant, is in *free fall;* that is, the spring is exerting zero force on the block.   ■

EXAMPLE 3.2   *The Simple Pendulum*

The so-called simple pendulum consists of a small plum bob of mass m swinging at the end of a light inextensible string of length l, Figure 3.7. The motion is along a circular arc defined by the angle θ, as shown. The restoring force is the component of the weight $m\mathbf{g}$ acting in the direction of increasing θ along the path of motion: $F_s = -mg \sin \theta$. If we treat the bob as a particle, the differential equation of motion is, therefore,

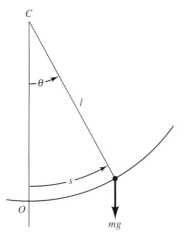

Figure 3.7   The simple pendulum.

$$m\ddot{s} = -mg \sin \theta$$

Now $s = l\theta$, and, for small θ, $\sin \theta = \theta$ to a fair approximation. So, after canceling the m's and rearranging terms, we can write the differential equation of motion either in terms of θ or s as follows:

$$\ddot{\theta} + \frac{g}{l}\theta = 0 \qquad \ddot{s} + \frac{g}{l}s = 0$$

Although the motion is along a curved path rather than a straight line, the differential equation is mathematically identical to that of the linear harmonic oscillator, Equation 3.3, with the quantity g/l replacing k/m. Thus, to the extent that the approximation $\sin \theta = \theta$ is valid, we can conclude that the motion is simple harmonic with angular frequency

$$\omega_0 = \sqrt{\frac{g}{l}}$$

and period

$$T_0 = \frac{2\pi}{\omega_0} = 2\pi\sqrt{\frac{l}{g}}$$

It is interesting to note that the above formula gives a period of very nearly 2 s, or a half-period of 1 s, when the length l is 1 m. More accurately, for a half-period of 1 s, known as the "seconds pendulum," the precise length is obtained by setting $T_0 = 2$ s and solving for l. This gives $l = g/\pi^2$, numerically, when g is expressed in m/s^2. At sea level at a latitude of 45°, the value of the acceleration of gravity is $g = 9.8062$ m/s^2. Accordingly, the length of a seconds pendulum at that location is $9.8062/9.8696 = 0.9936$ m. ∎

3.3   ENERGY CONSIDERATIONS IN HARMONIC MOTION

Consider a particle under the action of a linear restoring force $F_x = -kx$. Let us calculate the work done by an external force F_{ext} in moving the particle from the equilibrium position ($x = 0$) to some position x. We have $F_{ext} = -F_x = kx$, so

$$W = \int_0^x F_{ext}dx = \int_0^s kx\,dx = \frac{k}{2}x^2$$

In the case of a spring obeying Hooke's law, the work is stored in the spring as potential energy: $W = V(x)$ where

$$V(x) = \frac{1}{2}kx^2 \tag{3.14}$$

Thus, $F_x = -dV/dx = -kx$ as required by the definition of V. The total energy, when the particle is undergoing harmonic motion, is given by the sum of the kinetic and potential energies, namely

$$E = \frac{1}{2}m\dot{x}^2 + \frac{1}{2}kx^2 \tag{3.15}$$

This equation epitomizes the harmonic oscillator in a rather fundamental way: The kinetic energy is quadratic in the velocity variable, and the potential energy is quadratic in the displacement variable. The total energy is constant if there are no other forces except the restoring force acting on the particle.

The motion of the particle can be found by starting with the energy equation (3.15). Solving for the velocity gives

$$\frac{dx}{dt} = \pm\left(\frac{2E}{m} - \frac{k}{m}x^2\right)^{1/2} \tag{3.16}$$

which can be integrated to give t as a function of x as follows:

$$t = \int \frac{dx}{\pm[(2E/m) - (k/m)x^2]^{1/2}} = \mp(m/k)^{1/2} \cos^{-1}(x/A) + C \tag{3.17}$$

in which C is a constant of integration and A is the amplitude given by

$$A = \left(\frac{2E}{k}\right)^{1/2} \tag{3.18}$$

Upon solving the integrated equation for x as a function of t, we find the same relationship as that of the previous section, with the addition that we now have an explicit value for the amplitude. We can also obtain the amplitude directly from the energy equation (3.15) by finding the turning points of the motion where $\dot{x} = 0$: The value of x must lie between $(2E/k)^{1/2}$ and $-(2E/k)^{1/2}$ in order for $\dot{x}$ to be real. This is illustrated in Figure 3.8.

We also see from the energy equation that the maximum value of the speed, which we shall call v_{max}, occurs at $x = 0$. Accordingly, we can write

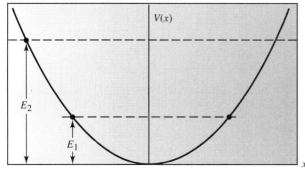

Figure 3.8 Graph of the parabolic potential energy function of the harmonic oscillator. The turning points defining the amplitude are indicated for two different values of the total energy.

$$E = \frac{1}{2}mv_{max}^2 = \frac{1}{2}kA^2 \tag{3.19}$$

As the particle oscillates, the kinetic and potential energies continually change. The constant total energy is entirely in the form of kinetic energy at the center where $x = 0$ and $\dot{x} = \pm v_{max}$, and it is all potential energy at the extrema where $\dot{x} = 0$ and $x = \pm A$.

EXAMPLE 3.3   *The Energy Function of the Simple Pendulum*

The potential energy of the simple pendulum is given by the expression

$$V = mgh$$

where h is the vertical distance from the reference level (which we choose to be the level of the equilibrium position). For a displacement through an angle θ, Figure 3.7, we see that $h = l - l \cos \theta$, so

$$V(\theta) = mgl(1 - \cos \theta)$$

Now the series expansion for the cosine is $\cos \theta = 1 - \theta^2/2! + \theta^4/4! - \cdots$, so for small θ we have approximately $\cos \theta = 1 - \theta^2/2$. This gives

$$V(\theta) = \frac{1}{2}mgl\,\theta^2$$

or, equivalently, since $s = l\theta$,

$$V(s) = \frac{1}{2}\frac{mg}{l}s^2$$

Thus, to a first approximation, the potential energy function is quadratic in the displacement variable. In terms of s, the total energy is given by

$$E = \frac{1}{2}m\dot{s}^2 + \frac{1}{2}\frac{mg}{l}s^2$$

in accordance with the general statement concerning the energy of the harmonic oscillator discussed above. ∎

EXAMPLE 3.4

Calculate the average kinetic, potential, and total energies of the harmonic oscillator.

Solution:

$$\langle K \rangle = \frac{1}{T_0}\int_0^{T_0} K(t)dt = \frac{1}{T_0}\int_0^{T_0} \frac{1}{2}m\dot{x}^2\,dt$$

but

$$x = A \sin (\omega_0 t + \phi_0)$$

$$\dot{x} = \omega_0 A \cos (\omega_0 t + \phi_0)$$

Setting $\phi_0 = 0$ and letting $u = \omega_0 t = (2\pi/T_0) \cdot t$, we obtain

$$\langle K \rangle = \frac{1}{T_0} \left[\frac{1}{2} m \omega_0^2 A^2 \int_0^{T_0} \cos^2 (\omega_0 t) \, dt \right]$$

$$= \frac{1}{2\pi} \left[\frac{1}{2} m \omega_0^2 A^2 \int_0^{2\pi} \cos^2 u \, du \right]$$

We can make use of the fact that

$$\frac{1}{2\pi} \int_0^{2\pi} (\sin^2 u + \cos^2 u) \, du = \frac{1}{2\pi} \int_0^{2\pi} du = 1$$

to obtain

$$\frac{1}{2\pi} \int_0^{2\pi} \cos^2 u \, du = \frac{1}{2}$$

since the areas under the $\cos^2$ and $\sin^2$ terms throughout one cycle are identical. Thus,

$$\langle K \rangle = \frac{1}{4} m \omega_0^2 A^2$$

The calculation of the average potential energy proceeds along similar lines.

$$V = \frac{1}{2} k x^2 = \frac{1}{2} k A^2 \sin^2 \omega_0 t$$

$$\langle V \rangle = \frac{1}{2} k A^2 \frac{1}{T_0} \int_0^{T_0} \sin^2 \omega_0 t \, dt$$

$$= \frac{1}{2} k A^2 \frac{1}{2\pi} \int_0^{2\pi} \sin^2 u \, du$$

$$= \frac{1}{4} k A^2$$

Now since $k/m = \omega_0^2$ or $k = m\omega_0^2$, we obtain

$$\langle V \rangle = \frac{1}{4} k A^2 = \frac{1}{4} m \omega_0^2 A^2 = \langle K \rangle$$

$$\langle E \rangle = \langle K \rangle + \langle V \rangle = \frac{1}{2} m \omega_0^2 A^2 = \frac{1}{2} k A^2 = E$$

The average kinetic energies and potential energies are equal; therefore, the average energy of the oscillator is equal to its total instantaneous energy. ∎

3.4   DAMPED HARMONIC MOTION

The foregoing analysis of the harmonic oscillator is somewhat idealized in that we have failed to take into account frictional forces. These are always present in a mechanical system to some extent. Analogously, there is always a certain amount of resistance in an electrical circuit. For a specific model let us consider an object of mass m that is supported by a light spring of stiffness k. We shall assume that there is a viscous retarding force varying *linearly* with the velocity, such as is produced by air drag at low speeds.[2] The forces are indicated in Figure 3.9.

If x is the displacement from equilibrium, then the restoring force is $-kx$, and the retarding force is $-c\dot{x}$ *where c is a constant of proportionality. The differential equation of motion is, therefore, $m\ddot{x} = -kx - c\dot{x}$ or

$$m\ddot{x} + c\dot{x} + kx = 0 \tag{3.20}$$

As with the undamped case, we divide Equation 3.20 by m to obtain

$$\ddot{x} + \frac{c}{m}\dot{x} + \frac{k}{m}x = 0 \tag{3.21}$$

The presence of the velocity-dependent term $c\dot{x}/m$ complicates the problem; simple sine or cosine solutions do not work as can be verified by trying them. We will introduce a method of solution that works rather well for second-order differential equations with

---

[2]Nonlinear drag is more realistic in many situations; however, the equations of motion are very much more difficult to solve and will not be treated here.

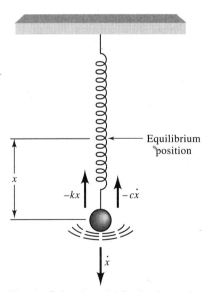

Figure 3.9   A model for the damped harmonic oscillator.

constant coefficients. Let D be the differential operator d/dt. We "operate" on x with a quadratic function of D chosen in such a way that we generate Equation 3.21

$$\left[D^2 + \frac{c}{m}D + \frac{k}{m} \right] x = 0 \tag{3.22}$$

We interpret this equation as an "operation" by the term in brackets on x. The operation by D^2 means first operate on x with D and then operate on the result of that operation with D again. This procedure yields $\ddot{x}$, the first term in Equation 3.21. The operator equation (Equation 3.22) is therefore equivalent to the differential equation (Equation 3.21). The simplification that we get by writing the equation this way arises when we factor the operator term, using the binomial theorem, to obtain

$$\left[D + \frac{c}{2m} - \sqrt{\frac{c^2}{4m^2} - \frac{k}{m}} \right]\left[D + \frac{c}{2m} + \sqrt{\frac{c^2}{4m^2} - \frac{k}{m}} \right] x = 0 \tag{3.22a}$$

The operation in Equation 3.22a is identical to that of Equation 3.22, but we have reduced the operation from second-order to a product of two first-order ones. Since the order of operation is arbitrary, the general solution is a sum of solutions obtained by setting the result of each first-order operation on x equal to zero. Thus, we obtain

$$x(t) = A_1 e^{-(\gamma - q)t} + A_2 e^{-(\gamma + q)t} \tag{3.23}$$

where

$$\gamma = \frac{c}{2m} \tag{3.24}$$

$$q = +\left(\frac{c^2}{4m^2} - \frac{k}{m} \right)^{1/2}$$

The student can verify that this is a solution by direct substitution into Equation 3.21. A problem that we will soon encounter, though, is that the above exponents may be real or complex since the factor q could be imaginary. We will see what this means in just a minute.

There are three possible scenarios:

I. q real > 0 *overdamping*
II. q real $= 0$ *critical damping*
III. q imaginary   *underdamping*

  I. *Overdamped.* Both exponents in Equation 3.23 are real. The constants A_1 and A_2 are determined by the initial conditions. The motion is an exponential decay with two different decay constants $(\gamma - q)$ and $(\gamma + q)$. A mass, given some initial displacement and released from rest, returns slowly to equilibrium, prevented from oscillating by the strong damping force. This situation is depicted in Figure 3.10.

  II. *Critical damping.* Here $q = 0$. The two exponents in Equation 3.23 are each equal to γ. The two constants A_1 and A_2 are no longer independent. Their sum forms a single constant A. The solution degenerates to a single exponential decay function.

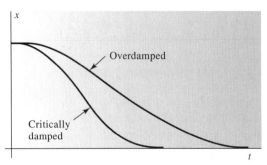

Figure 3.10 Displacement versus time for critically damped and overdamped oscillators released from rest after an initial displacement.

A completely general solution requires two different functions and independent constants in order to satisfy the boundary conditions specified by an initial position and velocity. To find a solution with two independent constants, we return to Equation 3.22a

$$(D + \gamma)(D + \gamma)x = 0$$

Switching the order of operation does not work here, since the operators are the same. We have to carry out the entire operation on x before setting the result to zero. To do this we make the substitution $u = (D + \gamma)x$, which gives

$$(D + \gamma)u = 0$$
$$u = Ae^{-\gamma t}$$

Equating this to $(D + \gamma)x$, the final solution is obtained as follows

$$Ae^{-\gamma t} = (D + \gamma)x$$
$$A = e^{\gamma t}(D + \gamma)x = D(xe^{\gamma t}) \tag{3.25}$$
$$\therefore xe^{\gamma t} = At + B$$
$$x(t) = Ate^{-\gamma t} + Be^{-\gamma t}$$

The solution consists of two different functions, $te^{-\gamma t}$ and $e^{-\gamma t}$, and two constants of integration A and B, as required. As in case I, if a mass is released from rest after an initial displacement, the motion is nonoscillatory, returning asymptotically to equilibrium. This case is also shown in Figure 3.10. Critical damping is highly desirable in many systems, such as the mechanical suspension systems of motor vehicles.

III. *Underdamping.* If the constant c in Equation 3.21 is small enough such that $c^2/4m^2 - k/m < 0$, the factor q in Equation 3.24 is imaginary. A mass, initially displaced and then released from rest, will oscillate, not unlike the situation described earlier for no damping force at all. The only difference is the presence of the real factor $-\gamma$ in the exponent of the solution that will cause the oscillatory motion to die away ultimately. Let us now reverse the factors under the square root sign in Equation 3.24 and write q as $i\omega_d$. Thus,

$$\omega_d = \sqrt{\frac{k}{m} - \frac{c^2}{4m^2}} = \sqrt{\omega_0^2 - \gamma^2} \tag{3.26}$$

where $\omega_0^2 = k/m$ is the angular frequency of a freely running, undamped harmonic oscillator. We now rewrite the general solution represented by Equation 3.23 in terms of the factors described here

$$
\begin{aligned}
x(t) &= C_+ e^{-(\gamma - i\omega_d)t} + C_- e^{-(\gamma + i\omega_d)t} \\
&= e^{-\gamma t}(C_+ e^{i\omega_d t} + C_- e^{-i\omega_d t})
\end{aligned}
\tag{3.27}
$$

where the constants of integration are C_+ and C_-. The solution contains a sum of imaginary exponentials. But the solution must be real—it is supposed to describe the real world! This reality demands that C_+ and C_- be complex conjugates of each other, a condition that will ultimately allow us to express the solution in terms of sines and/or cosines. Thus,

$$
\begin{aligned}
x^*(t) &= e^{-\gamma t}(C_+^* e^{-i\omega_d t} + C_-^* e^{+i\omega_d t}) = x(t) \\
\therefore\ C_+^* &= C_- = C \\
C_-^* &= C_+ = C^* \\
\therefore\ x(t) &= e^{-\gamma t}(C^* e^{+i\omega_d t} + C e^{-i\omega_d t})
\end{aligned}
\tag{3.28}
$$

It looks as though we have a solution that now has only a single constant of integration. In fact, C is a complex number. It is composed of two constants. We can express C and C^* in terms of two real constants A and θ_0 in the following way

$$C_- = C = \frac{A}{2} e^{-i\theta_0}$$

$$C_+ = C^* = \frac{A}{2} e^{+i\theta_0}$$

where we will soon see that A is the maximum displacement and θ_0 is the initial phase angle of the motion. Thus, Equation 3.28 becomes

$$x(t) = e^{-\gamma t}\left(\frac{A}{2} e^{+i(\omega_d t + \theta_0)} + \frac{A}{2} e^{-i(\omega_d t + \theta_0)} \right)$$

We will now apply Euler's identity[3] to the above expressions thus obtaining

$$\frac{A}{2} e^{+i(\omega_d t + \theta_0)} = \frac{A}{2}\cos(\omega_d t + \theta_0) + i\frac{A}{2}\sin(\omega_d t + \theta_0)$$

$$\frac{A}{2} e^{-i(\omega_d t + \theta_0)} = \frac{A}{2}\cos(\omega_d t + \theta_0) - i\frac{A}{2}\sin(\omega_d t + \theta_0) \tag{3.29}$$

$$\therefore\ x(t) = e^{-\gamma t}(A\cos(\omega_d t + \theta_0))$$

Following our discussion in Section 3.2 concerning the rotating vector construct,

---

[3] Euler's identity relates imaginary exponentials to sines and cosines. It is given by the expression $e^{iu} = \cos u + i\sin u$. This equality is demonstrated in Appendix D.

we see that we can express the solution equally well as a sine function

$$x(t) = e^{-\gamma t}(A \sin (\omega_d + \phi_0)) \qquad (3.29a)$$

The constants A, θ_0, and ϕ_0 have the same interpretation as those of Section 3.2. In fact, we see that the solution for the underdamped oscillator is nearly identical to that of the undamped oscillator. There are two differences: (1) the presence of the real exponential factor $e^{-\gamma t}$, which leads to a gradual death of the oscillations, and (2) the underdamped oscillator's angular frequency is ω_d, not ω_0, due to the presence of the damping force. The damped oscillator vibrates a little more slowly than does the undamped oscillator. The period of the underdamped oscillator is given by

$$T_d = \frac{2\pi}{\omega_d} = \frac{2\pi}{(\omega_0^2 - \gamma^2)^{1/2}} \qquad (3.30)$$

A plot of the motion is shown in Figure 3.11. Equation 3.29 shows that the two curves given by $x = Ae^{-\gamma t}$ and $x = -Ae^{-\gamma t}$ form an envelope of the curve of motion because the cosine factor takes on values between $+1$ and -1, including $+1$ and -1, at which points the curve of motion touches the envelope. Accordingly, the points of contact are separated by a time interval of one-half period $T_d/2$. These points, however, are not quite the maxima and minima of the displacement. It is left to the student to show that the actual maxima and minima are also separated in time by the same amount. In one complete period the amplitude diminishes by a factor $e^{-\gamma T_d}$; also, in a time $\gamma^{-1} = 2m/c$ the amplitude decays by a factor $e^{-1} = 0.3679$.

In summary, our analysis of the freely running harmonic oscillator has shown that the presence of damping of the linear type causes the oscillator, given an initial motion, to return eventually to a state of rest at the equilibrium position. The return to equilibrium is either oscillatory, or not, depending on the amount of damping. The critical condition, given by $c^2 = 4mk$, characterizes the limiting case of the nonoscillatory mode

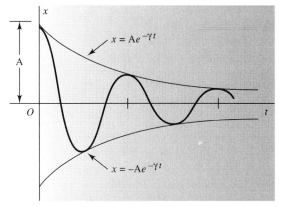

Figure 3.11 Graph of displacement versus time for the underdamped harmonic oscillator.

of return.

Mechanical Suspensions

Consider a mechanical system with linear damping. For fixed values of the damping constant c and stiffness k, the criterion determining the type of damping is determined by the value of the mass. If the system is critically damped for a certain critical mass m_{crit}, we have $c^2 = 4km_{crit}$. The discriminant is then

$$c^2 - 4km = 4k(m_{crit} - m)$$

so the system will be overdamped, or underdamped, respectively, depending on whether the mass is less than or greater than m_{crit}. Thus, in an automobile suspension system (springs and shock absorbers) increasing the load (passengers) causes the system to tend to be underdamped or oscillatory, while decreasing the load results in a stiffer, nonoscillatory condition.

Energy Considerations

The total energy of the damped harmonic oscillator is given by the sum of the kinetic and potential energies

$$E = \frac{1}{2}m\dot{x}^2 + \frac{1}{2}kx^2$$

This is constant for the undamped oscillator, as stated previously. Let us differentiate the above expression with respect to t:

$$\frac{dE}{dt} = m\dot{x}\ddot{x} + kx\dot{x} = (m\ddot{x} + kx)\dot{x}$$

Now the differential equation of motion is $m\ddot{x} + c\dot{x} + kx = 0$, or $m\ddot{x} + kx = -c\dot{x}$. Thus, we can write

$$\frac{dE}{dt} = -c\dot{x}^2$$

for the time rate of change of total energy. We see that it is given by the product of the damping force and the velocity. Since this is always either zero or negative, the total energy continually decreases and, like the amplitude, eventually becomes negligibly small. The energy is dissipated as frictional heat by virtue of the viscous resistance to the motion.

Quality Factor

The rate of energy loss of a weakly damped harmonic oscillator is best characterized by a single parameter Q called the *quality factor* of the oscillator. It is defined to be 2π times the energy stored in the oscillator divided by the energy lost in a single period of oscillation T_d. If the oscillator is weakly damped, the energy lost per cycle is small and Q is, therefore, large. We will calculate Q in terms of parameters already derived and show that this is true.

The average rate of energy dissipation for the damped oscillator is given by the expression derived above, $\dot{E} = -c\dot{x}^2$, so we need to calculate $\dot{x}$. We will assume that the oscillator is released from rest after an initial displacement equal to A. Thus, $\theta_0 = 0$ in Equation 3.29, and we have

$$x = Ae^{-\gamma t} \cos \omega_d t$$

$$\dot{x} = -Ae^{-\gamma t} (\gamma \cos \omega_d t + \omega_d \sin \omega_d t)$$

The energy lost in a single cycle is

$$\Delta E = \int_0^{T_d} \dot{E}\,dt = \frac{1}{\omega_d} \int_0^{2\pi} \dot{E}\,d(\omega_d t) = \frac{1}{\omega_d} \int_0^{2\pi} \dot{E}\,d\phi$$

$$= \frac{-cA^2}{\omega_d} \int_0^{2\pi} e^{-2\gamma t} [\gamma^2 \cos^2 \phi + 2\gamma\omega_d \cos \phi \sin \phi + \omega_d^2 \sin^2 \phi]\,d\phi$$

$$\approx \frac{-cA^2 T_d}{2\pi} e^{-2\gamma t} \int_0^{2\pi} [\gamma^2 \cos^2 \phi + 2\gamma\omega_d \cos \phi \sin \phi + \omega_d^2 \sin^2 \phi]\,d\phi$$

$$= \frac{-cA^2 T_d}{2\pi} e^{-2\gamma t}\pi\,(\gamma^2 + \omega_d^2) = -\left(\frac{c}{2m}\right) m\omega_0^2 A^2 e^{-2\gamma t} T_d$$

where we have used the following facts in the derivation above: (1) We extracted the exponential factor from the integral, since, in the case of weak damping, its value does not change very much during a single cycle of oscillation. (2) The integral of both $\sin^2$ and $\cos^2$ over a period of oscillation is π, and the integral of the (sin)(cos) product is 0. (3) $\omega_0^2 = \omega_d^2 + \gamma^2$ and $\gamma = c/2m$. (4) Furthermore, if we identify the damping factor c/m with a time constant τ, such that $c/m = 1/\tau$, then we have

$$\Delta E = \left(\frac{1}{2} mA^2\omega_0^2 e^{-t/\tau}\right) \frac{T_d}{\tau}$$

$$\frac{\Delta E}{E}(t) = \frac{T_d}{\tau}$$

where the energy stored in the oscillator (See Example 3.3a) at any time t has been identified as

$$E(t) = \frac{1}{2} m\omega_0^2 A^2 e^{-t/\tau}$$

Clearly, the energy remaining in the oscillator during any cycle dies away exponentially with time constant τ. We, therefore, see that the quality factor Q is just 2π times the inverse of the ratios given in the expression above or

$$Q = \frac{2\pi}{(T_d/\tau)} = \frac{2\pi}{(2\pi/\omega_d\tau)} = \omega_d\tau = \frac{\omega_d}{2\gamma} \tag{3.30a}$$

For weak damping the period of oscillation T_d is much less than the time constant τ,

TABLE 3.1 *Values of Q for Several Physical Systems.*

Earth (for earthquake)	250–1400
Piano string	3000
Crystal in digital watch	10^4
Microwave cavity	10^4
Excited atom	10^7
Neutron star	10^{12}
Excited Fe^{57} nucleus	3×10^{12}

which characterizes the energy loss rate of the oscillator. Q is large under such circumstances. Some values of Q are given in Table 3.1 for several different kinds of oscillators.

EXAMPLE 3.5

An automobile suspension system is critically damped, and its period of free oscillation with no damping is 1 s. If the system is initially displaced by an amount x_0 and released with zero initial velocity, find the displacement at $t = 1$ s.

Solution:
For critical damping we have $\gamma = c/2m = (k/m)^{1/2} = \omega_0 = 2\pi/T_0$. Hence $\gamma = 2\pi$ s^{-1} in our case, since $T_0 = 1$ s. Now the general expression for the displacement in the critically damped case, Equation 3.25, is $x(t) = (At + B)e^{-\gamma t}$, so, for $t = 0$, $x_0 = B$. Differentiating, we have $\dot{x}(t) = (A - \gamma B - \gamma At)e^{-\gamma t}$, which gives $\dot{x}_0 = A - \gamma B = 0$, so $A = \gamma B = \gamma x_0$ in our problem. Accordingly,

$$x(t) = x_0(1 + \gamma t)e^{-\gamma t} = x_0(1 + 2\pi t)e^{-2\pi t}$$

is the displacement as a function of time. For $t = 1$ s, we obtain $x_0(1 + 2\pi)e^{-2\pi} = x_0(7.28)e^{-6.28} = 0.0136 x_0$. The system has practically returned to equilibrium.

■

EXAMPLE 3.6

The frequency of a damped harmonic oscillator is one-half the frequency of the same oscillator with no damping. Find the ratio of the maxima of successive oscillations.

Solution:
We have $\omega_d = \frac{1}{2}\omega_0 = (\omega_0^2 - \gamma^2)^{1/2}$, which gives $\omega_0^2/4 = \omega_0^2 - \gamma^2$, so $\gamma = \omega_0(3/4)^{1/2}$. Consequently,

$$\gamma T_d = \omega_0(3/4)^{1/2}[2\pi/(\omega_0/2)] = 10.88$$

Thus, the amplitude ratio is

$$e^{-\gamma T_d} = e^{-10.88} = 0.00002$$

This is a *highly damped* oscillator.

■

EXAMPLE 3.7

Given: The terminal speed of a baseball in free fall is 30 m/s. Assuming a linear air drag, calculate the effect of air resistance on a simple pendulum using a baseball as the plum bob.

Solution:

In Chapter 2 we found the terminal speed for the case of linear air drag to be given by $v_t = mg/c_1$ where c_1 is the linear drag coefficient. This gives

$$\gamma = \frac{c_1}{2m} = \frac{(mg/v_t)}{2m} = \frac{g}{2v_t} = \frac{9.8\text{ms}^{-2}}{60\text{ ms}^{-1}} = 0.163 \text{ s}^{-1}$$

for the exponential damping constant. Consequently, the baseball pendulum's amplitude drops off by a factor e^{-1} in a time $\gamma^{-1} = 6.13$ s. Note that this is independent of the length of the pendulum. Earlier, in Example 3.2, we showed that the angular frequency of oscillation of the simple pendulum of length l is given by $\omega_0 = (g/l)^{1/2}$ for *small* amplitude. Therefore, from Equation 3.30, the period of our pendulum is

$$T_d = 2\pi(\omega_0^2 - \gamma^2)^{-1/2} = 2\pi\left(\frac{g}{l} - 0.0265 \text{ s}^{-2}\right)^{-1/2}$$

In particular, for a baseball "seconds pendulum" for which the half-period is 1 s in the absence of damping, we have $g/l = \pi^2$, so the half-period with damping in our case is

$$\frac{T_d}{2} = \pi(\pi^2 - 0.0265)^{-1/2} \text{ s} = 1.00134 \text{ s}$$

Our solution somewhat exaggerates the effect of air resistance, because the drag function for a baseball is more nearly quadratic in the velocity, rather than linear, except at very low velocities as discussed in Section 2.4.   ■

EXAMPLE 3.8

A spherical ball of radius .00265 m and mass 5×10^{-4} kg is attached to a spring of force constant $k = .05$ N/m underwater. The mass is set to oscillation under the action of the spring. The coefficient of viscosity η for water is 10^{-3} Ns/m^2. (a) Find the number of oscillations that the ball will execute in the time it takes for the amplitude of the oscillation to drop by a factor of 2 from its initial value. (b) Calculate the Q of the oscillator.

Solution:

Stokes law for objects moving in a viscous medium can be used to find c, the constant of proportionality of the $\dot{x}$ term, in the equation of motion for the damped oscillator. The relationship is

$$c = 6\pi\eta r = 5 \cdot 10^{-5} \text{ Ns/m}^2$$

The energy of the oscillator dies away exponentially with time constant τ, and the amplitude dies away as $A = A_0 e^{-t/2\tau}$. Thus,

$$\frac{A}{A_0} = \frac{1}{2} = e^{-t/2\tau}$$

$$\therefore t = 2\tau \ln 2$$

Consequently, the number of oscillations during this time is

$$n = \omega_d t/2\pi$$
$$= \pi(\omega_d \tau) \ln 2$$
$$= \pi Q \ln 2$$

Since $\omega_0^2 = k/m = 100 \text{ s}^{-2}$, $\tau = m/c = 10 \text{ s}$, and $\gamma = 1/2\tau = .05 \text{ s}^{-1}$, we obtain

$$Q = (\omega_0^2 - \gamma^2)^{1/2}\tau = (100 - .0025)^{1/2} \, 10 = 100$$
$$n = Q \ln 2/\pi = 22$$

If we had asked how many oscillations would occur in the time it takes for the amplitude to drop to $e^{-1/2}$ or about 0.606 times its initial value, the answer would have been Q. Clearly Q is a measure of the rate at which an oscillator loses energy. ∎

3.5 FORCED HARMONIC MOTION. RESONANCE

In this section we shall study the motion of a damped harmonic oscillator that is subjected to a periodic driving force by an external agent. Suppose a force of the form $F_0 \cos (\omega t)$ is exerted upon such an oscillator. The equation of motion is

$$m\ddot{x} = -kx - c\dot{x} + F_0 \cos (\omega t) \tag{3.31}$$

The most striking feature of such an oscillator is the way in which it responds as a function of the driving frequency even when the driving force is of fixed amplitude. A remarkable phenomenon occurs when the driving frequency is close in value to the natural frequency ω_0 of the oscillator. It is called *resonance*. Anyone who has ever pushed a child on a swing knows that the amplitude of oscillation can be made quite large if even the smallest push is made at just the right time. Small, periodic forces exerted on oscillators at frequencies well above or below the natural frequency are much less effective; the amplitude remains small. We will initiate our discussion of forced harmonic motion with a qualitative description of the behavior that we might expect. Then we will carry out a detailed analysis of the equation of motion (Equation 3.31), with our eyes peeled for the appearance of the phenomenon of resonance.

We already know that the undamped harmonic oscillator, subjected to any sort of disturbance that displaces it from its equilibrium position, will oscillate at its natural frequency, $\omega_0 = \sqrt{(k/m)}$. The dissipative forces inevitably present in any real system will change slightly the frequency of the oscillator from ω_0 to ω_d and cause the free oscillation to die out. This motion is represented by a solution to the *homogeneous* differential equation (Equation 3.20, which is Equation 3.31 without the driving force present). A periodic driving force will do two things to the oscillator: (1) it will initiate a "free" oscillation at its natural frequency, and (2) it will force the oscillator to vibrate eventually at the driving frequency ω. For a short time the actual motion is a linear superposition of oscillations at these two frequencies, but with one dying away and the other persisting. The motion that dies away is called the *transient*. The final, surviving motion, an oscillation at the driving frequency, is called the *steady-state* motion. It represents a solution to the inhomogeneous equation (Equation 3.31). Here we will focus only upon the steady-state motion, whose anticipated features we describe below. To aid in the descriptive process, we will assume for the moment that the damping term $-c\dot{x}$ is vanishingly small. Unfortunately, this approximation leads to the physical absurdity that the transient term never dies out—a rather paradoxical situation for a phenomenon described by the word *transient*! We will just ignore this difficulty and focus totally upon the steady-state description, in hopes that the simplicity gained by this approximation will give us insight that will help when we finally solve the problem of the driven, damped oscillator.

In the absence of damping, Equation 3.30a can be written as

$$m\ddot{x} + kx = F_0 \cos (\omega t) \qquad (3.31a)$$

The most dramatic feature of the resulting motion of this driven, undamped oscillator will be a catastrophically large response at $\omega = \omega_0$. This we shall soon see, but what response might we anticipate at both extremely low ($\omega \ll \omega_0$) and high ($\omega \gg \omega_0$) frequencies? At low frequencies, we might expect the inertial term $m\ddot{x}$ to be negligible compared to the spring force $-kx$. The spring should appear to be quite stiff, compressing and relaxing very slowly, with the oscillator moving pretty much in phase with the driving force. Thus, we might guess that

$$x \approx A \cos (\omega t)$$

$$A = \frac{F_0}{k}$$

At high frequencies the acceleration should be large, so we might guess that $m\ddot{x}$ should dominate the spring force $-kx$. The response, in this case, is controlled by the mass of the oscillator. Its displacement should be small and $180°$ out of phase with the driving force, since the acceleration of a harmonic oscillator is $180°$ out of phase with the displacement. The veracity of these preliminary considerations will emerge during the process of obtaining an actual solution.

First, let us solve Equation 3.31a representing the driven, undamped oscillator. In keeping with our previous descriptions of harmonic motion, we try a solution of the form

$$x(t) = A \cos (\omega t - \phi)$$

Thus, we assume that the steady-state motion is harmonic and that in the steady-state it ought to respond at the driving frequency ω. We note, though, that its response might differ in phase from that of the driving force by an amount ϕ. ϕ is not the result of some initial condition! (It does not make any sense to talk about initial conditions for a steady-state solution.) To see if this assumed solution works, we substitute it into Equation 3.31a obtaining

$$-m\omega^2 A \cos(\omega t - \phi) + kA \cos(\omega t - \phi) = F_0 \cos \omega t$$

This works if ϕ can take on only two values, 0 and π. Let us see what is implied by this requirement. Solving the above equation for $\phi = 0$ and π, respectively, yields

$$A = \frac{F_0/m}{(\omega_0^2 - \omega^2)} \qquad \phi = 0 \qquad \omega < \omega_0$$

$$= \frac{F_0/m}{(\omega^2 - \omega_0^2)} \qquad \phi = \pi \qquad \omega > \omega_0$$

We plot the amplitude A and phase angle ϕ as a function of ω in Figure 3.12. Indeed, as can be seen from the plots, as ω passes through ω_0, the amplitude becomes catastrophically large, and, perhaps even more surprisingly, the displacement shifts discontinuously from being in phase with the driving force to being 180° out of phase. True, these results are not physically possible. However, they are idealizations of real situations. As we shall soon see, if we throw in just a little damping, at ω close to ω_0 the amplitude becomes large but finite. The phase shift "smooths out"; it is no longer discontinuous, although the shift is still quite abrupt.

Note: The behavior of the system mimics our description of the low-frequency and high-frequency limits.

The 0° and 180° phase differences between the displacement and driving force can be simply and vividly demonstrated. Hold the lightest end of a pencil or a pair of scissors

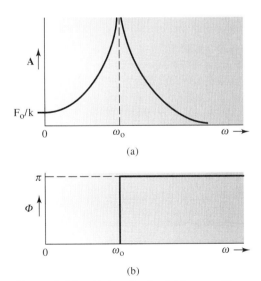

(a)

(b)

Figure 3.12 (a) Amplitude of driven oscillator versus ω with no damping. (b) Phase lag of the displacement relative to the driving force versus ω.

(closed) or a spoon delicately between forefinger and thumb, squeezing just hard enough that it does not drop. To demonstrate the 0° phase difference, slowly move hand back and forth horizontally in a direction parallel to the line formed between forefinger and thumb. The bottom of this makeshift pendulum will swing back and forth in phase with the hand motion and with a larger amplitude than the hand motion. To see the 180° phase shift, move hand back and forth rather rapidly (high frequency). The bottom of the pendulum will hardly move at all, but what little motion it does undergo will be 180° out of phase with the hand motion.

The Driven, Damped Harmonic Oscillator

We will now seek the steady-state solution to Equation 3.31, representing the driven, damped harmonic oscillator. It is fairly straightforward to solve this equation directly, but it is algebraically simpler to use complex exponentials instead of sines and/or cosines. First, we will represent the driving force as

$$F = F_0 e^{i\omega t}$$

so that Equation 3.31 becomes

$$m\ddot{x} + c\dot{x} + kx = F_0 e^{i\omega t} \tag{3.32}$$

The variable x is now complex, as is the applied force F. Remember, though, that by Euler's identity the real part of F is $F_0 \cos(\omega t)$. If we solve Equation 3.32 for x, its real part will be a solution to Equation 3.31. In fact, when we find a solution to the above complex equation (Equation 3.32), we can be sure that the real parts of both sides will be equal (as will the imaginary parts). It is the real parts that are equivalent to Equation 3.31 and, thus, the real, physical situation.

For the steady-state solution, let us, therefore, try the complex exponential

$$x(t) = A e^{i(\omega t - \phi)} \tag{3.33}$$

where the amplitude A and phase difference ϕ are constants to be determined. If this "guess" is correct, we must have

$$m\frac{d^2}{dt^2}A e^{i(\omega t - \phi)} + c\frac{d}{dt}A e^{i(\omega t - \phi)} + kA e^{i(\omega t - \phi)} = F_0 e^{i\omega t}$$

hold for all values of t. Upon performing the indicated operations and canceling the common factor $e^{i\omega t}$, we find

$$-m\omega^2 A + i\omega cA + kA = F_0 e^{i\phi} = F_0(\cos\phi + i\sin\phi) \tag{3.34}$$

Equating the real and imaginary parts yields the two equations

$$A(k - m\omega^2) = F_0 \cos\phi \tag{3.35}$$

$$c\omega A = F_0 \sin\phi$$

Upon dividing the second by the first and using the identity $\tan\phi = \sin\phi/\cos\phi$, we obtain the following relation for the phase angle

$$\tan \phi = \frac{c\omega}{k - m\omega^2} \tag{3.36}$$

By squaring both sides of Equations 3.35 and adding and employing the identity $\sin^2 \phi + \cos^2 \phi = 1$, we find

$$A^2(k - m\omega^2)^2 + c^2\omega^2 A^2 = F_0^2$$

We can then solve for A, the amplitude of the steady-state oscillation, as a function of the driving frequency

$$A(\omega) = \frac{F_0}{[(k - m\omega^2)^2 + c^2\omega^2]^{1/2}} \tag{3.37}$$

In terms of our previous abbreviations $\omega_0^2 = k/m$ and $\gamma = c/2m$, we can write the expressions in another form as follows:

$$\tan \phi = \frac{2\gamma\omega}{\omega_0^2 - \omega^2} \tag{3.38}$$

$$A(\omega) = \frac{F_0/m}{[(\omega_0^2 - \omega^2)^2 + 4\gamma^2\omega^2]^{1/2}} \tag{3.39}$$

A plot of the above amplitude A and phase difference ϕ versus driving frequency ω (see Figure 3.13) reveals a fetching similarity to the plots of Figure 3.12 for the case of

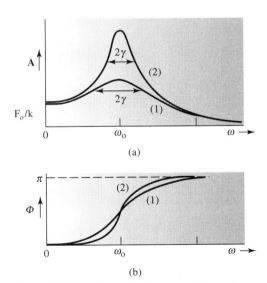

(a)

(b)

Figure 3.13 (a) Amplitude versus driving frequency for two values of the damping constant (1) $c = \frac{1}{2} m\omega_0$. (2) $c = \frac{1}{4} m\omega_0$. (b) Phase shift versus driving frequency for same values of c as in (a).

the undamped oscillator. As can be seen from the plots, as the damping term approaches 0, the resonant peak gets larger and narrower, and the phase shift sharpens up, ultimately approaching infinity and discontinuity, respectively, at ω_0. What is not so obvious from these plots is that the amplitude resonant frequency is not ω_0 when damping is present (although the phase shift always passes through $\pi/2$ at ω_0)! Amplitude resonance occurs at some other value ω_r, which can be calculated by differentiating $A(\omega)$ and setting the result equal to zero. Upon solving the resultant equation for ω, we obtain

$$\omega_r^2 = \omega_0^2 - 2\gamma^2 \tag{3.40}$$

ω_r approaches ω_0 as γ, the damping term, goes to zero. Note also that, since the angular frequency of the freely running damped oscillator is given by $\omega_d = (\omega_0^2 - \gamma^2)^{1/2}$, we have

$$\omega_r^2 = \omega_d^2 - \gamma^2 \tag{3.41}$$

When the damping is weak, the resonant frequency ω_r, the freely running, damped oscillator frequency ω_d, and the natural frequency ω_0 of the undamped oscillator are essentially identical, but only under this condition.

At the extreme of strong damping, no amplitude resonance occurs if $\gamma > \omega_0/\sqrt{2}$, because the amplitude then becomes a monotonically decreasing function of ω. To see this, consider the limiting case $\gamma^2 = \omega_0^2/2$. Equation 3.39 then gives

$$A(\omega) = \frac{F_0/m}{[(\omega_0^2 - \omega^2)^2 + 2\omega_0^2\omega^2]^{1/2}} = \frac{F_0/m}{(\omega_0^4 + \omega^4)^{1/2}}$$

which clearly decreases with increasing values of ω starting with $\omega = 0$.

Amplitude of Oscillation at the Resonance Peak

The steady-state amplitude at the resonant frequency, which we shall call A_{max}, is obtained from Equations 3.39 and 3.40. The result is

$$A_{max} = \frac{F_0/m}{2\gamma\sqrt{\omega_0^2 - \gamma^2}} = \frac{F_0}{c\omega_d} \tag{3.42}$$

In the case of weak damping, we can neglect γ^2 and write

$$A_{max} \simeq \frac{F_0}{2\gamma m\omega_0} = \frac{F_0}{c\omega_0} \tag{3.42a}$$

Thus, the amplitude of the induced oscillation at the resonant condition becomes very large if the damping constant c is very small, and conversely. In mechanical systems it may, or may not, be desirable to have large resonant amplitudes. In the case of electric motors, for example, rubber or spring mounts are used to minimize the transmission of vibration. The stiffness of these mounts is chosen so as to ensure that the resulting resonant frequency is far from the running frequency of the motor.

Sharpness of the Resonance. Quality Factor

The sharpness of the resonance peak is frequently of interest. Let us consider the case of weak damping $\gamma \ll \omega_0$. Then in the expression for steady-state amplitude (Equation 3.39), we can make the following substitutions:

$$\omega_0^2 - \omega^2 = (\omega_0 + \omega)(\omega_0 - \omega)$$
$$\approx 2\omega_0(\omega_0 - \omega)$$
$$\gamma\omega \approx \gamma\omega_0$$

These, together with the expression for A_{max}, allow us to write the amplitude equation in the following approximate form:

$$A(\omega) \approx \frac{A_{max}\gamma}{\sqrt{(\omega_0 - \omega)^2 + \gamma^2}} \tag{3.43}$$

The above equation shows that when $|\omega_0 - \omega| = \gamma$, or equivalently, if

$$\omega = \omega_0 \pm \gamma$$

then

$$A^2 = \frac{1}{2}A_{max}^2$$

This means that γ is a measure of the width of the resonance curve. Thus, 2γ is the frequency difference between the points for which the energy is down by a factor of $\frac{1}{2}$ from the energy at resonance, because the energy is proportional to A^2. This is illustrated in Figure 3.13a.

The quality factor Q, defined in Equation 3.30a, which characterizes the rate of energy loss in the undriven, damped harmonic oscillator, also characterizes the sharpness of the resonance peak for the driven oscillator. In the case of weak damping, Q can be expressed as

$$Q = \frac{\omega_d}{2\gamma} \approx \frac{\omega_0}{2\gamma} \tag{3.44}$$

Thus, the total width $\Delta\omega$ at the half-energy points is approximately

$$\Delta\omega = 2\gamma \approx \frac{\omega_0}{Q}$$

or, since $\omega = 2\pi f$,

$$\frac{\Delta\omega}{\omega_0} = \frac{\Delta f}{f_0} \approx \frac{1}{Q} \tag{3.45}$$

giving the fractional width of the resonance peak.

This last expression for Q, so innocuous looking, represents a key feature of feedback and control in electrical systems. Many electrical systems require the existence of a well-defined and precisely maintained frequency. High Q (of order 10^5) quartz oscillators, vibrating at their resonant frequency, are commonly employed as the control element in feedback circuits to provide frequency stability. A high Q results in a sharp resonance. If the frequency of the circuit under control by the quartz oscillator starts to wander or drift by some amount δf away from the resonance peak, feedback circuitry, exploiting the sharpness of the resonance, drives the circuit vigorously back toward the resonant frequency. The higher the Q of the oscillator and thus the narrower δf, the more stable the output of the frequency of the circuit.

The Phase Difference ϕ

The difference in phase ϕ between the applied driving force and the steady-state response is given by Equation 3.38, namely

$$\phi = \tan^{-1}\left[\frac{2\gamma\omega}{(\omega_0^2 - \omega^2)}\right]$$

The phase difference is plotted in Figure 3.13(b). We saw that for the driven, undamped oscillator, ϕ was $0°$ at $\omega < \omega_0$ and $180°$ for $\omega > \omega_0$. It can now be seen that these values are the low- and high-frequency limits of the real motion. Furthermore, ϕ changed discontinuously at $\omega = \omega_0$. This, too, is an idealization of the real motion where the transition between the two limits is smooth, although for very small damping it will be quite abrupt, changing essentially from one limit to the other as ω passes through a region within $\pm \gamma$ about ω_0.

At low driving frequencies $\omega \ll \omega_0$, we see that $\phi \to 0$ and the response is nearly in phase with the driving force. That this is reasonable can be seen upon examination of the amplitude of the oscillation (Equation 3.39). In the low-frequency limit, it becomes

$$A(\omega \to 0) \approx \frac{F_0/m}{\omega_0^2} = \frac{F_0/m}{k/m} = \frac{F_0}{k}$$

In other words, just as we claimed during our preliminary discussion of the driven oscillator, the spring, and not the mass or the friction, controls the response; the mass is slowly pushed back and forth by a force acting against the retarding force of the spring.

At resonance the response can be enormous. Physically, how can this be? Perhaps some insight can be gained by thinking about pushing a child on a swing. How is it done? Clearly, anyone who has experience pushing a swing did not stand behind the child and push when the swing is on the backswing. They would push in the same direction that the swing is *moving,* essentially in phase with its velocity, regardless of its position. To push a small child, we usually stand somewhat to the side and give a very small shove forward as the swing passes through the equilibrium position, when its speed was a maximum and the displacement was zero! In fact, this is the optimum way to achieve a resonance condition, in which a rather gentle force, judiciously applied, can lead to a large amplitude of oscillation. The maximum amplitude at resonance is given by Equation 3.42 and in the case of weak damping by Equation 3.42a, $A_{max} \approx F_0/c\omega_0$. But from the expression above for the amplitude as $\omega \to 0$, we have $A(\omega \to 0) \approx F_0/m\omega_0^2$. Hence, the ratio is

$$\frac{A_{max}}{A(\omega \to 0)} = \frac{F_0/(c\omega_0)}{F_0/(m\omega_0^2)} = \omega_0 \frac{m}{c} = \omega_0\tau = Q$$

The result is simply the Q of the oscillator. Imagine what would happen to the child on the swing if there were no frictional losses! We would continue to pump little bits of energy into the swing on a cycle-by-cycle basis, and with no energy loss per cycle, the amplitude would soon grow to a catastrophic dimension.

Now let us look at the phase difference. At $\omega = \omega_0$, $\phi = \pi/2$. Hence, the displacement "lags" or is behind the driving force by 90°. In view of our discussion above, this should make sense. The optimum way to dump energy into the oscillator is when it swings through zero at maximum velocity; that is, the power input $\mathbf{F} \cdot \mathbf{v}$ is a maximum. For example, the displacement of the oscillator is given by taking the real part of Equation 3.33

$$x(t) = A(\omega) \; Re(e^{i\omega t}) = A(\omega) \cos(\omega t - \phi) \tag{3.46}$$

and at resonance, for small damping, this becomes

$$x(t) = A(\omega_0) \cos(\omega_0 t - \pi/2)$$
$$= A(\omega_0) \sin \omega_0 t$$

The velocity, in general, is

$$\dot{x}(t) = -\omega A(\omega) \sin(\omega t - \phi) \tag{3.47}$$

which at resonance becomes

$$\dot{x}(t) = \omega_0 A(\omega_0) \cos(\omega_0 t) \tag{3.48}$$

Since the driving force, at resonance, is given by

$$F = F_0 Re(e^{i\omega_0 t}) = F_0 \cos(\omega_0 t)$$

we can see that the driving force is, indeed, in phase with the velocity of the oscillator, or 90° ahead of the displacement.

Finally, for large values of ω, $\omega \gg \omega_0$, $\phi \to \pi$, and the displacement is 180° out of phase with the driving force. The amplitude of the displacement becomes

$$A(\omega \gg \omega_0) \approx \frac{F_0}{m\omega^2}$$

In this case, the amplitude falls off as $1/\omega^2$. The mass responds essentially like a free object, being rapidly shaken back and forth by the applied force. The main effect of the spring is to cause the displacement to lag behind the driving force by 180°.

Electrical–Mechanical Analogs

When an electric current flows in a circuit comprised of inductive, capacitive, and resistive elements, there is a precise analogy with a moving mechanical system of masses and springs with frictional forces of the type studied previously. Thus, if a current $i = dq/dt$ (q being the charge) flows through an inductance L, the potential differ-

TABLE 3.2   Electrical–Mechanical Analogs.

	Mechanical		**Electrical**
x	Displacement	q	Charge
$\dot{x}$	Velocity	$\dot{q} = i$	Current
m	Mass	L	Inductance
k	Stiffness	C^{-1}	Reciprocal of capacitance
c	Damping resistance	R	Resistance
F	Force	V	Potential difference

ence across the inductance is $L\ddot{q}$, and the stored energy is $\frac{1}{2}L\dot{q}^2$. Hence, inductance and charge are analogous to mass and displacement, respectively, and potential difference is analogous to force. Similarly, if a capacitance C carries a charge q, the potential difference is $C^{-1}q$, and the stored energy is $\frac{1}{2}C^{-1}q^2$. Consequently, we see that the reciprocal of C is analogous to the stiffness constant of a spring. Finally, for an electric current i flowing through a resistance R, the potential difference is $iR = \dot{q}R$, and the rate of energy dissipation is $i^2R = \dot{q}^2R$ in analogy with the quantity $c\dot{x}^2$ for a mechanical system. Table 3.2 summarizes the situation.

EXAMPLE 3.9

The exponential damping factor γ of a spring suspension system is one tenth the critical value. If the undamped frequency is ω_0, find (a) the resonant frequency, (b) the quality factor, (c) the phase angle ϕ when the system is driven at a frequency $\omega = \omega_0/2$, and (d) the steady-state amplitude at this frequency.

Solution:
(a) We have $\gamma = \gamma_{crit}/10 = \omega_0/10$, so from Equation 3.40

$$\omega_r = [\omega_0^2 - 2(\omega_0/10)^2]^{1/2} = \omega_0(0.98)^{1/2} = 0.99\omega_0$$

(b) The system can be regarded as weakly damped, so, from Equation 3.44

$$Q \simeq \frac{\omega_0}{2\gamma} = \frac{\omega_0}{2(\omega_0/10)} = 5$$

(c) From Equation 3.38 we have

$$\phi = \tan^{-1}\left(\frac{2\gamma\omega}{\omega_0^2 - \omega^2}\right) = \tan^{-1}\left[\frac{2(\omega_0/10)(\omega_0/2)}{\omega_0^2 - (\omega_0/2)^2}\right]$$

$$= \tan^{-1}0.133 = 7.6°$$

(d) From Equation 3.39 we first calculate the value of the resonance denominator

$$D(\omega = \omega_0/2) = [(\omega_0^2 - \omega_0^2/4)^2 + 4(\omega_0/10)^2(\omega_0/2)^2]^{1/2}$$

$$= [(9/16) + (1/100)]^{1/2}\omega_0^2 = 0.7566\,\omega_0^2$$

From this, the amplitude is

$$A(\omega = \omega_0/2) = \frac{F_0/m}{0.7566\ \omega_0^2} = 1.322\frac{F_0}{m\omega_0^2}$$

Notice that the factor $(F_0/m\omega_0^2) = F_0/k$ is the steady-state amplitude for zero driving frequency. ∎

3.6 THE NONLINEAR OSCILLATOR. METHOD OF SUCCESSIVE APPROXIMATIONS

When a system is displaced from its equilibrium position, the restoring force may vary in a manner other than in direct proportion to the displacement. For example, a spring may not obey Hooke's law exactly; also, in many physical cases the restoring force function is inherently nonlinear, as is the case with the simple pendulum discussed in the example to follow.

In the nonlinear case the restoring force can be expressed as

$$F(x) = -kx + \epsilon(x) \tag{3.49}$$

in which the function $\epsilon(x)$ represents the departure from linearity. It is necessarily quadratic, or higher order, in the displacement variable x. The differential equation of motion under such a force, assuming no external forces are acting, can be written in the form

$$m\ddot{x} + kx = \epsilon(x) = \epsilon_2 x^2 + \epsilon_3 x^3 + \cdots \tag{3.50}$$

Here we have expanded $\epsilon(x)$ as a power series.

The above type of equation usually requires some method of approximation to arrive at a solution. To illustrate one method we shall take a particular case in which only the cubic term in $\epsilon(x)$ is of importance. Then we have

$$m\ddot{x} + kx = \epsilon_3 x^3 \tag{3.51}$$

Upon division by m and introduction of the abbreviations $\omega_0^2 = k/m$ and $\epsilon_3/m = \lambda$, we can write

$$\ddot{x} + \omega_0^2 x = \lambda x^3 \tag{3.52}$$

We shall find the solution by *the method of successive approximations*.

Now we now that for $\lambda = 0$ a solution is $x = A \cos \omega_0 t$. Suppose we try a *first* approximation of the same form

$$x = A \cos \omega t \tag{3.53}$$

where, as we shall see, ω is not quite equal to ω_0. Inserting our trial solution into the differential equation gives

$$-A\omega^2 \cos \omega t + A\omega_0^2 \cos \omega t = \lambda A^3 \cos^3 \omega t = \lambda A^3 \left(\frac{3}{4} \cos \omega t + \frac{1}{4} \cos 3\omega t\right)$$

In the last step we have used the trigonometric identity $\cos^3 u = \frac{3}{4} \cos u + \frac{1}{4} \cos 3u$, which is easily derived by use of the relation $\cos^3 u = [(e^{iu} + e^{-iu})/2]^3$. Upon transposing and collecting terms we get

$$(-\omega^2 + \omega_0^2 - \frac{3}{4} \lambda A^2)A \cos \omega t - \frac{1}{4} \lambda A^3 \cos 3\omega t = 0 \qquad (3.54)$$

Excluding the trivial case $A = 0$, we see that our trial solution does not exactly satisfy the differential equation. However, an approximation to the value of ω, *which is valid for small* λ, is obtained by setting the quantity in parentheses equal to zero. This yields

$$\omega^2 = \omega_0^2 - \frac{3}{4} \lambda A^2 \qquad (3.55)$$

for the frequency of our freely running nonlinear oscillator. As we can see, it is a function of the amplitude A.

   To obtain a better solution we must take into account the dangling term involving the third harmonic, $\cos 3\omega t$. Accordingly, we shall take a *second* trial solution of the form

$$x = A \cos \omega t + B \cos 3\omega t \qquad (3.56)$$

Putting this into the differential equation we find, after collecting terms,

$$\left(-\omega^2 + \omega_0^2 - \frac{3}{4} \lambda A^2\right)A \cos \omega t + \left(-9B\omega^2 + \omega_0^2 B - \frac{1}{4} \lambda A^3\right) \cos 3\omega t$$
$$+ \text{(terms involving } B\lambda \text{ and higher multiples of } \omega t) = 0$$

Setting the first quantity in parentheses equal to zero gives the same value for ω found above. Equating the second to zero gives a value for the coefficient B, namely,

$$B = \frac{\frac{1}{4} \lambda A^3}{-9\omega^2 + \omega_0^2} = \frac{\lambda A^3}{-32\omega_0^2 + 27\lambda A^2} \approx -\frac{\lambda A^3}{32\omega_0^2} \qquad (3.57)$$

where we have assumed that the term in the denominator involving λA^2 is small enough to neglect. Our second approximation can be expressed as

$$x = A \cos \omega t - \frac{\lambda A^3}{32\omega_0^2} \cos 3\omega t \qquad (3.58)$$

$$\omega = \omega_0 \left(1 - \frac{3\lambda A^2}{4\omega_0^2}\right)^{1/2} \qquad (3.59)$$

We shall stop at this point, but the process could be repeated to find yet a third approximation, and so on.

   The above analysis, although it is admittedly very crude, brings out two essential features of free oscillation under a nonlinear restoring force; that is, the period of oscillation is a function of the amplitude of vibration, and the oscillation is not strictly sinusoidal but can be considered as the superposition of a mixture of harmonics. It can be shown that the vibration of a nonlinear system driven by a purely sinusoidal driving force will also be distorted; that is, it will contain harmonics. The loudspeaker of a stereo system, for example, may introduce distortion (harmonics) over and above that introduced by the electronic amplifying system.

EXAMPLE 3.10   *The Simple Pendulum as a Nonlinear Oscillator*

In Example 3.2 we treated the simple pendulum as a linear harmonic oscillator by using the approximation $\sin \theta \simeq \theta$. Actually, the sine can be expanded as a power series

$$\sin \theta = \theta - \frac{\theta^3}{3!} + \frac{\theta^5}{5!} - \cdots$$

so the differential equation for the simple pendulum, $\ddot{\theta} + (g/l) \sin \theta = 0$, may be written in the form of Equation 3.50, and, by retaining only the linear and the cubic terms in the expansion for the sine, the differential equation becomes

$$\ddot{\theta} + \omega_0^2 \theta = \frac{\omega_0^2}{3!} \theta^3$$

in which $\omega_0^2 = g/l$. This is mathematically identical to Equation 3.52 with the constant $\lambda = \omega_0^2/3! = \omega_0^2/6$. The improved expression for the angular frequency, Equation 3.59, then gives

$$\omega = \omega_0 \left[1 - \frac{3(\omega_0^2/6)A^2}{4\omega_0^2} \right]^{1/2} = \omega_0 \left(1 - \frac{A^2}{8} \right)^{1/2}$$

and

$$T = \frac{2\pi}{\omega} = 2\pi \sqrt{\frac{l}{g}} \left(1 - \frac{A^2}{8} \right)^{-1/2} = T_0 \left(1 - \frac{A^2}{8} \right)^{-1/2}$$

for the period of the simple pendulum. Here A is the amplitude of oscillation expressed in radians. Our method of approximation shows that the period for nonzero amplitude is longer by the factor $(1 - A^2/8)^{-1/2}$ than that calculated earlier assuming $\sin \theta = \theta$. For instance, if the pendulum is swinging with an amplitude of $90° = \pi/2$ radians (a fairly large amplitude) the factor is $(1 - \pi^2/32)^{-1/2} = 1.2025$, so the period is about 20% longer than the period for small amplitude. This is considerably greater than the increase due to damping of the baseball pendulum, treated in Example 3.6.   ■

*3.7   NONSINUSOIDAL DRIVING FORCE. FOURIER SERIES

In order to determine the motion of a harmonic oscillator that is driven by an external periodic force that is *other* than "pure" sinusoidal, it is necessary to employ a somewhat

---

*As in Chapter 1, sections marked with an asterisk may be omitted without loss of continuity.

more involved method than that of the previous sections. In this more general case it is convenient to use the *principle of superposition*. The principle is applicable to any system governed by a linear differential equation. In our application the principle states that if the external driving force acting on a damped harmonic oscillator is given by a superposition of force functions

$$F_{ext} = \sum_n F_n(t) \tag{3.60}$$

such that the differential equation

$$m\ddot{x}_n + c\dot{x}_n + kx_n = F_n(t)$$

is individually satisfied by the functions $x_n(t)$, then the solution of the differential equation of motion

$$m\ddot{x} + c\dot{x} + kx = F_{ext} \tag{3.61}$$

is given by the superposition

$$x(t) = \sum_n x_n(t) \tag{3.62}$$

The validity of the principle is easily verified by substitution:

$$m\ddot{x} + c\dot{x} + kx = \sum_n (m\ddot{x}_n + c\dot{x}_n + kx_n) \sum_n F_n(t) = F_{ext}$$

In particular, when the driving force is periodic, that is, if for any value of the time t

$$F_{ext}(t) = F_{ext}(t + T)$$

where T is the period, then the force function can be expressed as a superposition of harmonic terms according to *Fourier's theorem*. This theorem states that any periodic function $f(t)$ can be expanded as a sum as follows:

$$f(t) = \frac{1}{2}a_0 + \sum_{n=1}^{\infty} [a_n \cos(n\omega t) + b_n \sin(n\omega t)] \tag{3.63}$$

The coefficients are given by the following formulas (derived in Appendix G):

$$a_n = \frac{2}{T} \int_{-T/2}^{T/2} f(t) \cos(n\omega t) \, dt \quad n = 0, 1, 2, \ldots$$

$$b_n = \frac{2}{T} \int_{-T/2}^{T/2} f(t) \sin(n\omega t) \, dt \quad n = 1, 2, \ldots \tag{3.64}$$

Here T is the period and $\omega = 2\pi/T$ is the fundamental frequency. If the function $f(t)$ is an *even* function, that is, if $f(t) = f(-t)$, then the coefficients $b_n = 0$ for all n. The series expansion is then known as a *Fourier cosine series*. Similarly, if we have an *odd* function so that $f(t) = -f(-t)$, then the a_n all vanish, and the series is called a *Fourier sine series*. By use of the relation $e^{iu} = \cos u + i \sin u$, it is straightforward to verify that Equations 3.63 and 3.64 may also be expressed in complex exponential form as follows:

$$f(t) = \sum_n c_n e^{in\omega t} \qquad n = 0, \pm 1, \pm 2, \ldots \qquad (3.65)$$

$$c_n = \frac{1}{T} \int_{-T/2}^{T/2} f(t) e^{-in\omega t} dt \qquad (3.66)$$

Thus, to find the steady-state motion of our harmonic oscillator subject to a given periodic driving force, we express the force as a Fourier series of the form of Equation 3.63 or 3.65, using equation 3.64 or 3.66 to determine the Fourier coefficients a_n and b_n, or c_n. For each value of n, corresponding to a given harmonic $n\omega$ of the fundamental driving frequency ω, there is a response function $x_n(t)$. This function is the steady-state solution of the driven oscillator treated in Section 3.5. The superposition of all the $x_n(t)$ gives the actual motion. In the event that one of the harmonics of the driving frequency coincides, or nearly coincides, with the resonance frequency ω_r, then the response at that harmonic will dominate the motion. As a result, if the damping constant γ is very small, the resulting oscillation may be very nearly sinusoidal even if a highly nonsinusoidal driving force is applied.

EXAMPLE 3.11 *Periodic Pulse*

To illustrate the above theory we shall analyze the motion of a harmonic oscillator that is driven by an external force consisting of a succession of rectangular pulses:

$$F_{ext}(t) = F_0 \qquad NT - \frac{1}{2}\Delta T \le t \le NT + \frac{1}{2}\Delta T$$

$$F_{ext}(t) = 0 \qquad \text{otherwise}$$

where $N = 0, \pm 1, \pm 2, \ldots$, T is the time from one pulse to the next, and ΔT is the width of each pulse as shown in Figure 3.14. In this case $F_{ext}(t)$ is an even function

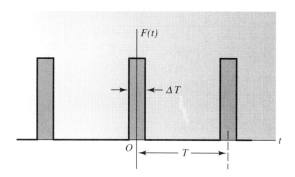

Figure 3.14 Rectangular-pulse driving force.

of t, so it can be expressed as a Fourier cosine series. Equation 3.64 gives the coefficients

$$
\begin{aligned}
a_n &= \frac{2}{T} \int_{-\Delta T/2}^{+\Delta T/2} F_0 \cos(n\omega t)\, dt \\[2mm]
&= \frac{2}{T} F_0 \left[\frac{\sin(n\omega t)}{n\omega} \right]_{-\Delta T/2}^{+\Delta T/2} \\[2mm]
&= F_0 \frac{2\,\sin(n\pi\Delta T/T)}{n\pi}
\end{aligned}
$$

where in the last step we use the fact that $\omega = 2\pi/T$. We see also that

$$
a_0 = \frac{2}{T} \int_{-\Delta T/2}^{+\Delta T/2} F_0\, dt = F_0 \frac{2\Delta T}{T}.
$$

Thus, for our periodic pulse force we can write

$$
\begin{aligned}
F_{ext}(t) = F_0 \Bigg[\frac{\Delta T}{T} &+ \frac{2}{\pi} \sin\left(\pi\,\frac{\Delta T}{T}\right) \cos(\omega t) \\
&+ \frac{2}{2\pi} \sin\left(2\pi\frac{\Delta T}{T}\right) \cos(2\omega t) + \frac{2}{3\pi} \sin\left(3\pi\frac{\Delta T}{T}\right) \cos(3\omega t) + \cdots \Bigg]
\end{aligned}
\tag{3.68}
$$

The first term in the above series expansion is just the *average* value of the external force: $F_{avg} = F_0(\Delta T/T)$. The second term is the Fourier component at the fundamental frequency ω. The remaining terms are harmonics of the fundamental: 2ω, 3ω, and so on.

Referring to Equations 3.38 and 3.39, we can now write the final expression for the motion of our pulse-driven oscillator. It is given by the superposition principle

$$
x(t) = \sum_n x_n(t) = \sum_n A_n \cos(n\omega t - \phi_n)
\tag{3.69}
$$

in which the respective amplitudes are

$$
A_n = \frac{a_n/m}{D_n(\omega)} = \frac{(F_0/m)(2/n\pi)\,\sin(n\pi\Delta T/T)}{[(\omega_0^2 - n^2\omega^2)^2 + 4\gamma^2 n^2\omega^2]^{1/2}}
\tag{3.70}
$$

and the phase angles

$$
\phi_n = \tan^{-1}\left(\frac{2\gamma n\omega}{\omega_0^2 - n^2\omega^2} \right)
\tag{3.71}
$$

Here m is the mass, γ is the decay constant, and ω_0 is the frequency of the freely running oscillator with no damping.

As a specific numerical example let us consider the spring suspension system of Example 3.9 under the action of a periodic pulse for which the pulse width is one tenth the pulse period: $\Delta T/T = 0.1$. As before, we shall take the damping constant to be one-tenth critical: $\gamma = 0.1\,\omega_0$ and the pulse frequency to be one half the undamped frequency of the system: $\omega = \omega_0/2$. The Fourier series for the driving force (Equation 3.68) is then

$$F_{ext}(t) = F_0\left[0.1 + \frac{2}{\pi} \sin(0.1\pi) \cos(\omega t) + \frac{2}{2\pi} \sin(0.2\pi) \cos(2\omega t)\right.$$

$$\left. + \frac{2}{3\pi} \sin(0.3\pi) \cos(3\omega t) + \cdots\right]$$

$$= F_0[0.1 + 0.197 \cos(\omega t) + 0.187 \cos(2\omega t)$$

$$+ 0.172 \cos(3\omega t) + \cdots]$$

The resonance denominators in Equation 3.70 are given by

$$D_n = \left[\left(\omega_0^2 - n^2\frac{\omega_0^2}{4}\right)^2 + 4(0.1)^2\omega_0^2 n^2\frac{\omega_0^2}{4}\right]^{1/2} = \left[\left(1 - \frac{n^2}{4}\right)^2 + 0.01n^2\right]^{1/2}\omega_0^2$$

Thus,

$$D_0 = \omega_0^2 \qquad D_1 = 0.757\omega_0^2 \qquad D_2 = 0.2\omega_0^2 \qquad D_3 = 1.285\omega_0^2$$

The phase angles, Equation 3.71, are

$$\phi_n = \tan^{-1}\left(\frac{0.2n\omega_0^2/2}{\omega_0^2 - n^2\omega_0^2/4}\right) = \tan^{-1}\left(\frac{0.4n}{4 - n^2}\right)$$

which gives

$$\phi_0 = 0 \qquad \phi_1 = \tan^{-1}(0.133) = 0.132$$

$$\phi_2 = \tan^{-1}\infty = \pi/2 \qquad \phi_3 = \tan^{-1}(-0.24) = -0.236$$

The steady-state motion of the system is, therefore, given by the following series (Equation 3.69):

$$x(t) = \frac{F_0}{m\omega_0^2}[0.1 + 0.26 \cos(\omega t - 0.132) + 0.935 \sin(2\omega t)$$

$$+ 0.134 \cos(3\omega t + 0.236) + \cdots]$$

The dominant term is the one involving the second harmonic $2\omega = \omega_0$ because ω_0 is close to the resonant frequency. Note also the phase of this term: $\cos(2\omega t - \pi/2) = \sin(2\omega t)$. ∎

PROBLEMS

3.1 A guitar string vibrates harmonically with a frequency of 512 Hz (one octave above middle C on the musical scale). If the amplitude of oscillation of the center point of the string is 0.002 m (2 mm), what is the maximum speed and the maximum acceleration at that point?

3.2 A piston executes simple harmonic motion with an amplitude of 0.1 m. If it passes through the center of its motion with a speed of 0.5 m/s, what is the period of oscillation?

3.3 A particle undergoes simple harmonic motion with a frequency of 10 Hz. Find the displacement x at any time t for the following initial condition:

$$t = 0 \qquad x = 0.25 \text{ m} \qquad \dot{x} = 0.1 \text{ m/s}$$

3.4 Verify the relations among the four quantities C, D, ϕ_0, and A given by Equation 3.12a.

3.5 A particle undergoing simple harmonic motion has a velocity $\dot{x}_1$ when the displacement is x_1 and a velocity $\dot{x}_2$ when the displacement is x_2. Find the angular frequency and the amplitude of the motion in terms of the given quantities.

3.6 On the surface of the moon, the acceleration of gravity is about one sixth that on the earth. What is the half-period of a simple pendulum of length 1 m on the moon?

3.7 Two springs having stiffness k_1 and k_2, respectively, are used in a vertical position to support a single object of mass m. Show that the angular frequency of oscillation is $[(k_1 + k_2)/m]^{1/2}$, if the springs are tied in parallel, and $[k_1 k_2/(k_1 + k_2)m]^{1/2}$, if the springs are tied in series.

3.8 A spring of stiffness k supports a box of mass M in which is placed a block of mass m. If the system is pulled downward a distance d from the equilibrium position and then released, find the force of reaction between the block and the bottom of the box as a function of time. For what value of d will the block just begin to leave the bottom of the box at the top of the vertical oscillations? Neglect any air resistance.

3.9 Show that the ratio of two successive maxima in the displacement of a damped harmonic oscillator is constant. [*Note:* The maxima do not occur at the points of contact of the displacement curve with the curve $Ae^{-\gamma t}$.]

3.10 The frequency f_d of a damped harmonic oscillator is 100 Hz, and the ratio of the amplitude of two successive maxima is one half. (a) What is the undamped frequency f_0 of this oscillator? (b) What is the resonant frequency f_r?

3.11 Given: The amplitude of a damped harmonic oscillator drops to $1/e$ of its initial value after n complete cycles. Show that the ratio of period of oscillation to the period of the same oscillator with no damping is given by

$$\frac{T_d}{T_0} = \left(1 + \frac{1}{4\pi^2 n^2}\right)^{1/2} \simeq 1 + \frac{1}{8\pi^2 n^2}$$

where the approximation in the last expression is valid if n is large. (See approximation formulas in Appendix D.)

3.12 Work all parts of Example 3.9 for the case in which the exponential damping factor γ is one half the critical value and the driving frequency is equal to $2\omega_0$.

3.13 For a lightly damped harmonic oscillator $\gamma \ll \omega_0$, show that the driving frequency for which the steady-state amplitude is one half the steady-state amplitude at the resonant frequency is given by $\omega \simeq \omega_0 \pm \gamma\sqrt{3}$.

3.14 If a series LCR circuit is connected across the terminals of an electric generator that produces a voltage $V = V_0 e^{i\omega t}$, the flow of electrical charge q through the circuit is given by the following second-order differential equation:

$$L\frac{d^2 q}{dt^2} + R\frac{dq}{dt} + \frac{1}{C}q = V_0 e^{i\omega t}$$

(a) Verify the correspondence shown in Table 3.2 between the parameters of a driven mechanical oscillator and the above driven electrical oscillator.

(b) Calculate the Q of the electrical circuit in terms of the coefficients of the above differential equation.

(c) Show that, in the case of small damping, Q can be written as $Q = R_0/R$ where $R_0 = \sqrt{L/C}$ is the *characteristic impedance* of the circuit.

3.15 A damped harmonic oscillator is driven by an external force of the form

$$F_{\text{ext}} = F_0 \sin\omega t$$

Show that the steady-state solution is given by

$$x(t) = A(\omega) \sin(\omega t - \phi)$$

where $A(\omega)$ and ϕ are identical to the expressions given by Equations 3.37 and 3.38.

3.16 Solve the differential equation of motion of the damped harmonic oscillator driven by a damped harmonic force:

$$F_{ext}(t) = F_0 e^{-\alpha t} \cos \omega t$$

[*Hint:* $e^{-\alpha t} \cos \omega t = \text{Re}(e^{-\alpha t + i\omega t}) = \text{Re}(e^{\beta t})$, where $\beta = -\alpha + i\omega$. Assume a solution of the form $Ae^{\beta t - i\phi}$.]

3.17 A simple pendulum of length l oscillates with an amplitude of 45°. (a) What is the period? (b) If this pendulum is used as a laboratory experiment to determine the value of g, find the error included in the use of the elementary formula $T_0 = 2\pi(l/g)^{1/2}$. (c) Find the approximate amount of third-harmonic content in the oscillation of the pendulum.

3.18 Verify Equations 3.65 and 3.66 in the text.

3.19 Show that the Fourier series for a periodic square wave is

$$f(t) = \frac{4}{\pi}\left[\sin(\omega t) + \frac{1}{3} \sin(3\omega t) + \frac{1}{5} \sin(5\omega t) + \cdots \right]$$

where

$$\begin{aligned} f(t) &= +1 && \text{for } 0 < \omega t < \pi,\ 2\pi < \omega t < 3\pi, \text{ and so on} \\ f(t) &= -1 && \text{for } \pi < \omega t < 2\pi,\ 3\pi < \omega t < 4\pi, \text{ and so on} \end{aligned}$$

3.20 Use the above result to find the steady-state motion of a damped harmonic oscillator that is driven by a periodic square-wave force of amplitude F_0. In particular, find the relative amplitudes of the first three terms $A_1 A_3$, and A_5 of the response function $x(t)$ in the case that the third harmonic 3ω of the driving frequency coincides with the frequency ω_0 of the undamped oscillator. Let the quality factor $Q = 100$.

COMPUTER/CALCULATOR APPLICATIONS

3.1 The exact equation of motion for a simple pendulum of length L is given by (see Example 3.2)

$$\ddot{\theta} + \omega_0^2 \sin \theta = 0$$

where $\omega_0^2 = g/L$. Find $\theta(t)$ by numerically integrating this equation of motion. Let $L = 1.00$ m. Let the initial conditions be $\theta_0 = \pi/2$ rad and $\dot{\theta}_0 = 0$ rad/s.
(a) Plot $\theta(t)$ from $t = 0$ to 4 s. Also, plot the solution obtained by using the small angle approximation ($\sin \theta \approx \theta$) on the same graph.
(b) Repeat (a) for $\theta_0 = 3.10$ rad.
(c) Plot the period of the pendulum as a function of the amplitude θ_0 from 0 to 3.10 rad. At

what amplitude does the period deviate by more than 2% from $\sqrt{g/L}$? (*Hint: A computer spreadsheet program, such as Quattro Pro, is ideal for this problem.*)

3.2 Assume that the damping force for the damped harmonic oscillator is proportional to the square of its velocity; that is, it is given by $-c_2\dot{x}|\dot{x}|$. The equation of motion for such an oscillator is, thus,

$$\ddot{x} + 2\gamma\dot{x}|\dot{x}| + \omega_0^2 x = 0$$

where $\gamma = c_2/2m$ and $\omega_0^2 = k/m$. Find $x(t)$ by numerically integrating the above equation of motion. Let $\gamma = 0.20$ m^{-1} s^{-1} and $\omega_0 = 2.00$ rad/s. Let the initial conditions be $x(0) = 1.00$ m and $\dot{x}(0) = 0$ m/s.

(a) Plot $x(t)$ from $t = 0$ to 20 s. Also, on the same graph, plot the solution for the damped harmonic oscillator where the damping force is linearly proportional to the velocity, that is, it is given by $-c_1\dot{x}$. Again, let $\gamma = c_1/2m = 0.20$ s^{-1} and $\omega_0 = 2.00$ rad/s.

(b) For the case of linear damping, plot the log of the absolute value of the successive extrema versus their time of occurrence. Find the slope of this plot, and use it to estimate γ. (This method works well for the case of weak damping.)

(c) Find the value of γ that results in critical damping for the linear case. Plot this solution from $t = 0$ to 5 s. Can you find a well-defined value of γ that results in critical damping for the quadratic case? If not, what value of γ is required to limit the first negative excursion of the oscillator to less than 2% of the initial amplitude? (*Hint: Use a computer spreadsheet program, such as Quattro Pro, to solve this fairly simple numerical problem.*)

4

GENERAL MOTION OF A PARTICLE IN THREE DIMENSIONS

"Sir Isaac Newton, and his followers, have also a very odd opinion concerning the work of God. According to their doctrine, God almighty needs to wind up his watch from time to time; otherwise it would cease to move. He had not, it seems, sufficient foresight to make it a perpetual motion. Nay, the machine of God's making, is so imperfect, according to these gentlemen, that He is obliged to clean it now and then by an extraordinary concourse, and even to mend it, as a clockmaker mends his work; who must consequently be so much the more unskillful a workman, as He is often obliged to mend his work and set it right. According to

my opinion, the same force and vigour [energy] remains always in the world, and only passes from one part to another, agreeable to the laws of nature, and the beautiful pre-established order—"

> *(Gottfried Wilhelm Liebniz—Letter to Caroline, Princess of Wales, 1715; The Leibniz-Clarke Correspondence, Manchester, Manchester Univ. Press, 1956)*

4.1   INTRODUCTION. GENERAL PRINCIPLES

We now wish to examine the general case of the motion of a particle in three dimensions. The vector form of the equation of motion for such a particle is

$$\mathbf{F} = \frac{d\mathbf{p}}{dt} \tag{4.1}$$

in which $\mathbf{p} = m\mathbf{v}$ is the linear momentum of the particle. This vector equation is equivalent to three scalar equations in Cartesian coordinates.

$$\begin{aligned}
F_x &= m\ddot{x} \\
F_y &= m\ddot{y} \\
F_z &= m\ddot{z}
\end{aligned} \tag{4.2}$$

The three force components may be explicit or implicit functions of the coordinates, their time and spatial derivatives, and possibly time itself. There is no general method for obtaining an analytic solution to the above equations of motion. In problems of even the mildest complexity, we might have to resort to the use of applied numerical techniques. However, there are many problems that can be solved using relatively simple analytical methods. It may be true that such problems are sometimes overly simplistic in their representation of reality. However, they ultimately serve as the basis of models of real physical systems, and so it is well worth the effort that we take here to develop the analytical skills necessary to solve such idealistic problems. Even these may prove capable of taxing our analytic ability.

It is rare that one knows the explicit way in which $\mathbf{F}$ depends on time. Therefore, we will not worry about this situation, but, instead, we will focus on the more normal situation in which $\mathbf{F}$ is known as an explicit function of spatial coordinates and their derivatives. The simplest situation is one in which $\mathbf{F}$ is known to be a function of spatial coordinates only. We will devote most of our effort toward solving such problems. There are many, only slightly more complex situations, in which $\mathbf{F}$ is a known function of coordinate derivatives as well. Such cases include projectile motion with air resistance and the motion of a charged particle in a static electromagnetic field. We will solve problems such as these, too. Finally, $\mathbf{F}$ may be an implicit function of time, as in situations where the coordinate and coordinate derivative dependency is nonstatic. A prime example of such a situation involves the motion of a charged particle in a time-varying electromagnetic field. We will not solve problems such as these. For now, we will begin our study of three-dimensional motion with a development of several powerful analytical techniques that can be applied when $\mathbf{F}$ is a known function of $\mathbf{r}$ and/or $\dot{\mathbf{r}}$.

The Work Principle

Work done on a particle causes it to gain or lose kinetic energy. The work concept was introduced in Chapter 2 for the case of motion of a particle in one dimension. We would like to generalize the results obtained there to the case of three-dimensional motion. To do so, we first take the dot product of both sides of Equation 4.1 with the velocity $\mathbf{v}$

$$\mathbf{F} \cdot \mathbf{v} = \frac{d\mathbf{p}}{dt} \cdot \mathbf{v} = \frac{d(m\mathbf{v})}{dt} \cdot \mathbf{v} \qquad (4.3)$$

Since $d(\mathbf{v} \cdot \mathbf{v})/dt = 2\mathbf{v} \cdot \dot{\mathbf{v}}$ and assuming that the mass is constant, independent of the velocity of the particle, we may write Equation 4.3 as

$$\mathbf{F} \cdot \mathbf{v} = \frac{d}{dt} \left(\frac{1}{2} m\mathbf{v} \cdot \mathbf{v} \right) = \frac{dT}{dt} \qquad (4.4)$$

in which T is the kinetic energy, $mv^2/2$. *Since* $\mathbf{v} = d\mathbf{r}/dt$, we can rewrite Equation 4.4 and then integrate the result to obtain

$$\mathbf{F} \cdot \frac{d\mathbf{r}}{dt} = \frac{dT}{dt}$$

$$\therefore \int \mathbf{F} \cdot d\mathbf{r} = \int dT = T_f - T_i = \Delta T \qquad (4.5)$$

The left-hand side of this equation is a *line integral,* or the integral of $F_r \cdot dr$, the component of $\mathbf{F}$ parallel to the particle's displacement vector $\mathbf{dr}$. The integral is carried out along the trajectory of the particle from some initial point in space A to some final point B. This situation is pictured in Figure 4.1. The line integral represents the work done on the particle by the force $\mathbf{F}$ as the particle moves along its trajectory from A to B. The right-hand side of the equation is the net change in the kinetic energy of the particle. It is important to note that $\mathbf{F}$ is the net sum of all vector forces acting on the particle. Hence, the equation states that the work done on a particle by the net force acting on it, in moving from one position in space to another, is equal to the difference in the kinetic energy of the particle at those two positions.

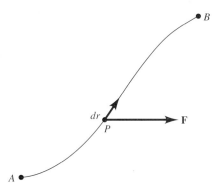

Figure 4.1 The work done by a force $\mathbf{F}$ is the line integral $_A\int^B \mathbf{F} \cdot d\mathbf{r}.$

Conservative Forces and Force Fields

In Chapter 2 we also introduced the concept of potential energy. We stated there that if the force acting on a particle was *conservative,* it could be derived from a scalar potential energy function $V(x)$, as the derivative of that function, $F_x = -dV(x)/dx$. This condition led us to the notion that the work done by such a force in moving a particle from point A to point B along the x-axis was $\int F_x \, dx = -\Delta V = V(A) - V(B)$, or equal to minus the change in the potential energy of the particle. Thus, we no longer required a detailed knowledge of the motion of the particle from A to B in order to calculate the work done on it by a conservative force. We needed to know only that it started at point A and ended up at point B. The work done depended only upon the potential energy function evaluated at the endpoints of the motion. Moreover, since the work done was also equal to the change in kinetic energy of the particle, $\Delta T = T(B) - T(A)$, we were able to establish a general conservation of total energy principle, namely, $E_{tot} = V(A) + T(A) = V(B) + T(B) =$ constant throughout the motion of the particle.

This principle was based on the condition that the force acting on the particle was conservative. Indeed, the very name implies that something is being conserved as the particle moves under the action of such a force. We would like to generalize this concept for a particle moving in three dimensions, and, more importantly, we would like to define just what is meant by the word *conservative.* Clearly, we would like to have some prescription that would tell us whether or not a particular force was conservative and, thus, whether or not a potential energy function exists for the particle. Then we could invoke the powerful conservation of energy principle in solving the motion of a particle.

In searching for such a prescription, we will first describe an example of a nonconservative force that, in fact, is a well-defined function of position but cannot be derived from a potential energy function. This should give us a hint of the critical characteristic that a force must have if it is to be conservative. Consider the two-dimensional force field depicted in Figure 4.2. The term *force field* simply means that if a small test par-

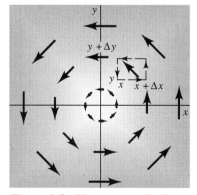

Figure 4.2   Non-conservative force field whose force components are $F_x = -by$ and $F_y = +bx$.

ticle[1] were to be placed at any point (x_1, y_1) on the xy plane, it would experience a force
F. Thus, we can think of the xy plane as permeated or "mapped out" with the potential
for generating a force.

This situation can be mathematically described by assigning a vector **F** to every
point in the xy plane. The field is, therefore, a vector field, represented by the function
F(x, y). Its components are $F_x = -by$ and $F_y = +bx$, where b is some constant. The
arrows in the figure represent the vector $\mathbf{F} = -\mathbf{i}by + \mathbf{j}bx$ evaluated at each point on
which the center of the arrow is located. You can see by looking at the figure that there
seems to be a general counterclockwise "circulation" of the force vectors around the
origin. The magnitude of the vectors increases with increasing distance from the origin.
If we were to turn a small test particle loose in such a "field," the particle would tend to
circulate around counterclockwise, gaining kinetic energy all the while.

This situation, at first glance, does not appear to be so unusual. After all, when you
drop a ball in a gravitational force field, it falls and gains kinetic energy, with an accompanying loss of an equal amount of potential energy. The question here is, can we even
define a potential energy function for this circulating particle such that it would lose an
amount of "potential energy" equal to the kinetic energy it gained, thus preserving its
overall energy, as it travels from one point to another? That is not the case here. If we
were to calculate the work done on this particle in tracing out some path that came back
on itself (such as the rectangular path indicated by the dashed line in Figure 4.2), we
would obtain a nonzero result! In traversing such a loop, over and over again, our particle would continue to gain kinetic energy equal to the nonzero value of work done
per loop. But if the particle could be assigned a potential energy dependent only upon
its (x, y) position, then its change in potential energy upon traversing the closed loop
would be zero. It should be clear that there is no way in which we could assign a unique
value of potential energy for this particle at any particular point on the xy plane. Any
value assigned would depend on the previous history of the particle. For example, how
many loops has the particle already made before arriving at its current position.

We can further expose the nonuniqueness of any proposed potential energy function
by examining the work done on the particle as it travels between two points A and B,
but along two different paths. First, we let the particle move from (x, y) to $(x + \Delta x,
y + \Delta y)$ by traveling in the $+x$ direction to $(x + \Delta x, y)$ and then in the $+y$ direction to
$(x + \Delta x, y + \Delta y)$. Then we let the particle travel first along the $+y$ direction from (x, y)
to $(x, y + \Delta y)$ and then along the $+x$ direction to $(x + \Delta x, y + \Delta y)$. We will see that a
different amount of work is done depending upon which path we let the particle take. If
this is true, then the work done cannot be set equal to the difference between the values
of some scalar potential energy function evaluated at the two endpoints of the motion
since such a difference would give a unique, *path-independent* result. The difference in
work done along these two paths is equal to $2b\Delta x\Delta y$ (see Equation 4.6). This difference

---

[1] A test particle is one whose mass is small enough that its presence does not alter its environment. Concep-
tually, we might imagine it placed at some point in space in order to serve as a "test probe" for the presence
of suspected forces at that point by "observing" its acceleration. We further imagine that its presence does
not disturb the sources of those forces.

is just equal to the value of the closed-loop work integral. Hence, the statement that the work done in going from one point to another in this force field is path-dependent is equivalent to the statement that the closed-loop work integral is nonzero. The particular force field represented in Figure 4.2 demands that we know the complete history of the particle in order to calculate the work done and, therefore, its kinetic energy gain. The potential energy concept, from which the force could presumably be derived, is rendered meaningless in this particular context.

The only way in which we could assign a unique value to the potential energy would be if the closed-loop work integral vanished. In such cases, the work done along a path from A to B would be path-independent and would equal both the potential energy loss as well as the kinetic energy gain. The total energy of the particle would be a constant, independent of its location in such a force field! It, therefore, behooves us to find the constraint that a particular force must obey if its closed-loop work integral is to vanish.

In order to find the desired constraint, let us calculate the work done in taking a test particle counterclockwise around the rectangular loop of area $\Delta x\,\Delta y$ from the point (x, y) and back again as indicated in Figure 4.2. We get the following result:

$$
\begin{aligned}
W &= \oint \mathbf{F} \cdot d\mathbf{r} \\
&= \int_x^{x+\Delta x} F_x(y)dx + \int_y^{y+\Delta y} F_y(x + \Delta x)dy \\
&\quad + \int_{x+\Delta x}^x F_x(y + \Delta y)dx + \int_{y+\Delta y}^y F_y(x)dy \\
&= \int_y^{y+\Delta y} (F_y(x + \Delta x) - F_y(x))dy \qquad\qquad (4.6)\\
&\quad + \int_x^{x+\Delta x} (F_x(y) - F_x(y + \Delta y))dx \\
&= (b(x + \Delta x) - bx)\,\Delta y + (b(y + \Delta y) - by)\,\Delta x \\
&= 2b\Delta x\Delta y
\end{aligned}
$$

The work done is nonzero and is proportional to the area of the loop, $\Delta A = \Delta x \cdot \Delta y$, which was chosen in an arbitrary fashion. If we divide the work done by the area of the loop and take limits as $\Delta A \to 0$, we obtain the value $2b$. The result is dependent on the precise nature of this particular nonconservative force field.

If we were to reverse the direction of one of the force components, say let $F_x = +by$ (thus "destroying" the circulation of the force field but everywhere preserving its magnitude), then the work done per unit area in traversing the closed loop would vanish. The resulting force field is conservative. It is shown in Figure 4.3. Clearly, the value of the closed loop integral depends upon the precise way in which the vector force $\mathbf{F}$ changes its direction as well as its magnitude as we move around on the xy plane.

There is obviously some sort of constraint that $\mathbf{F}$ must obey if the closed-loop work integral is to vanish. We can derive this condition of constraint by evaluating the forces at $x + \Delta x$ and $y + \Delta y$ using a Taylor's expansion and then inserting the resultant expansion into the closed-loop work integral of Equation 4.6. The result follows:

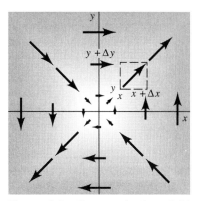

Figure 4.3 Conservative force field whose components are $F_x = by$ and $F_y = bx$.

$$F_x(y + \Delta y) = F_x(y) + \frac{\partial F_x}{\partial y} \Delta y \tag{4.7}$$

$$F_y(x + \Delta x) = F_y(x) + \frac{\partial F_y}{\partial x} \Delta x$$

$$\oint \mathbf{F} \cdot d\mathbf{r} = \int_y^{y+\Delta y} \left(\frac{\partial F_y}{\partial x} \Delta x \right) dy - \int_x^{x+\Delta x} \left(\frac{\partial F_x}{\partial y} \Delta y \right) dx \tag{4.8}$$

$$= \left(\frac{\partial F_y}{\partial x} - \frac{\partial F_x}{\partial y} \right) \Delta x \Delta y = 2b \Delta x \Delta y$$

This last equation contains the term $(\partial F_y/\partial x - \partial F_x/\partial y)$, whose zero or nonzero value represents the test we are looking for. If this term were identically equal to zero instead of $2b$, then the closed-loop work integral would vanish, which would ensure the existence of a potential energy function from which the force could be derived.

This condition is a rather simplified version of a very general mathematical theorem called *Stoke's theorem*.[2] It is written as

$$\oint \mathbf{F} \cdot d\mathbf{r} = \int_s \text{curl } \mathbf{F} \cdot \hat{\mathbf{n}} \, da \tag{4.9}$$

$$\text{curl } \mathbf{F} = \hat{\imath} \left(\frac{\partial F_z}{\partial y} - \frac{\partial F_y}{\partial z} \right) + \hat{\jmath} \left(\frac{\partial F_x}{\partial z} - \frac{\partial F_z}{\partial x} \right) + \hat{k} \left(\frac{\partial F_y}{\partial x} - \frac{\partial F_x}{\partial y} \right)$$

---

[2] See any advanced calculus textbook (e.g., S. I. Grossman and W. R. Derrick, *Advanced Engineering Mathematics,* Harper Collins, New York, 1988) or any advanced electricity and magnetism textbook (e.g., J. R. Reitz, F. J. Milford, and R. W. Christy, *Foundations of Electromagnetic Theory,* Addison-Wesley, New York, 1992).

The theorem states that the closed-loop line integral of *any* vector function $\mathbf{F}$ is equal to curl $\mathbf{F} \cdot \mathbf{n} \, da$ integrated over a surface S surrounded by the closed loop. $\mathbf{n}$ is a unit vector normal to the surface area integration element da. Its direction is in the sense that a right-hand screw would advance if the screw were turned in the same rotational sense as the direction of traversal around the closed loop. In Figure 4.2, $\mathbf{n}$ would be directed out of the paper. The surface would be the rectangular area enclosed by the dashed rectangular loop. Thus, a vanishing curl $\mathbf{F}$ ensures that the line integral of $\mathbf{F}$ around a closed path is zero and, thus, that $\mathbf{F}$ is a conservative force.

4.2   THE POTENTIAL ENERGY FUNCTION IN THREE-DIMENSIONAL MOTION. THE DEL OPERATOR

Assume that we have a test particle subject to some force whose curl vanishes. Then all the components of curl $\mathbf{F}$ in Equation 4.9 vanish. We can make certain that the curl vanishes if we derive $\mathbf{F}$ from a potential energy function $V(x, y, z)$ according to

$$F_x = -\frac{\partial V}{\partial x} \qquad F_y = -\frac{\partial V}{\partial y} \qquad F_z = -\frac{\partial V}{\partial z} \qquad (4.10)$$

For example, the z component of curl $\mathbf{F}$ becomes

$$\frac{\partial F_x}{\partial y} = -\frac{\partial^2 V}{\partial y \partial x} \qquad \frac{\partial F_y}{\partial x} = -\frac{\partial^2 V}{\partial x \partial y} = -\frac{\partial^2 V}{\partial y \partial x} \qquad \therefore \; \frac{\partial F_y}{\partial x} - \frac{\partial F_x}{\partial y} = 0$$

This last step follows if we assume that V is everywhere continuous and differentiable. We reach the same conclusion for the other components of curl $\mathbf{F}$. It is true that one might wonder whether or not there are other reasons why curl $\mathbf{F}$ might vanish other than $\mathbf{F}$ being derivable from a potential energy function. However, curl $\mathbf{F} = 0$ is a necessary and sufficient condition for the existence of $V(x, y, z)$ such that Equation 4.10 holds.[3]

We can now express a conservative force $\mathbf{F}$ vectorially as

$$\mathbf{F} = -\mathbf{i}\frac{\partial V}{\partial x} - \mathbf{j}\frac{\partial V}{\partial y} - \mathbf{k}\frac{\partial V}{\partial z} \qquad (4.11)$$

This equation can be more succinctly written as

$$\mathbf{F} = -\nabla V \qquad (4.12)$$

where we have introduced the vector operator *del*

$$\nabla = \mathbf{i}\frac{\partial}{\partial x} + \mathbf{j}\frac{\partial}{\partial y} + \mathbf{k}\frac{\partial}{\partial z} \qquad (4.13)$$

The expression ∇V is also called the *gradient of V* and is sometimes written grad V. Mathematically, the gradient of a function is a vector that represents the maximum spatial derivative of the function in direction and magnitude. Physically, the negative gradient of the potential energy function gives the direction and magnitude of the force that acts on a particle located in a field created by other particles. The meaning of the

---

[3] See, for example, S. I. Grossman, op cit. Also, an interesting discussion of conservancy criteria when the force field contains singularities has been given by Feng in *Amer. J. Phys.* **37,** 616 (1969).

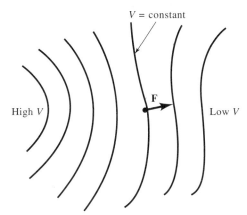

Figure 4.4 A force field represented by equipotential contour curves.

negative sign is that the particle is urged to move in the direction of *decreasing* potential energy rather than in the opposite direction. An illustration of the gradient is shown in Figure 4.4. Here the potential energy function is plotted out in the form of contour lines representing the curves of constant potential energy. The force at any point is always normal to the equipotential curve or surface passing through the point in question.

We can express curl **F** using the del operator. Look at the components of curl **F** in Equation 4.9. They are the components of the vector $\nabla \times \mathbf{F}$. Thus, $\nabla \times \mathbf{F} = \text{curl } \mathbf{F}$. The condition that a force be conservative can be written compactly as

$$\nabla \times \mathbf{F} = \mathbf{i}\left(\frac{\partial F_z}{\partial y} - \frac{\partial F_y}{\partial z}\right) + \mathbf{j}\left(\frac{\partial F_x}{\partial z} - \frac{\partial F_z}{\partial x}\right) + \mathbf{k}\left(\frac{\partial F_y}{\partial x} - \frac{\partial F_x}{\partial y}\right) = 0 \quad (4.14)$$

Furthermore, if $\nabla \times \mathbf{F} = 0$, then **F** can be derived from a scalar function V by the operation $\mathbf{F} = -\nabla V$, since $\nabla \times \nabla V \equiv 0$, or the curl of any gradient is identically 0.

We are now able to generalize the conservation of energy principle to three dimensions. The work done by a conservative force in moving a particle from point A to point B can be written as

$$\int_A^B \mathbf{F} \cdot d\mathbf{r} = -\int_A^B \nabla V(\mathbf{r}) \cdot d\mathbf{r} = -\int_{A_x}^{B_x} \frac{\partial V}{\partial x} dx - \int_{A_y}^{B_y} \frac{\partial V}{\partial y} dy - \int_{A_z}^{B_z} \frac{\partial V}{\partial z} dz \quad (4.15)$$

$$= -\int_A^B dV(\mathbf{r}) = -\Delta V = V(A) - V(B)$$

The last step illustrates the fact that $\nabla V \cdot d\mathbf{r}$ is an *exact* differential equal to dV. The work done by any force is always equal to the change in kinetic energy, so

$$\int_A^B \mathbf{F} \cdot d\mathbf{r} = \Delta T = -\Delta V$$

$$\therefore \Delta(T + V) = 0 \quad (4.16)$$

$$\therefore T(A) + V(A) = T(B) + V(B) = E = \text{constant}$$

and we have arrived at our desired law of conservation of total energy.

If $\mathbf{F}'$ is a nonconservative force, it cannot be set equal to $-\nabla V$. The work increment $\mathbf{F}' \cdot d\mathbf{r}$ is not an exact differential and cannot be equated to $-dV$. In those cases where both conservative forces $\mathbf{F}$ and nonconservative forces $\mathbf{F}'$ are present, the total work increment is $(\mathbf{F} + \mathbf{F}') \cdot d\mathbf{r} = -dV + \mathbf{F}' \cdot d\mathbf{r} = dT$ and the generalized form of the work energy theorem becomes

$$\int_A^B \mathbf{F}' \cdot d\mathbf{r} = \Delta(T + V) = \Delta E$$

The total energy E does not remain a constant throughout the motion of the particle but increases or decreases depending upon the nature of the nonconservative force $\mathbf{F}'$. In the case of dissipative forces like friction or air resistance, the direction of $\mathbf{F}'$ is always *opposite* the motion; hence, $\mathbf{F}' \cdot d\mathbf{r}$ is negative, and the total energy of the particle decreases as it moves through space.

EXAMPLE 4.1

Given the potential energy function

$$V(\mathbf{r}) = \alpha x^2 + \beta xy + \gamma z + C,$$

in which α, β, γ, and C are constants, find the force function.

Solution:
Applying the del operator, we have

$$\mathbf{F} = -\nabla V = -\left(\mathbf{i}\frac{\partial V}{\partial x} + \mathbf{j}\frac{\partial V}{\partial y} + \mathbf{k}\frac{\partial V}{\partial z} \right)$$

$$= -\mathbf{i}(2x\alpha + y\beta) - \mathbf{j}(x\beta) - \mathbf{k}\gamma$$

Notice that the constant C does not appear in the force function; as in the one-dimensional case the value of C is arbitrary. It is equal to the potential energy at the origin in this case: $C = V(0)$. ∎

EXAMPLE 4.2

Suppose a particle of mass m is moving in the above force field, and at time $t = 0$ the particle passes through the origin with speed v_0. What will the speed of the particle be if and when it passes through the point $\mathbf{r} = \mathbf{i} + 2\mathbf{j} + \mathbf{k}$?

Solution:
To answer the given question we need only to use the fact that the force is conservative (we know that a potential energy function exists) and so the total energy is constant: $T + V = E = $ constant. In our case

$$E = \frac{1}{2}mv^2 + V(\mathbf{r}) = \frac{1}{2}mv_0^2 + V(0)$$

so

$$v^2 = v_0^2 + \frac{2}{m}[V(0) - V(\mathbf{r})]$$

$$= v_0^2 + \frac{2}{m}[C - (\alpha x^2 + \beta xy + \gamma z + C)]$$

$$= v_0^2 - \frac{2}{m}(\alpha + 2\beta + \gamma)$$

at the given point. ∎

EXAMPLE 4.3

Is the force field $\mathbf{F} = \mathbf{i}xy + \mathbf{j}xz + \mathbf{k}yz$ conservative? The curl of $\mathbf{F}$ is

$$\nabla \times \mathbf{F} = \begin{vmatrix} \mathbf{i} & \mathbf{j} & \mathbf{k} \\ \partial/\partial x & \partial/\partial y & \partial/\partial z \\ xy & xz & yz \end{vmatrix} = \mathbf{i}(z - x) + \mathbf{j}0 + \mathbf{k}(z - x)$$

The final expression is not zero for all values of the coordinates; hence, the field is *not* conservative. ∎

EXAMPLE 4.4

For what values of the constants a, b, and c is the force $\mathbf{F} = \mathbf{i}(ax + by^2) + \mathbf{j}cxy$ conservative? Taking the curl, we have

$$\nabla \times \mathbf{F} = \begin{vmatrix} \mathbf{i} & \mathbf{j} & \mathbf{k} \\ \partial/\partial x & \partial/\partial y & \partial/\partial z \\ ax + by^2 & cxy & 0 \end{vmatrix} = \mathbf{k}(c - 2b)y$$

This shows that the force is conservative, provided $c = 2b$. The value of a is immaterial. ∎

EXAMPLE 4.5

Show that the inverse-square law of force in three dimensions $\mathbf{F} = (-k/r^2)\mathbf{e}_r$ is conservative by the use of the curl. Use spherical coordinates. The curl is given in Appendix F as

$$\nabla \times \mathbf{F} = \frac{1}{r^2 \sin \theta} \begin{vmatrix} \mathbf{e}_r & \mathbf{e}_\theta r & \mathbf{e}_\phi r \sin \theta \\ \dfrac{\partial}{\partial r} & \dfrac{\partial}{\partial \theta} & \dfrac{\partial}{\partial \phi} \\ F_r & rF_\theta & rF_\phi \sin \theta \end{vmatrix}$$

We have $F_r = -k/r^2$, $F_\theta = 0$, $F_\phi = 0$. The curl then reduces to

$$\nabla \times \mathbf{F} = \frac{\mathbf{e}_\theta}{r \sin \theta} \frac{\partial}{\partial \phi}\left(\frac{-k}{r^2}\right) - \frac{\mathbf{e}_\phi}{r} \frac{\partial}{\partial \theta}\left(\frac{-k}{r^2}\right) = 0$$

which, of course, vanishes since both partial derivatives are zero. Thus, the force in question is conservative. ∎

4.3 FORCES OF THE SEPARABLE TYPE. PROJECTILE MOTION

It is often the case that a Cartesian coordinate system can be chosen such that the components of a force field involve the respective coordinates alone, that is,

$$\mathbf{F} = \mathbf{i}F_x(x) + \mathbf{j}F_y(y) + \mathbf{k}F_z(z) \tag{4.17}$$

Forces of this type are said to be *separable*. It is readily verified that the curl of such a force is identically zero:

$$\nabla \times \mathbf{F} = \begin{vmatrix} \mathbf{i} & \mathbf{j} & \mathbf{k} \\ \partial/\partial x & \partial/\partial y & \partial/\partial z \\ F_x(x) & F_y(y) & F_z(z) \end{vmatrix} = 0$$

The x component is $\partial F_z(z)/\partial y - \partial F_y(y)/\partial z$, and similarly for the other components. Hence, the field is conservative because each partial derivative is of the mixed type and therefore vanishes identically, since the coordinates x, y, and z are independent variables. The integration of the differential equations of motion is then very simple because each component equation is of the type $m\ddot{x} = F_x(x)$. In this case the equations can be solved by the methods described under rectilinear motion in Chapter 2.

In the event that the force components involve the time and the time derivatives of the respective coordinates, then it is no longer true that the force is necessarily conservative. Nevertheless, if the force is separable, then the component equations of motion are of the form $m\ddot{x} = F_x(x, \dot{x}, t)$ and may be solved by the methods used in Chapter 2. Some examples of separable forces, both conservative and nonconservative, will be discussed here and in the sections to follow.

Motion of a Projectile in a Uniform Gravitational Field

One of the famous classical problems of particle dynamics is the motion of a projectile. We shall study this problem in some detail because it illustrates most of the general principles that have been cited in the foregoing sections.

No Air Resistance

First, for simplicity, we consider the case of a projectile moving with no air resistance. In this idealized situation there is only one force acting, namely the force of gravity. Choosing the z-axis to be vertical, we have the differential equation of motion

$$m\frac{d^2\mathbf{r}}{dt^2} = -mg\mathbf{k}$$

If we further idealize the problem and assume that the acceleration of gravity g is constant, then the force function is clearly of the separable type and is also conservative, since it is a special case of that expressed by Equation 4.17. We shall particularize the problem further by choosing the initial speed to be v_0 and the initial position to be at the origin at time $t = 0$. The energy equation (Equation 4.9) then reads

$$\frac{1}{2}m(\dot{x}^2 + \dot{y}^2 + \dot{z}^2) + mgz = \frac{1}{2}mv_0^2$$

or, equivalently,

$$v^2 = v_0^2 - 2gz \tag{4.18}$$

thus giving the speed as a function of height. This is all the information we can obtain directly from the energy equation.

In order to proceed further, we must go back to the differential equation of motion. This can be written

$$\frac{d}{dt}\left(\frac{d\mathbf{r}}{dt}\right) = -g\mathbf{k} \tag{4.19}$$

It can be integrated directly. A single integration gives the velocity as

$$\frac{d\mathbf{r}}{dt} = -gt\mathbf{k} + \mathbf{v}_0 \tag{4.19a}$$

in which the constant of integration $\mathbf{v}_0$ is the initial velocity. Another integration yields the position vector

$$\mathbf{r} = -\frac{1}{2}gt^2\mathbf{k} + \mathbf{v}_0 t + \mathbf{r}_0 \tag{4.19b}$$

The constant of integration $\mathbf{r}_0$ is zero in this case, since the initial position of the projectile is taken to be the origin. In components

$$x = \dot{x}_0 t$$
$$y = \dot{y}_0 t \tag{4.19c}$$
$$z = \dot{z}_0 t - \frac{1}{2}gt^2$$

Here $\dot{x}_0$, $\dot{y}_0$, and $\dot{z}_0$ are the components of the initial velocity $\mathbf{v}_0$. We have thus solved the problem of determining the position of the projectile as a function of time.

Concerning the path or trajectory of the projectile, we notice that if the time t is eliminated from the x and y equations, the result is

$$y = bx$$

in which the constant b is given by

$$b = \frac{\dot{y}_0}{\dot{x}_0}$$

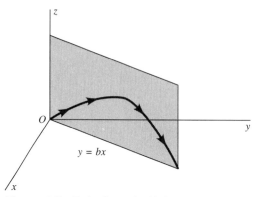

Figure 4.5   Path of a projectile in three dimensions.

Thus, the path lies entirely in a plane. In particular, if $\dot{y}_0 = 0$, then the path lies in the xz plane. Next, if we eliminate t between the x and z equations, we find the equation of the path to be of the form

$$z = Ax - Bx^2$$

where $A = \dot{z}_0/\dot{x}_0$ and $B = g/2\dot{x}_0^2$. Hence, the path is a parabola lying in the plane $y = bx$. This is shown in Figure 4.5.

Linear Air Resistance

We now consider the motion of a projectile for the more realistic situation in which there is a retarding force due to air resistance. In this case the motion is not conservative. The total energy continually diminishes as a result of frictional loss.

For simplicity, let us assume that the law of air resistance is linear so that the resisting force varies directly with the velocity $\mathbf{v}$. It will be convenient to write the constant of proportionality as $m\gamma$ where m is the mass of the projectile. Thus, we have two forces acting on the projectile, namely the air resistance $-m\gamma\mathbf{v}$ and the force of gravity, which is equal to $-mg\mathbf{k}$, as before. The differential equation of motion is then

$$m\frac{d^2\mathbf{r}}{dt^2} = -m\gamma\mathbf{v} - mg\mathbf{k}$$

or, upon canceling the m's, we have

$$\frac{d^2\mathbf{r}}{dt^2} = -\gamma\mathbf{v} - g\mathbf{k} \tag{4.20}$$

The integration of the above equation is conveniently accomplished by expressing it in component form as follows:

$$\begin{aligned} \ddot{x} &= -\gamma\dot{x} \\ \ddot{y} &= -\gamma\dot{y} \\ \ddot{z} &= -\gamma\dot{z} - g \end{aligned} \tag{4.20a}$$

We now see that the equations are separated. Hence, each can be solved individually by the methods of Chapter 2. Using our results from Section 2.5, we can write down the solutions immediately, by setting $\gamma = c_1/m$, c_1 being the linear drag coefficient. The results are

$$\dot{x} = \dot{x}_0 e^{-\gamma t}$$
$$\dot{y} = \dot{y}_0 e^{-\gamma t} \qquad\qquad (4.20b)$$
$$\dot{z} = \dot{z}_0 e^{-\gamma t} - \frac{g}{\gamma}(1 - e^{-\gamma t})$$

for the velocity components, and

$$x = \frac{\dot{x}_0}{\gamma}(1 - e^{-\gamma t})$$
$$y = \frac{\dot{y}_0}{\gamma}(1 - e^{-\gamma t}) \qquad\qquad (4.20c)$$
$$z = \left(\frac{\dot{z}_0}{\gamma} + \frac{g}{\gamma^2}\right)(1 - e^{-\gamma t}) - \frac{g}{\gamma}t$$

for the positional coordinates. Here, as before, the initial velocity components are $\dot{x}_0$, $\dot{y}_0$, and $\dot{z}_0$, and the initial position of the projectile is taken as the origin.

The above solution of the motion of a projectile with linear air resistance can be written vectorially in the following way:

$$\mathbf{r} = \left(\frac{\mathbf{v}_0}{\gamma} + \frac{\mathbf{k}g}{\gamma^2}\right)(1 - e^{-\gamma t}) - \mathbf{k}\frac{gt}{\gamma} \qquad\qquad (4.20d)$$

That it is a solution of the vector differential equation of motion is easily verified by differentiation.

As in the case of zero air resistance, the path of the projectile lies in a vertical plane $y = bx$ with $b = \dot{y}_0/\dot{x}_0$. The path in this plane is not a parabola, however, but rather a curve that lies below the corresponding parabolic trajectory. This is illustrated in Figure 4.6. Inspection of the x and y equations shows that, for large t, the values of x and y approach the limiting values

$$x \to \frac{\dot{x}_0}{\gamma} \qquad y \to \frac{\dot{y}_0}{\gamma}$$

This means that the complete trajectory of the projectile, if it did not hit anything, would have a vertical asymptote as shown in Figure 4.6.

In the actual motion of a projectile through the atmosphere, the law of resistance is by no means linear, but it is a very complicated function of the velocity. An accurate calculation of the trajectory can be done by means of numerical integration methods aided by the use of high-speed computers. (See reference cited under Example 2.9.)

Horizontal Range

The horizontal range of a projectile with linear air drag is found by setting $z = 0$ in the third of Equations 4.20c and then eliminating t among the three equations. We shall set

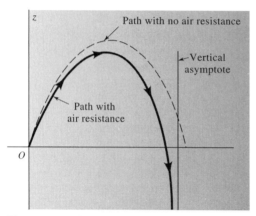

Figure 4.6   Comparison of the paths of a projectile with and without air resistance. Motion is shown in the plane $y = bx$.

$\dot{y}_0 = 0$ so that the trajectory lies in the xz plane. From the first of Equations 4.20c, we have $1 - \gamma x/\dot{x}_0 = e^{-\gamma t}$, so $t = -\gamma^{-1} \ln (1 - \gamma x/\dot{x}_0)$. Thus, the horizontal range x_h is given by the implicit expression

$$\left(\frac{\dot{z}_0}{\gamma} + \frac{g}{\gamma^2} \right) \frac{\gamma x_h}{\dot{x}_0} + \frac{g}{\gamma^2} \ln\left(1 - \frac{\gamma x_h}{\dot{x}_0} \right) = 0 \tag{4.21}$$

This is a transcendental equation and must be solved by some approximation method to find x_h. We can expand the logarithmic term by use of the series

$$\ln(1 - u) = -u - \frac{u^2}{2} - \frac{u^3}{3} - \cdots$$

which is valid for $|u| < 1$. With $u = \gamma x_h/\dot{x}_0$ it is left as a problem to show that this leads to the following expression for the horizontal range:

$$x_h = \frac{2\dot{x}_0\dot{z}_0}{g} - \frac{8\dot{x}_0\dot{z}_0^2}{3g^2}\gamma + \cdots \tag{4.22}$$

If the projectile is fired at angle of elevation α with initial speed v_0, then $\dot{x}_0 = v_0 \cos \alpha$, $\dot{z}_0 = v_0 \sin \alpha$, and $2\dot{x}_0\dot{z}_0 = 2v_0^2 \sin \alpha \cos \alpha = v_0^2 \sin 2\alpha$. An equivalent expression is then

$$x_h = \frac{v_0^2 \sin 2\alpha}{g} - \frac{4v_0^3 \sin 2\alpha \sin \alpha}{3g^2}\gamma + \cdots \tag{4.22a}$$

The first term on the right is the range in the absence of air resistance. The remainder is the decrease due to air resistance.

EXAMPLE 4.6 *Horizontal Range of a Golf Ball*

For objects of baseball or golf ball size traveling at normal speeds, the air drag is more nearly quadratic in v, rather than linear, as pointed out in Section 2.5. However, the approximate expression found above can be used to find the range for flat trajectories by "linearizing" the force function given by Equation 2.22, which may be written in three dimensions as

$$\mathbf{F}(\mathbf{v}) = -\mathbf{v}(c_1 + c_2|\mathbf{v}|)$$

To linearize it we set $|\mathbf{v}|$ equal to the initial speed v_0 and so the constant γ is given by

$$\gamma = \frac{c_1 + c_2 v_0}{m}$$

(A better approximation would be to take the average speed, but that is not a given quantity.) Although this method exaggerates the effect of air drag, it allows a quick ballpark estimate to be found easily.

For a golf ball of diameter $D = 0.042$ m and mass $m = 0.046$ kg, we find that c_1 is negligible and so

$$\gamma = \frac{c_2 v_0}{m} = \frac{0.22 D^2 v_0}{m}$$

$$= \frac{0.22(0.042)^2 v_0}{0.046} = 0.0084\ v_0$$

numerically, where v_0 is in m/s. For a chip shot with, say, $v_0 = 20$ m/s, we find $\gamma = 0.0084 \times 20 = 0.17$ s^{-1}. The horizontal range is then, for $\alpha = 30°$,

$$x_h = \frac{(20)^2 \sin 60°}{9.8}m - \frac{4(20)^3 \sin 60° \sin 30° \times 0.17}{3(9.8)^2}m$$

$$= 35.3\ m - 8.2\ m = 27.1\ m$$

Our estimate thus gives a reduction of about one fourth due to air drag on the ball. ∎

4.4 THE HARMONIC OSCILLATOR IN TWO AND THREE DIMENSIONS

Consider the motion of a particle that is subject to a linear restoring force, which is always directed toward a fixed point, the origin of our coordinate system. Such a force can be represented by the expression

$$\mathbf{F} = -k\mathbf{r}$$

Accordingly, the differential equation of motion is simply expressed as

$$m\frac{d^2\mathbf{r}}{dt^2} = -k\mathbf{r} \tag{4.23}$$

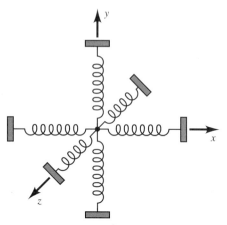

Figure 4.7   Model of a three-dimensional
harmonic oscillator.

The situation can be represented approximately by a particle attached to a set of elastic
springs as shown in Figure 4.7. This is the three-dimensional generalization of the linear
oscillator studied earlier. Equation 4.23 is the differential equation of the *linear isotropic
oscillator.*

The Two-Dimensional Isotropic Oscillator

In the case of motion in a single plane, Equation 4.23 is equivalent to the two component
equations

$$m\ddot{x} = -kx \tag{4.23a}$$
$$m\ddot{y} = -ky$$

These are separated, and we can immediately write down the solutions in the form

$$x = A \cos(\omega t + \alpha) \qquad y = B \cos(\omega t + \beta) \tag{4.23b}$$

in which

$$\omega = \left(\frac{k}{m}\right)^{1/2}$$

The constants of integration A, B, α, and β are determined from the initial conditions in
any given case.

In order to find the equation of the path, we eliminate the time t between the two
equations. To do this, let us write the second equation in the form

$$y = B \cos(\omega t + \alpha + \Delta)$$

where

$$\Delta = \beta - \alpha$$

Then

$$y = B[\cos(\omega t + \alpha) \cos \Delta - \sin(\omega t + \alpha) \sin \Delta]$$

From the first of Equations 4.23b, we then have

$$\frac{y}{B} = \frac{x}{A} \cos \Delta - \left(1 - \frac{x^2}{A^2}\right)^{1/2} \sin \Delta$$

or, upon squaring and transposing terms, we obtain

$$\frac{x^2}{A^2} - xy\frac{2 \cos \Delta}{AB} + \frac{y^2}{B^2} = \sin^2 \Delta \qquad (4.24)$$

which is a quadratic equation in x and y. Now the general quadratic

$$ax^2 + bxy + cy^2 + dx + ey = f$$

represents an ellipse, a parabola, or a hyperbola, depending on whether the discriminant

$$b^2 - 4ac$$

is negative, zero, or positive, respectively. In our case the discriminant is equal to $-(2 \sin \Delta/AB)^2$, which is negative, so the path is an ellipse as shown in Figure 4.8.

In particular, if the phase difference Δ is equal to $\pi/2$, then the equation of the path reduces to the equation

$$\frac{x^2}{A^2} + \frac{y^2}{B^2} = 1$$

which is the equation of an ellipse whose axes coincide with the coordinate axes. On the other hand, if the phase difference is 0 or π, then the equation of the path reduces to that of a straight line, namely,

$$y = \pm\frac{B}{A}x$$

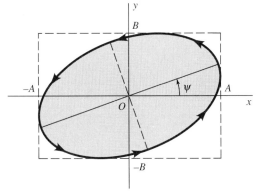

Figure 4.8 Elliptical path of a two-dimensional isotropic oscillator.

The positive sign is taken if $\Delta = 0$, and the negative sign, if $\Delta = \pi$. In the general case it is possible to show that the axis of the elliptical path is inclined to the x-axis by the angle ψ where

$$\tan 2\psi = \frac{2AB \cos \Delta}{A^2 - B^2} \tag{4.25}$$

The derivation is left as an exercise.

The Three-Dimensional Isotropic Harmonic Oscillator

In the case of three-dimensional motion, the differential equation of motion is equivalent to the three equations

$$m\ddot{x} = -kx \qquad m\ddot{y} = -ky \qquad m\ddot{z} = -kz \tag{4.26}$$

which are separated. Hence, the solutions may be written in the form of Equations 4.23b or, alternately, we may write

$$
\begin{aligned}
x &= A_1 \sin \omega t + B_1 \cos \omega t \\
y &= A_2 \sin \omega t + B_2 \cos \omega t \\
z &= A_3 \sin \omega t + B_3 \cos \omega t
\end{aligned}
$$

The six constants of integration are determined from the initial position and velocity of the particle. Now Equations 4.26a can be expressed vectorially as

$$\mathbf{r} = \mathbf{A} \sin \omega t + \mathbf{B} \cos \omega t \tag{4.26b}$$

in which the components of $\mathbf{A}$ are A_1, A_2, and A_3, and similarly for $\mathbf{B}$. It is clear that the motion takes place entirely in a single plane, which is common to the two constant vectors $\mathbf{A}$ and $\mathbf{B}$, and that the path of the particle in that plane is an ellipse, as in the two-dimensional case. Hence, the analysis concerning the shape of the elliptical path under the two-dimensional case also applies to the three-dimensional case.

Nonisotropic Oscillator

The above discussion considered the motion of the isotropic oscillator, wherein the restoring force is independent of the direction of the displacement. If the magnitudes of the components of the restoring force depend on the direction of the displacement, we have the case of the *nonisotropic oscillator.* For a suitable choice of axes, the differential equations for the nonisotropic case can be written

$$
\begin{aligned}
m\ddot{x} &= -k_1 x \\
m\ddot{y} &= -k_2 y \\
m\ddot{z} &= -k_3 z
\end{aligned} \tag{4.27}
$$

Here we have a case of *three* different frequencies of oscillation: $\omega_1 = \sqrt{k_1/m}$, $\omega_2 = \sqrt{k_2/m}$, $\omega_3 = \sqrt{k_3/m}$, and the motion is given by the solutions

$$x = A \cos(\omega_1 t + \alpha)$$
$$y = B \cos(\omega_2 t + \beta) \tag{4.28}$$
$$z = C \cos(\omega_3 t + \gamma)$$

Again, the six constants of integration in the above equations are determined from the initial conditions. The resulting oscillation of the particle lies entirely within a rectangular box (whose sides are $2A$, $2B$, and $2C$) centered on the origin. In the event that ω_1, ω_2, and ω_3 are commensurate, that is, if

$$\frac{\omega_1}{n_1} = \frac{\omega_2}{n_2} = \frac{\omega_3}{n_3} \tag{4.29}$$

where n_1, n_2, and n_3 are integers, the path, called a *Lissajous* figure, will be closed, because after a time $2\pi n_1/\omega_1 = 2\pi n_2/\omega_2 = 2\pi n_3/\omega_3$ the particle will return to its initial position and the motion will be repeated. (In Equation 4.29 it is assumed that any common integral factor is canceled out.) On the other hand, if the ω's are *not* commensurate, the path is not closed. In this case the path may be said to fill completely the rectangular box mentioned above, at least in the sense that if we wait long enough, the particle will come arbitrarily close to any given point.

The net restoring force exerted on a given atom in a solid crystalline substance is approximately linear in the displacement in many cases. The resulting frequencies of oscillation usually lie in the infrared region of the spectrum: 10^{12} to 10^{14} vibrations per second.

Energy Considerations

In the preceding chapter we showed that the potential energy function of the one-dimensional harmonic oscillator is quadratic in the displacement, $V(x) = \frac{1}{2}kx^2$. For the general three-dimensional case it is easy to verify that

$$V(x,\ y,\ z) = \frac{1}{2}k_1 x^2 + \frac{1}{2}k_2 y^2 + \frac{1}{2}k_3 z^2$$

since $F_x = -\partial V/\partial x = -k_1 x$, and similarly for F_y and F_z. If $k_1 = k_2 = k_3 = k$, we have the isotropic case, and

$$V(x,\ y,\ z) = \frac{1}{2}k(x^2 + y^2 + z^2) = \frac{1}{2}kr^2$$

The total energy in the isotropic case is then given by the simple expression

$$\frac{1}{2}mv^2 + \frac{1}{2}kr^2 = E$$

which is similar to that of the one-dimensional case discussed in the previous chapter.

EXAMPLE 4.7

A particle of mass m moves in two dimensions under the following potential energy function:

$$V(\mathbf{r}) = \frac{1}{2}k(x^2 + 4y^2)$$

Find the resulting motion, given the initial condition at $t = 0$: $x = a$, $y = 0$, $\dot{x} = 0$, $\dot{y} = v_0$.

Solution:

This is a nonisotropic oscillator potential. The force function is

$$\mathbf{F} = -\nabla V = -\mathbf{i}kx - \mathbf{j}4ky = m\ddot{\mathbf{r}}$$

The component differential equations of motion are then

$$m\ddot{x} + kx = 0 \qquad m\ddot{y} + 4ky = 0$$

The x-motion has angular frequency $\omega = (k/m)^{1/2}$, while the y-motion has angular frequency just twice that, namely, $\omega_y = (4k/m)^{1/2} = 2\omega$. We shall write the general solution in the form

$$x = A_1 \cos \omega t + B_1 \sin \omega t$$
$$y = A_2 \cos 2\omega t + B_2 \sin 2\omega t$$

To use the initial condition we must first differentiate with respect to t to find the general expression for the velocity components

$$\dot{x} = -A_1\omega \sin \omega t + B_1\omega \cos \omega t$$
$$\dot{y} = -2A_2\omega \sin 2\omega t + 2B_2\omega \cos 2\omega t$$

Thus, at $t = 0$, we see that the above equations for the components of position and velocity reduce to

$$a = A_1 \qquad 0 = A_2 \qquad 0 = B_1\omega \qquad v_0 = 2B_2\omega$$

These equations give directly the values of the amplitude coefficients, $A_1 = a$, $A_2 = B_1 = 0$, $B_2 = v_0/2\omega$, so the final equations for the motion are

$$x = a \cos \omega t$$
$$y = \frac{v_0}{2\omega} \sin 2\omega t$$

The path is a Lissajous figure having the shape of a figure eight as shown in Figure 4.9.   ■

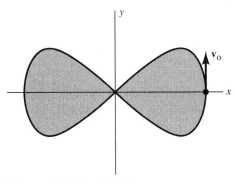

Figure 4.9 Lissajous figure.

4.5 MOTION OF CHARGED PARTICLES IN ELECTRIC AND MAGNETIC FIELDS

When an electrically charged particle is in the vicinity of other electric charges, it experiences a force. This force $\mathbf{F}$ is said to be due to the electric field $\mathbf{E}$, which arises from these other charges. We write

$$\mathbf{F} = q\mathbf{E}$$

where q is the electric charge carried by the particle in question.[4] The equation of motion of the particle is then

$$m\frac{d^2\mathbf{r}}{dt^2} = q\mathbf{E} \tag{4.30}$$

or, in component form,

$$\begin{aligned} m\ddot{x} &= qE_x \\ m\ddot{y} &= qE_y \\ m\ddot{z} &= qE_z \end{aligned} \tag{4.30a}$$

The field components are, in general, functions of the position coordinates x, y, and z. In the case of time-varying fields (that is, if the charges producing $\mathbf{E}$ are moving) the components, of course, also involve t.

Let us consider a simple case, namely that of a uniform constant electric field. We can choose one of the axes, say the z-axis, to be in the direction of the field. Then $E_x = E_y = 0$, and $E = E_z$. The differential equations of motion of a particle of charge q moving in this field are then

$$\ddot{x} = 0 \qquad \ddot{y} = 0 \qquad \ddot{z} = \frac{qE}{m} = \text{constant}$$

---

[4] In SI units F is in newtons, q in coulombs, and E in volts per meter. In cgs units F is in dynes, q in electrostatic units, and E in statvolts per centimeter.

These are of exactly the same form as those for a projectile in a uniform gravitational field. The path is therefore a parabola, if $\dot{x}$ and $\dot{y}$ are not both zero initially. Otherwise, the path is a straight line, as with a body falling vertically.

It is shown in textbooks dealing with electromagnetic theory[5] that

$$\nabla \times \mathbf{E} = 0$$

if $\mathbf{E}$ is due to static charges. This means that motion in such a field is conservative, and that there exists a potential function Φ such that $\mathbf{E} = -\nabla\Phi$. The potential energy of a particle of charge q in such a field is then $q\Phi$, and the total energy is constant and is equal to $\frac{1}{2}mv^2 + q\Phi$.

In the presence of a static magnetic field $\mathbf{B}$ (called the magnetic induction), the force acting on a moving particle is conveniently expressed by means of the cross product, namely,

$$\mathbf{F} = q(\mathbf{v} \times \mathbf{B}) \tag{4.31}$$

where $\mathbf{v}$ is the velocity, and q is the charge.[6] The differential equation of motion of a particle moving in a purely magnetic field is then

$$m\frac{d^2\mathbf{r}}{dt^2} = q(\mathbf{v} \times \mathbf{B}) \tag{4.32}$$

Equation 4.32 states that the acceleration of the particle is always at right angles to the direction of motion. This means that the tangential component of the acceleration ($\dot{v}$) is zero, and so the particle moves with constant speed. This is true even if $\mathbf{B}$ is a varying function of the position $\mathbf{r}$ as long as it does not vary with time.

EXAMPLE 4.8

Let us examine the motion of a charged particle in a uniform constant magnetic field. Suppose we choose the z-axis to be in the direction of the field; that is, we shall write

$$\mathbf{B} = \mathbf{k}B$$

The differential equation of motion now reads

$$m\frac{d^2\mathbf{r}}{dt^2} = q(\mathbf{v} \times \mathbf{k}B) = qB\begin{vmatrix} \mathbf{i} & \mathbf{j} & \mathbf{k} \\ \dot{x} & \dot{y} & \dot{z} \\ 0 & 0 & 1 \end{vmatrix}$$

$$m(\mathbf{i}\ddot{x} + \mathbf{j}\ddot{y} + \mathbf{k}\ddot{z}) = qB(\mathbf{i}\dot{y} - \mathbf{j}\dot{x})$$

---

[5] For example, Reitz, Milford, and Christy, op cit.

[6] Equation 4.31 is valid for SI units: F is in newtons, q in coulombs, v in meters per second, and B in webers per square meter. In cgs units we must write $F = (q/c)(\mathbf{v} \times \mathbf{B})$, where F is in dynes, q in electrostatic units, c is the speed of light (3×10^{10} cm per second), and B is in gauss. (See Reitz, Milford and Christy, footnote 5.)

Equating components, we have

$$m\ddot{x} = qB\dot{y}$$
$$m\ddot{y} = -qB\dot{x} \tag{4.33}$$
$$\ddot{z} = 0$$

Here, for the first time, we meet a set of differential equations of motion that are *not* of the separated type. The solution is relatively simple, however, for we can integrate at once with respect to t to obtain

$$m\dot{x} = qBy + c_1$$
$$m\dot{y} = -qBx + c_2$$
$$\dot{z} = \text{constant} = \dot{z}_0$$

or

$$\dot{x} = \omega y + C_1 \qquad \dot{y} = -\omega x + C_2 \qquad \dot{z} = \dot{z}_0 \tag{4.34}$$

where we have used the abbreviation $\omega = qB/m$. The c's are constants of integration, and $C_1 = c_1/m$, $C_2 = c_2/m$. Upon inserting the expression for $\dot{y}$ from the second part of Equation 4.34 into the first part of Equation 4.33, we obtain the following separated equation for x:

$$\ddot{x} + \omega^2 x = \omega^2 a \tag{4.35}$$

where $a = C_2/\omega$. The solution is clearly

$$x = a + A\cos(\omega t + \theta_0) \tag{4.36}$$

where A and θ_0 are constants of integration. Now, if we differentiate with respect to t, we have

$$\dot{x} = -A\omega\sin(\omega t + \theta_0) \tag{4.37}$$

The above expression for $\dot{x}$ may be substituted for the left-hand side of the first of Equations 4.34 and the resulting equation solved for y. The result is

$$y = b - A\sin(\omega t + \theta_0) \tag{4.38}$$

where $b = -C_1/\omega$. To find the form of the path of motion, we eliminate t between Equation 4.36 and Equation 4.38 to get

$$(x - a)^2 + (y - b)^2 = A^2 \tag{4.39}$$

Thus, the projection of the path of motion on the xy plane is a circle of radius A centered at the point (a, b). Since, from the third of Equations 4.34, the speed in the z direction is constant, we conclude that the path is a *helix*. The axis of the winding path is in the direction of the magnetic field, as shown in Figure 4.10. From Equation 4.38 we have

$$\dot{y} = -A\omega\cos(\omega t + \theta_0) \tag{4.40}$$

Upon eliminating t between Equation 4.37 and Equation 4.40, we find

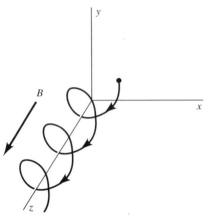

Figure 4.10   Helical path of a particle moving in a magnetic field.

$$\dot{x}^2 + \dot{y}^2 = A^2\omega^2 = A^2\left(\frac{qB}{m}\right)^2 \qquad (4.41)$$

Letting $v_1 = (\dot{x}^2 + \dot{y}^2)^{1/2}$, we see that the radius A of the helix is given by

$$A = \frac{v_1}{\omega} = v_1\frac{m}{qB} \qquad (4.42)$$

If there is no component of the velocity in the z direction, the path is a circle of radius A. It is evident that A is directly proportional to the speed v_1 and that the angular frequency ω of motion in the circular path is independent of the speed. ω is known as the cyclotron frequency. The cyclotron, invented by Ernest Lawrence, depends for its operation on the fact that ω is independent of the speed of the charged particle. ■

4.6   CONSTRAINED MOTION OF A PARTICLE

When a moving particle is restricted geometrically in the sense that it must stay on a certain definite surface or curve, the motion is said to be *constrained*. A piece of ice sliding around a bowl, or a bead sliding on a wire, are examples of constrained motion. The constraint may be complete, as with the bead, or it may be one sided, as with the ice in the bowl. Constraints may be fixed, or they may be moving. In this chapter we shall study only fixed constraints.

The Energy Equation for Smooth Constraints

The total force acting on a particle moving under constraint can be expressed as the vector sum of the net external force **F** and the force of constraint **R**. The latter force is the reaction of the constraining agent upon the particle. The equation of motion may therefore be written

$$m\frac{d\mathbf{v}}{dt} = \mathbf{F} + \mathbf{R} \qquad (4.43)$$

If we take the dot product with the velocity **v,** we have

$$m\frac{d\mathbf{v}}{dt} \cdot \mathbf{v} = \mathbf{F} \cdot \mathbf{v} + \mathbf{R} \cdot \mathbf{v} \qquad (4.44)$$

Now in the case of a *smooth* constraint—for example, a frictionless surface—the reaction **R** is normal to the surface or curve while the velocity **v** is tangent to the surface. Hence, **R** is perpendicular to **v,** and the dot product **R · v** vanishes. Equation 4.44 then reduces to

$$\frac{d}{dt}\left(\frac{1}{2}m\mathbf{v} \cdot \mathbf{v}\right) = \mathbf{F} \cdot \mathbf{v}$$

Consequently, if **F** is conservative, we can integrate as in Section 4.2, and we find the same energy relation as Equation 4.9, namely,

$$\frac{1}{2}mv^2 + V(x, y, z) = \text{constant} = E$$

Thus, the particle, although remaining on the surface or curve, moves in such a way that the total energy is constant. We might, of course, have expected this to be the case for frictionless constraints.

EXAMPLE 4.9

A particle is placed on top of a smooth sphere of radius a. If the particle is slightly disturbed, at what point will it leave the sphere?

Solution:
The forces acting on the particle are the downward force of gravity and the reaction **R** of the spherical surface. The equation of motion is

$$m\frac{d\mathbf{v}}{dt} = m\mathbf{g} + \mathbf{R}$$

Let us choose coordinate axes as shown in Figure 4.11. The potential energy is then mgz, and the energy equation reads

$$\frac{1}{2}mv^2 + mgz = E \qquad (4.45)$$

From the initial conditions ($v = 0$ for $z = a$) we have $E = mga$, so, as the particle slides down, its speed is given by the equation

$$v^2 = 2g(a - z)$$

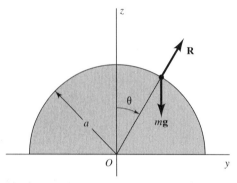

Figure 4.11   Particle sliding on a smooth sphere.

Now, if we take radial components of the equation of motion, we can write the force equation as

$$-\frac{mv^2}{a} = -mg\,\cos\theta + R = -mg\frac{z}{a} + R$$

Hence,

$$R = mg\frac{z}{a} - \frac{mv^2}{a} = mg\frac{z}{a} - \frac{m}{a}2g(a - z)$$

$$= \frac{mg}{a}(3z - 2a)$$

Thus, R vanishes when $z = \frac{2}{3}a$, at which point the particle will leave the sphere. This may be argued from the fact that the sign of R changes from positive to negative there.   ∎

EXAMPLE 4.10   *Constrained Motion on a Cycloid*

Consider a particle sliding under gravity in a smooth cycloidal trough, Figure 4.12, represented by the parametric equations

$$x = A(2\phi + \sin 2\phi)$$
$$z = A(1 - \cos 2\phi)$$

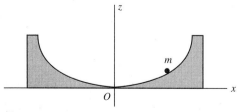

Figure 4.12   Particle sliding in a smooth cycloidal trough.

where ϕ is the parameter. Now the energy equation for the motion, assuming no y-motion, is

$$E = \frac{m}{2}v^2 + V(z) = \frac{m}{2}(\dot{x}^2 + \dot{z}^2) + mgz$$

Since $\dot{x} = 2A\dot{\phi}(1 + \cos 2\phi)$ and $\dot{z} = 2A\dot{\phi} \sin 2\phi$, we find the following expression for the energy in terms of ϕ:

$$E = 4mA^2\dot{\phi}^2(1 + \cos 2\phi) + mgA(1 - \cos 2\phi)$$

or, by use of the identities $1 + \cos 2\phi = 2 \cos^2 \phi$ and $1 - \cos 2\phi = 2 \sin^2 \phi$,

$$E = 8mA^2\dot{\phi}^2 \cos^2 \phi + 2mgA \sin^2 \phi$$

Let us introduce the variable s defined by $s = 4A \sin \phi$. The energy equation can then be written

$$E = \frac{m}{2}\dot{s}^2 + \frac{1}{2}\left(\frac{mg}{4A}\right)s^2$$

This is just the energy equation for harmonic motion in the single variable s. Thus, the particle undergoes periodic motion whose frequency is independent of the amplitude of oscillation, unlike the simple pendulum for which the frequency depends on the amplitude. The periodic motion in the present case is said to be *isochronous*. (The linear harmonic oscillator under Hooke's law is, of course, isochronous.)

The Dutch physicist and mathematician Christiaan Huygens discovered the above fact in connection with attempts to improve the accuracy of pendulum clocks. He also discovered the theory of evolutes and found that the evolute of a cycloid is also a cycloid. Hence, by providing cycloidal "cheeks" for a pendulum, the motion of the bob must follow a cycloidal path and the period is thus independent of the amplitude. Though ingenious, the invention never found extensive practical use. ∎

PROBLEMS

4.1 Find the force for each of the following potential energy functions:
(a) $V = cxyz + C$
(b) $V = \alpha x^2 + \beta y^2 + \gamma z^2 + C$
(c) $V = ce^{-(\alpha x + \beta y + \gamma z)}$
(d) $V = cr^n$ in spherical coordinates

4.2 By finding the curl, determine which of the following forces are conservative:
(a) $\mathbf{F} = \mathbf{i}x + \mathbf{j}y + \mathbf{k}z$
(b) $\mathbf{F} = \mathbf{i}y - \mathbf{j}x + \mathbf{k}z^2$
(c) $\mathbf{F} = \mathbf{i}y + \mathbf{j}x + \mathbf{k}z^3$
(d) $\mathbf{F} = -kr^{-n}\mathbf{e}_r$ in spherical coordinates

4.3 Find the value of the constant c such that each of the following forces is conservative:
(a) $\mathbf{F} = \mathbf{i}xy + \mathbf{j}cx^2 + \mathbf{k}z^3$
(b) $\mathbf{F} = \mathbf{i}(z/y) + c\mathbf{j}(xz/y^2) + \mathbf{k}(x/y)$

4.4 A particle of mass m moving in three dimensions under the potential energy function $V(x, y, z) = \alpha x + \beta y^2 + \gamma z^3$ has speed v_0 when it passes through the origin.
(a) What will its speed be if and when it passes through the point $(1, 1, 1)$?
(b) If the point $(1, 1, 1)$ is a turning point in the motion $(v = 0)$, what is v_0?
(c) What are the component differential equations of motion of the particle?

(*Note:* It is *not* necessary to solve the differential equations of motion in this problem.)

4.5 Consider the two force functions
(a) $\mathbf{F} = \mathbf{i}x + \mathbf{j}y$
(b) $\mathbf{F} = \mathbf{i}y - \mathbf{j}x$
Verify that (a) is conservative and that (b) is nonconservative by showing that the integral $\int \mathbf{F} \cdot d\mathbf{r}$ is independent of the path of integration for (a), but not for (b), by taking two paths in which the starting point is the origin $(0, 0)$, and the end point is $(1, 1)$. For one path take the line $x = y$. For the other path take the x-axis out to the piont $(1, 0)$ and then the line $x = 1$ up to the point $(1, 1)$.

4.6 Show that the variation of gravity with height can be accounted for approximately by the following potential energy function:

$$V = mgz\left(1 - \frac{z}{r_e}\right)$$

in which r_e is the radius of the earth. Find the force given by the above potential function. From this find the component differential equations of motion of a projectile under such a force. If the vertical component of the initial velocity is v_{0z}, how high does the projectile go? (Compare with Example 2.3.)

4.7 Particles of mud are thrown from the rim of a rolling wheel. If the forward speed of the wheel is v_0, and the radius of the wheel is b, show that the greatest height above the ground that the mud can go is

$$b + \frac{v_0^2}{2g} + \frac{gb^2}{2v_0^2}$$

At what point on the rolling wheel does this mud leave? (*Note:* It is necessary to assume that $v_0^2 \geq bg$.)

4.8 A gun is located at the bottom of a hill of constant slope ϕ. Show that the range of the gun measured up the slope of the hill is

$$\frac{2v_0^2 \cos \alpha \, \sin(\alpha - \phi)}{g \cos^2 \phi}$$

where α is the angle of elevation of the gun, and that the maximum value of the slope range is

$$\frac{v_0^2}{g(1 + \sin \phi)}$$

4.9 Write down the component form of the differential equations of motion of a projectile if the air resistance is proportional to the square of the speed. Are the equations separated? Show that the x and y components of the velocity are given by

$$\dot{x} = \dot{x}_0 e^{-\gamma s} \qquad \dot{y} = \dot{y}_0 e^{-\gamma s}$$

where s is the distance the projectile has traveled along the path of motion, and $\gamma = c_2/m$.

4.10 Fill in the steps leading to Equations 4.22 and 4.22a giving the horizontal range of a projectile that is subject to linear air drag.

4.11 The initial conditions for a two-dimensional isotropic oscillator are as follows: $t = 0$, $x = A$, $y = 4A$, $\dot{x} = 0$, $\dot{y} = 3\omega A$ where ω is the angular frequency. Find x and y as functions of t. Show that the motion takes place entirely within a rectangle of dimensions $2A$ and $10A$. Find the inclination ψ of the elliptical path relative to the x-axis. Make a sketch of the path.

4.12 A small lead ball of mass m is suspended by means of six light springs as shown in Figure 4.7. The stiffness constants are in the ratio $1:4:9$ so that the potential energy function can be expressed as

$$V = \frac{k}{2}(x^2 + 4y^2 + 9z^2)$$

At time $t = 0$ the ball is given a push in the $(1, 1, 1)$ direction imparting to it a speed v_0 at the origin. If $k = \pi^2 m$, numerically, find x, y, and z as functions of the time t. Does the ball ever retrace its path? If so, for what value of t does it first return to the origin with the same velocity that it had at $t = 0$?

4.13 Complete the derivation of Equation 4.25.

4.14 An atom is situated in a simple cubic crystal lattice. If the potential energy of interaction between any two atoms is of the form $cr^{-\alpha}$ where c and α are constants and r is the distance between the two atoms, show that the total energy of interaction of a given atom with its six nearest neighbors is approximately that of the three-dimensional harmonic oscillator potential

$$V \simeq A + B(x^2 + y^2 + z^2)$$

where A and B are constants. [*Note:* Assume that the six neighboring atoms are fixed and are located at the points $(\pm d, 0, 0)$, $(0, \pm d, 0)$, $(0, 0, \pm d)$, and that the displacement (x, y, z) of the given atom from the equilibrium position $(0, 0, 0)$ is small compared to d. Then $V = \Sigma cr_i^{-\alpha}$ where $r_1 = [(d - x)^2 + y^2 + z^2]^{1/2}$ with similar expressions for r_2, $r_3, \ldots, r_6$. See approximation formulas in Appendix D.]

4.15 An electron moves in a force field due to a uniform electric field $\mathbf{E}$ and a uniform magnetic field $\mathbf{B}$ which is at right angles to $\mathbf{E}$. Let $\mathbf{E} = \mathbf{j}E$ and $\mathbf{B} = \mathbf{k}B$. Take the initial position of the electron at the origin with initial velocity $\mathbf{v}_0 = \mathbf{i}v_0$ in the x direction. Find the resulting motion of the particle. Show that the path of motion is a cycloid:

$$\begin{aligned} x &= a \sin \omega t + bt \\ y &= a(1 - \cos \omega t) \\ z &= 0 \end{aligned}$$

Cycloidal motion of electrons is used in an electronic tube called a magnetron to produce the microwaves in a microwave oven.

4.16 A particle is placed on a smooth sphere of radius b at a distance $b/2$ above the central plane. As the particle slides down the side of the sphere, at what point will it leave?

4.17 A bead slides on a smooth rigid wire bent into the form of a circular loop of radius b. If the plane of the loop is vertical, and if the bead starts from rest at a point that is level with the center of the loop, find the speed of the bead at the bottom and the reaction of the wire on the bead at that point.

4.18 Show that the period of the particle sliding in the cycloidal trough of Example 4.10 is $4\pi (A/g)^{1/2}$.

COMPUTER/CALCULATOR APPLICATIONS

4.1 Mickey Mantle once hit a baseball that traveled a measured distance of 565 ft. The mass of a baseball is 0.3 kg. Assume that the force of air resistance is proportional to the square of the speed of the baseball, given approximately by $-c_2 v|v|$ where $c_2 \approx .001$ in SI units (see Chapter 2). Calculate the initial elevation angle θ_0 and initial velocity v_0 of the baseball, assuming that Mantle struck it in such a way that the initial angle was optimal for maximum range. Your answer should yield a maximum range to within 5 ft of 565 ft. (*Hint: Numerically integrate the equation of motion to find the trajectory of the baseball $y(x)$. Find the range of the baseball by locating the value of x where $y(x) \rightarrow 0$. To start your calculation, assume v_0 = 100 mph and θ_0 = 45°. Holding v_0 fixed, repeat the calculation for various angles. Find the maximum range and the corresponding value of θ_0 by interpolation. If the range is within 5 ft of the desired value, the problem is complete. Otherwise, estimate a new initial velocity required to give a maximal range of 565 ft at the optimal angle θ_0 computed from the above trial and repeat the process until the calculated maximum range is within the allowed margin of error.*)

Albert Einstein. (Stock Montage, Inc.)

5

NONINERTIAL REFERENCE SYSTEMS

"I was sitting in a chair at the patent office in Bern, when all of a sudden a thought occurred to me: If a person falls freely, he will not feel his own weight. I was startled. This simple thought made a deep impression on me. It impelled me toward a theory of gravitation."

(Albert Einstein, "The Happiest Thought of My Life"; see A. Pais, Inward Bound, New York, Oxford Univ. Press, 1986)

5.1   ACCELERATED COORDINATE SYSTEMS AND INERTIAL FORCES

In describing the motion of a particle, it is frequently convenient, and sometimes necessary, to employ a coordinate system that is not inertial. For example, a coordinate system fixed to the earth is the most convenient one to describe the motion of a projectile, even though the earth is accelerating and rotating.

We shall first consider the case of a coordinate system that undergoes pure translation. In Figure 5.1 $Oxyz$ are the primary coordinate axes (assumed fixed), and $O'x'y'z'$ are the moving axes. In the case of pure translation, the respective axes Ox and $O'x'$, and so on, remain parallel. The position vector of a particle P is denoted by $\mathbf{r}$ in the fixed system and by $\mathbf{r}'$ in the moving system. The displacement OO' of the moving origin is denoted by $\mathbf{R}_0$. Thus, from the triangle $OO'P$, we have

$$\mathbf{r} = \mathbf{R}_0 + \mathbf{r}' \tag{5.1}$$

Taking the first and second time derivatives gives

$$\mathbf{v} = \mathbf{V}_0 + \mathbf{v}' \tag{5.2}$$

$$\mathbf{a} = \mathbf{A}_0 + \mathbf{a}' \tag{5.3}$$

in which V_0 and $\mathbf{A}_0$ are, respectively, the velocity and acceleration of the moving system, and $\mathbf{v}'$ and $\mathbf{a}'$ are the velocity and acceleration of the particle *in* the moving system.

In particular, if the moving system is not accelerating, so that $\mathbf{A}_0 = 0$, then

$$\mathbf{a} = \mathbf{a}'$$

so the acceleration is the same in either system. Consequently, if the primary system is inertial, Newton's second law $\mathbf{F} = m\mathbf{a}$ becomes $\mathbf{F} = m\mathbf{a}'$ in the moving system; that is, the moving system is also an inertial system (provided it is not rotating). Thus, as far as Newtonian mechanics is concerned, we cannot specify a unique coordinate system; if Newton's laws hold in one system, they are also valid in any other system moving with uniform velocity relative to the first.

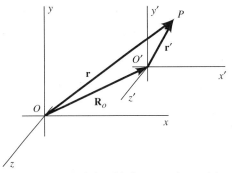

Figure 5.1   Relationship between the position vectors for two coordinate systems undergoing pure translation relative to one another.

On the other hand if the moving system is accelerating, then Newton's second law becomes

$$\mathbf{F} = m\mathbf{A}_0 + m\mathbf{a}'$$

or

$$\mathbf{F} - m\mathbf{A}_0 = m\mathbf{a}' \tag{5.4}$$

for the equation of motion in the accelerating system. If we wish, we can write Equation 5.4 in the form

$$\mathbf{F}' = m\mathbf{a}' \tag{5.5}$$

in which $\mathbf{F}' = \mathbf{F} + (-m\mathbf{A}_0)$. That is, an acceleration $\mathbf{A}_0$ of the reference system can be taken into account by adding an *inertial term* $-m\mathbf{A}_0$ to the force $\mathbf{F}$ and equating the result to the product of mass and acceleration in the moving system. Inertial terms in the equations of motion are sometimes called *inertial forces* or *fictitious forces*. Such "forces" are not due to interactions with other bodies, rather, they stem from the acceleration of the reference system. Whether or not one wishes to call them forces is purely a matter of terminology. In any case inertial terms are present if a noninertial coordinate system is used to describe the motion of a particle.

EXAMPLE 5.1

A block of wood rests on a rough horizontal table. If the table is accelerated in a horizontal direction, under what conditions will the block slip?

Solution:
Let μ_s be the coefficient of static friction between the block and the table top. Then the force of friction $\mathbf{F}$ has a maximum value of $\mu_s mg$, where m is the mass of the block. The condition for slipping is that the inertial force $-m\mathbf{A}_0$ exceeds the frictional force where $\mathbf{A}_0$ is the acceleration of the table. Hence, the condition for slipping is

$$|-m\mathbf{A}_0| > \mu_s mg$$

or

$$A_0 > \mu_s g \qquad \blacksquare$$

EXAMPLE 5.2

A pendulum is suspended from the ceiling of a railroad car, as shown in Figure 5.2a. Assume that the car is accelerating uniformly toward the right ($+x$ direction). A noninertial observer, the boy inside the car, sees the pendulum hanging at an angle θ, left of vertical. He believes it hangs this way because of the existence of an inertial force $\mathbf{F}'_x$, which acts upon all objects in his accelerated frame of reference (Figure 5.2b). An inertial observer, the girl outside the car, sees the same thing.

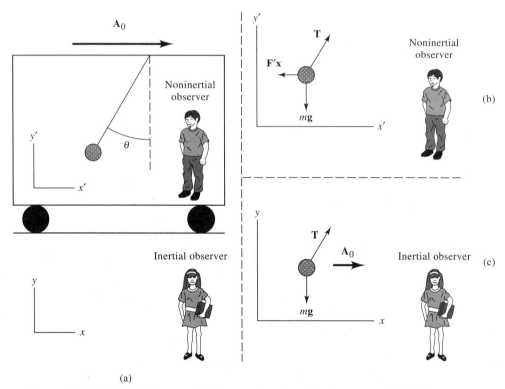

Figure 5.2   (a) Pendulum suspended in an accelerating railroad car as seen by (b) the non-inertial observer and (c) the inertial observer.

However, she knows there is no real force $\mathbf{F}'_x$ acting upon the pendulum. She knows that it hangs this way because a net force in the horizontal direction is required to accelerate it at the rate $\mathbf{A}_0$ that she observes (Figure 5.2c). Calculate the acceleration $\mathbf{A}_0$ of the car from the inertial observer's point of view. Show that, according to the noninertial observer, $\mathbf{F}'_x = -m\mathbf{A}_0$ is the force that causes the pendulum to hang at the angle θ.

Solution:
The inertial observer writes down Newton's second law for the hanging pendulum as

$$\sum \mathbf{F}_i = m\mathbf{a}$$
$$T \sin \theta = mA_0 \qquad T \cos \theta - mg = 0$$
$$\therefore A_0 = g \tan \theta$$

She concludes that the suspended pendulum hangs at the angle θ because the railroad car is accelerating in the horizontal direction and a horizontal force is needed to make it accelerate. This force is the x-component of the tension in the string. The acceleration of the car is proportional to the tangent of the angle of deflection. The pendulum, thus, serves as a linear accelerometer.

On the other hand the noninertial observer, unaware of the outside world (assume the railroad track is perfectly smooth—no vibration—and that the railroad car has no windows or other sensory clues for another reference point), observes that the pendulum just hangs there, tilted to the left of vertical. He concludes that

$$\sum \mathbf{F}'_i = m\mathbf{a}' = 0$$

$$T \sin \theta - F'_x = 0 \qquad T \cos \theta - mg = 0$$

$$\therefore F'_x = mg \tan \theta$$

All the forces acting on the pendulum are in balance and the reason that the pendulum hangs left of vertical is due to the force $\mathbf{F}'_x (= -m\mathbf{A}_0)$. In fact if this observer were to do some more experiments in the railroad car, such as drop balls or stones or whatever, he would see that they would also be deflected to the left of vertical. He would soon discover that the amount of the deflection would be independent of their mass. In other words he would conclude that there was a force, quite like a gravitational one (to be discussed in Chapter 6), pushing things to the left of the car with an acceleration $\mathbf{A}_0$ as well as the force pulling them down with an acceleration $\mathbf{g}$. ∎

EXAMPLE 5.3

Two astronauts are standing in a spaceship accelerating upward with an acceleration $\mathbf{A}_0$ as shown in Figure 5.3. Let the magnitude of $\mathbf{A}_0$ equal g. Astronaut #1 throws a ball directly toward astronaut #2 who is 10 m away on the other side of the ship. What must be the initial speed of the ball if it is to reach astronaut #2 before striking the floor? Assume astronaut #1 releases the ball at a height $h = 2$ m above the floor of the ship. Solve the problem from the perspective of both (a) a noninertial observer (inside the ship) and (b) an inertial observer (outside the ship).

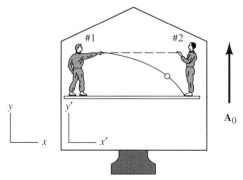

Figure 5.3 Two astronauts throwing a ball in a spaceship accelerating at $A_0 = g$.

Solution:

(a) The noninertial observer believes that a force $-m\mathbf{A_0}$ acts upon all objects in the ship. Thus, in the noninertial $(x'y')$ frame of reference, we conclude that the trajectory of the ball is a parabola, that is,

$$x'(t) = \dot{x}_0' t \qquad y'(t) = \dot{y}_0' - \frac{1}{2}A_0 t^2$$

$$\therefore y'(x') = \dot{y}_0' - \frac{1}{2}A_0\left(\frac{x'}{\dot{x}_0'}\right)^2$$

Setting $y'(x')$ equal to zero when $x' = 10$ m and solving for $\dot{x}_0'$ yields

$$\dot{x}_0' = \left(\frac{A_0}{2y_0'}\right)^{1/2} x'$$

$$= \left(\frac{9.8 \text{ ms}^{-2}}{4 \text{ m}}\right)^{1/2} (10 \text{ m}) = 15.6 \text{ ms}^{-1}$$

(b) The inertial observer sees the picture a little differently. It appears to him that the ball travels at constant velocity in a straight line after it is released and that the floor of the spaceship accelerates upward to intercept the ball. A plot of the vertical position of the ball and the floor of the spaceship is shown schematically in Figure 5.4. Both the ball and the rocket have the same initial upward speed $\dot{y}_0$ at the moment the ball is released by astronaut #1. The vertical positions of the ball and the floor coincide at a time that depends on the initial height of the ball

$$y_0 + \dot{y}_0 t = \dot{y}_0 t + \frac{1}{2}A_0 t^2$$

$$y_0 = \frac{1}{2}A_0 t^2$$

$$x = \dot{x}_0 t \qquad \text{or} \qquad t = \frac{x}{\dot{x}_0}$$

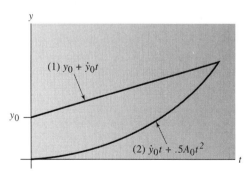

Figure 5.4   Vertical position of (1) a ball thrown in an accelerating rocket and (2) the floor of the rocket as seen by an inertial observer.

During this time, the ball has traveled a horizontal distance. Inserting this time into the relation above yields the required initial, horizontal speed of the ball

$$y_0 = \frac{1}{2}A_0\left(\frac{x}{\dot{x}_0}\right)^2$$

$$\dot{x}_0 = \left(\frac{A_0}{2y_0}\right)^{1/2} x$$

Thus, each observer calculates the same value for the initial horizontal velocity, as well they should.

The analysis seems less complex from the perspective of the noninertial observer. In fact the noninertial observer would physically experience the inertial force $-mA_0$. It would seem every bit as real as the gravitational force we experience here on earth. Our astronaut might even invent the concept of gravity to "explain" the dynamics of moving objects observed in the spaceship.   ■

5.2 ROTATING COORDINATE SYSTEMS. ANGULAR VELOCITY AS A VECTOR QUANTITY

To discuss the effects of rotation of the coordinate system it will be convenient to first consider the case of pure rotation. The origins of the two coordinate systems then coincide, Figure 5.5, and so $\mathbf{r} = \mathbf{r}'$, or explicitly

$$\mathbf{i}x + \mathbf{j}y + \mathbf{k}z = \mathbf{i}'x' + \mathbf{j}'y' + \mathbf{k}'z' \tag{5.6}$$

When we differentiate with respect to time to find the velocity, we must keep in mind the fact that the unit vectors $\mathbf{i}'$, $\mathbf{j}'$, and $\mathbf{k}'$ in the rotating system are *not* constant, whereas the primary unit vectors $\mathbf{i}$, $\mathbf{j}$, and $\mathbf{k}$ are considered constant. Thus, we can write

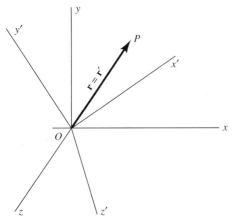

Figure 5.5 Rotating coordinate system (primed system).

$$\mathbf{i}\frac{dx}{dt} + \mathbf{j}\frac{dy}{dt} + \mathbf{k}\frac{dz}{dt} = \mathbf{i'}\frac{dx'}{dt} + \mathbf{j'}\frac{dy'}{dt} + \mathbf{k'}\frac{dz'}{dt} + x'\frac{d\mathbf{i'}}{dt} + y'\frac{d\mathbf{j'}}{dt} + z'\frac{d\mathbf{k'}}{dt}$$

The three terms on the left-hand side of the above equation clearly give the velocity vector $\mathbf{v}$ in the fixed system, and the first three terms on the right are the components of the velocity *in* the rotating system, which we shall call $\mathbf{v'}$, so the equation may be written

$$\mathbf{v} = \mathbf{v'} + x'\frac{d\mathbf{i'}}{dt} + y'\frac{d\mathbf{j'}}{dt} + z'\frac{d\mathbf{k'}}{dt} \tag{5.7}$$

The last three terms on the right represent the velocity due to rotation of the primed coordinate system. We must now determine how the time derivatives of the basis vectors are related to the rotation.

At any given instant the rotation of the primed system is specified by some axis of rotation and an angular speed about that axis. Let the direction of the axis be designated by a unit vector $\mathbf{n}$, and let the angular speed be ω. We shall call the product $\omega\mathbf{n}$ the *angular velocity vector* of the rotating system:

$$\boldsymbol{\omega} = \omega\mathbf{n} \tag{5.8}$$

The sense of the angular velocity vector is given by the right-hand rule, similar to the definition of the cross product, as shown in Figure 5.6.

In order to find the time derivatives $d\mathbf{i'}/dt$, $d\mathbf{j'}/dt$, and $d\mathbf{k'}/dt$, consider Figure 5.7. Here is shown the change $\Delta\mathbf{i'}$ in the unit vector $\mathbf{i'}$ due to a small rotation $\Delta\theta$ about the axis of rotation. (The vectors $\mathbf{j'}$ and $\mathbf{k'}$ are omitted for clarity.) From the figure we see that the magnitude of $\Delta\mathbf{i'}$ is given by the approximate relation

$$|\Delta\mathbf{i'}| \simeq (\sin\ \phi)\Delta\theta$$

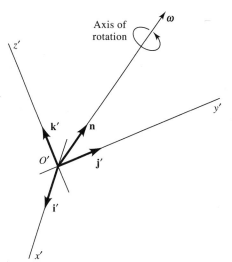

Figure 5.6   The angular velocity vector of a rotating coordinate system.

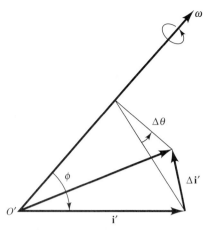

Figure 5.7   Change in the unit vector **i′** produced by a small rotation $\Delta\theta$.

where ϕ is the angle between **i′** and **ω**. Let Δt be the time interval for this change. Then we can write

$$\left|\frac{d\mathbf{i'}}{dt}\right| = \lim_{\Delta t \to 0} \left|\frac{\Delta \mathbf{i'}}{\Delta t}\right| = \sin\phi \frac{d\theta}{dt} = (\sin\phi)\omega$$

Now the direction of $\Delta\mathbf{i'}$ is perpendicular to *both* **ω** and **i′**; consequently, from the definition of the cross product, we can write the above equation in vector form

$$\frac{d\mathbf{i'}}{dt} = \boldsymbol{\omega} \times \mathbf{i'} \tag{5.9}$$

Similarly, we find $d\mathbf{j'}/dt = \boldsymbol{\omega} \times \mathbf{j'}$, and $d\mathbf{k'}/dt = \boldsymbol{\omega} \times \mathbf{k'}$.

We now apply the above result to the last three terms in Equation 5.7 as follows:

$$x'\frac{d\mathbf{i'}}{dt} + y'\frac{d\mathbf{j'}}{dt} + z'\frac{d\mathbf{k'}}{dt} = x'(\boldsymbol{\omega} \times \mathbf{i'}) + y'(\boldsymbol{\omega} \times \mathbf{j'}) + z'(\boldsymbol{\omega} \times \mathbf{k'})$$

$$= \boldsymbol{\omega} \times (\mathbf{i'}x' + \mathbf{j'}y' + \mathbf{k'}z')$$

$$= \boldsymbol{\omega} \times \mathbf{r'}$$

This is the velocity of P due to rotation of the primed coordinate system. Accordingly, Equation 5.7 can be shortened to read

$$\mathbf{v} = \mathbf{v'} + \boldsymbol{\omega} \times \mathbf{r'} \tag{5.10}$$

or, more explicitly

$$\left(\frac{d\mathbf{r}}{dt}\right)_{fixed} = \left(\frac{d\mathbf{r'}}{dt}\right)_{rot} + \boldsymbol{\omega} \times \mathbf{r'} = \left[\left(\frac{d}{dt}\right)_{rot} + \boldsymbol{\omega} \times\right]\mathbf{r'}$$

that is, the operation of differentiating the position vector with respect to time in the fixed system is equivalent to the operation of taking the time derivative in the rotating system plus the operation $\boldsymbol{\omega} \times$. A little reflection will show that the same applies to *any* vector $\mathbf{Q}$: $dQ/dt)_{fixed} = dQ/dt)_{rot} + \boldsymbol{\omega} \times \mathbf{Q}$. In particular, if that vector is the velocity, then we have

$$\left(\frac{d\mathbf{v}}{dt}\right)_{fixed} = \left(\frac{d\mathbf{v}}{dt}\right)_{rot} + \boldsymbol{\omega} \times \mathbf{v}$$

But $\mathbf{v} = \mathbf{v}' + \boldsymbol{\omega} \times \mathbf{r}'$, so

$$\left(\frac{d\mathbf{v}}{dt}\right)_{fixed} = \left(\frac{d}{dt}\right)_{rot} (\mathbf{v}' + \boldsymbol{\omega} \times \mathbf{r}') + \boldsymbol{\omega} \times (\mathbf{v}' + \boldsymbol{\omega} \times \mathbf{r}')$$

$$= \left(\frac{d\mathbf{v}'}{dt}\right)_{rot} + \left[\frac{d(\boldsymbol{\omega} \times \mathbf{r}')}{dt}\right]_{rot} + \boldsymbol{\omega} \times \mathbf{v}' + \boldsymbol{\omega} \times (\boldsymbol{\omega} \times \mathbf{r}')$$

$$= \left(\frac{d\mathbf{v}'}{dt}\right)_{rot} + \left(\frac{d\boldsymbol{\omega}}{dt}\right)_{rot} \times \mathbf{r}' + \boldsymbol{\omega} \times \left(\frac{d\mathbf{r}'}{dt}\right)_{rot}$$

$$+ \boldsymbol{\omega} \times \mathbf{v}' + \boldsymbol{\omega} \times (\boldsymbol{\omega} \times \mathbf{r}')$$

Now concerning the term involving the time derivative of $\boldsymbol{\omega}$, we have $(d\boldsymbol{\omega}/dt)_{fixed} = (d\boldsymbol{\omega}/dt)_{rot} + \boldsymbol{\omega} \times \boldsymbol{\omega}$. But the cross product of any vector with itself vanishes, so $(d\boldsymbol{\omega}/dt)_{fixed} = (d\boldsymbol{\omega}/dt)_{rot} = \dot{\boldsymbol{\omega}}$. Since $\mathbf{v}' = (d\mathbf{r}'/dt)_{rot}$ and $\mathbf{a}' = (d\mathbf{v}'/dt)_{rot}$, we can express the final result as follows:

$$\mathbf{a} = \mathbf{a}' + \dot{\boldsymbol{\omega}} \times \mathbf{r}' + 2\boldsymbol{\omega} \times \mathbf{v}' + \boldsymbol{\omega} \times (\boldsymbol{\omega} \times \mathbf{r}') \tag{5.11}$$

giving the acceleration in the fixed system in terms of the position, velocity, and acceleration in the rotating system.

In the general case in which the primed system is undergoing *both* translation and rotation (Figure 5.8), we must add the velocity of translation $\mathbf{V}_0$ to the right-hand side

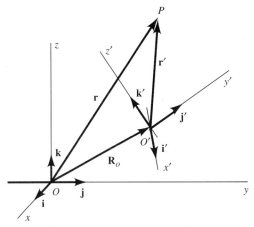

Figure 5.8   Geometry for the general case of translation and rotation of the moving coordinate system (primed system).

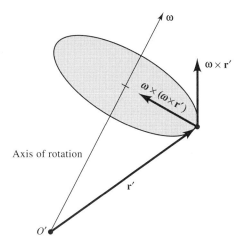

Figure 5.9 Illustrating the centripetal acceleration.

of Equation 5.10 and the acceleration A_0 of the moving system to the right-hand side of Equation 5.11. This gives the general equations for transforming from a fixed system to a moving and rotating system:

$$v = v' + \boldsymbol{\omega} \times r' + V_0 \tag{5.12}$$

$$a = a' + \dot{\boldsymbol{\omega}} \times r' + 2\boldsymbol{\omega} \times v' + \boldsymbol{\omega} \times (\boldsymbol{\omega} \times r') + A_0 \tag{5.13}$$

The term $2\boldsymbol{\omega} \times v'$ is known as the *Coriolis acceleration,* and the term $\boldsymbol{\omega} \times (\boldsymbol{\omega} \times r')$ is called the *centripetal acceleration.* The Coriolis acceleration appears whenever a particle moves in a rotating coordinate system (except when the velocity v' is parallel to the axis of rotation), and the centripetal acceleration is the result of the particle being carried around a circular path (for fixed r') in the rotating system. The centripetal acceleration is always directed toward the axis of rotation and is perpendicular to the axis as shown in Figure 5.9. The term $\dot{\boldsymbol{\omega}} \times r'$ is called the *transverse acceleration,* since it is perpendicular to the position vector r'. It appears as a result of any angular acceleration of the rotating system, that is, if the angular velocity vector is changing in either magnitude or direction, or both.

EXAMPLE 5.4

A wheel of radius b rolls along the ground with constant forward speed V_0. Find the acceleration, relative to the ground, of any point on the rim.

Solution:
Let us choose a coordinate system fixed to the rotating wheel, and let the moving origin be at the center with the x'-axis passing through the point in question, as shown in Figure 5.10. Then we have

$$r' = i'b \qquad a' = \ddot{r}' = 0 \qquad v' = \dot{r}' = 0$$

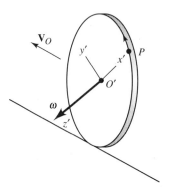

Figure 5.10 Rotating coordinates fixed to a rolling wheel.

The angular velocity vector is given by

$$\boldsymbol{\omega} = \mathbf{k}'\omega = \mathbf{k}'\frac{V_0}{b}$$

for the choice of coordinates shown. Hence, all terms in the expression for acceleration vanish except the centripetal term:

$$\mathbf{a} = \boldsymbol{\omega} \times (\boldsymbol{\omega} \times \mathbf{r}') = \mathbf{k}'\omega \times (\mathbf{k}'\omega \times \mathbf{i}'b)$$

$$= \frac{V_0^2}{b} \mathbf{k}' \times (\mathbf{k}' \times \mathbf{i}')$$

$$= \frac{V_0^2}{b} \mathbf{k}' \times \mathbf{j}'$$

$$= \frac{V_0^2}{b} (-\mathbf{i}')$$

Thus, **a** is of magnitude V_0^2/b and is always directed toward the center of the rolling wheel.   ∎

EXAMPLE 5.5

A bicycle travels with constant speed around a track of radius ρ. What is the acceleration of the highest point on one of its wheels? Let V_0 denote the speed of the bicycle and b the radius of the wheel.

Solution:

We choose a coordinate system with origin at the center of the wheel and with the x'-axis horizontal pointing toward the center of curvature C of the track. Rather than have the moving coordinate system rotate with the wheel, we choose a system in which the z'-axis remains vertical as shown in Figure 5.11. Thus, the $O'x'y'z'$ system rotates with angular velocity $\boldsymbol{\omega}$, which can be expressed as

$$\boldsymbol{\omega} = \mathbf{k}'\frac{V_0}{\rho}$$

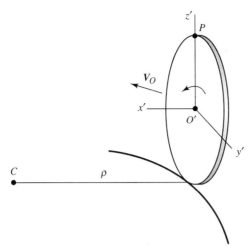

Figure 5.11 Wheel rolling on a curved track. The z' axis remains vertical as the wheel turns.

and the acceleration of the moving origin A_0 is given by

$$A_0 = i' \frac{V_0^2}{\rho}$$

Since each point on the wheel is moving in a circle of radius b with respect to the moving origin, the acceleration in the $O'x'y'z'$ system of any point on the wheel is directed toward O' and has magnitude V_0^2/b. Thus, in the moving system we have

$$\ddot{r}' = -k' \frac{V_0^2}{b}$$

for the point at the top of the wheel. Also, the velocity of this point in the moving system is given by

$$v' = -j'V_0$$

so the Coriolis acceleration is

$$2\omega \times v' = 2\left(\frac{V_0}{\rho}k'\right) \times (-j'V_0) = 2\frac{V_0^2}{\rho}i'$$

Since the angular velocity ω is constant, the transverse acceleration is zero. The centripetal acceleration is also zero because

$$\omega \times (\omega \times r') = \frac{V_0^2}{\rho^2}k' \times (k' \times bk') = 0$$

Thus, the net acceleration, relative to the ground, of the highest point on the wheel is

$$a = 3\frac{V_0^2}{\rho}i' - \frac{V_0^2}{b}k'$$

∎

5.3   DYNAMICS OF A PARTICLE IN A ROTATING COORDINATE SYSTEM

The fundamental equation of motion of a particle in an inertial frame of reference is

$$\mathbf{F} = m\mathbf{a}$$

where $\mathbf{F}$ is the vector sum of all real, physical forces acting on the particle. In view of Equation 5.13, we can write the equation of motion in a noninertial frame of reference as

$$\mathbf{F} - m\mathbf{A}_0 - 2m\boldsymbol{\omega} \times \mathbf{v}' - m\dot{\boldsymbol{\omega}} \times \mathbf{r}' - m\boldsymbol{\omega} \times (\boldsymbol{\omega} \times \mathbf{r}') = m\mathbf{a}' \quad (5.14)$$

All the terms from Equation 5.13, except $\mathbf{a}'$, have been multiplied by m and transposed to show them as inertial forces added to the real, physical forces $\mathbf{F}$. The $\mathbf{a}'$ term has been multiplied by m also, but left on the right-hand side. Thus, Equation 5.14 represents the dynamical equation of motion of a particle in a noninertial frame of reference subjected to both real, physical forces as well as those inertial forces that appear as a result of the acceleration of the noninertial frame of reference. The inertial forces have names corresponding to their respective accelerations, discussed in Section 5.2. The *Coriolis force* is

$$\mathbf{F}'_{Cor} = -2m\boldsymbol{\omega} \times \mathbf{v}'$$

The *transverse force* is

$$\mathbf{F}'_{trans} = -m\dot{\boldsymbol{\omega}} \times \mathbf{r}'$$

The *centrifugal force* is

$$\mathbf{F}'_{Cen} = -m\boldsymbol{\omega} \times (\boldsymbol{\omega} \times \mathbf{r}')$$

The remaining inertial force $-m\mathbf{A}_0$ appears whenever the (x', y', z') coordinate system is undergoing a translational acceleration, as discussed in Section 5.1.

A noninertial observer in an accelerated frame of reference who denotes the acceleration of a particle by the vector $\mathbf{a}'$ will be forced to include any or all of these inertial forces along with the real forces in order to calculate the correct motion of the particle. In other words, such an observer would write down the fundamental equation of motion as

$$\mathbf{F}' = m\mathbf{a}'$$

in which the sum of the vector forces $\mathbf{F}'$ acting on the particle is given by

$$\mathbf{F}' = \mathbf{F}_{physical} + \mathbf{F}'_{Cor} + \mathbf{F}'_{trans} + \mathbf{F}'_{centrif} - m\mathbf{A}_0$$

We have emphasized the real, physical nature of the force term $\mathbf{F}$ in Equation 5.14 by appending the subscript *physical* to it here. $\mathbf{F}$ (or $\mathbf{F}_{physical}$) forces are the only forces that a noninertial observer would claim are actually acting upon the particle. The inclusion of the remaining four inertial terms depend critically on the exact status of the noninertial frame of reference being used to describe the motion of the particle. They arise because of the inertial property of the matter whose motion is under investigation, rather than from the presence or action of any surrounding matter.

The Coriolis force is particularly interesting. It is present only if a particle is moving in a rotating coordinate system. Its direction is always perpendicular to the velocity vector of the particle in the moving system. The Coriolis force thus seems to deflect a moving particle at right angles to its direction of motion. (The Coriolis force has been rather fancifully called "the merry-go-round force." Try walking radially inward or outward on a moving merry-go-round to experience its effect.) This force is important in computing the trajectory of a projectile. Coriolis effects are responsible for the circulation of air around high- or low-pressure systems on the earth's surface. In the case of a high-pressure area,[1] as air spills down from the high, it flows outward and away, deflecting toward the right as it moves into the surrounding low, setting up a clockwise circulation pattern. In the southern hemisphere the reverse is true.

The transverse force is present only if there is an angular acceleration (or deceleration) of the rotating coordinate system. This force is always perpendicular to the radius vector $\mathbf{r}'$ in the rotating coordinate system.

The centrifugal force is the familiar one that arises from rotation about an axis. It is directed outward away from the axis of rotation and is perpendicular to that axis. These three inertial forces are illustrated in Figure 5.12 for the case of a mass m moving radially outward on a rotating platform, whose rate of rotation is decreasing ($\dot{\omega} < 0$). The z-axis is the axis of rotation, directed out of the paper. That is also the direction of the angular velocity vector $\boldsymbol{\omega}$. Since $\mathbf{r}'$, the radius vector denoting the position of m in the rotating system, is perpendicular to $\boldsymbol{\omega}$, the magnitude of the centrifugal force is $mr'\omega^2$. In general if the angle between $\boldsymbol{\omega}$ and $\mathbf{r}'$ is θ, then the magnitude of the centripetal force is $mr'\omega^2 \sin\theta$ where $r' \sin\theta$ is the shortest distance from the mass to the axis of protation.

---

[1] A high-pressure system is essentially a bump in the earth's atmosphere where more air is stacked up above some region on the earth's surface than it is for surrounding regions.

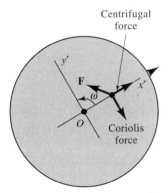

Figure 5.12 Forces on an insect crawling outward along a radial line on a rotating wheel.

EXAMPLE 5.6

A bug crawls outward with a constant speed v' along the spoke of a wheel that is rotating with constant angular velocity $\boldsymbol{\omega}$ about a vertical axis. Find all the apparent forces acting on the bug.

Solution:

First, let us choose a coordinate system fixed on the wheel, and let the x'-axis point along the spoke in question. Then we have

$$\dot{\mathbf{r}}' = \mathbf{i}\dot{x}' = \mathbf{i}v'$$
$$\ddot{\mathbf{r}}' = 0$$

for the velocity and acceleration of the bug as described in the rotating system. If we choose the z'-axis to be vertical, then

$$\boldsymbol{\omega} = \mathbf{k}'\omega$$

The various forces are then given by the following:

$$-2m\boldsymbol{\omega} \times \dot{\mathbf{r}}' = -2m\omega v'(\mathbf{k}' \times \mathbf{i}') = -2m\omega v'\mathbf{j}' \qquad Coriolis\ force$$
$$-m\dot{\boldsymbol{\omega}} \times \mathbf{r}' = 0 \qquad (\omega = \text{constant}) \qquad transverse\ force$$
$$-m\boldsymbol{\omega} \times (\boldsymbol{\omega} \times \mathbf{r}') = -m\omega^2[\mathbf{k}' \times (\mathbf{k}' \times \mathbf{i}'x')] \qquad centrifugal\ force$$
$$= -m\omega^2(\mathbf{k}' \times \mathbf{j}'x')$$
$$= m\omega^2 x'\mathbf{i}'$$

Thus, Equation 5.14 reads

$$\mathbf{F} - 2m\omega v'\mathbf{j}' + m\omega^2 x'\mathbf{i}' = 0$$

Here $\mathbf{F}$ is the real force exerted on the bug by the spoke. The forces are shown in Figure 5.13. ∎

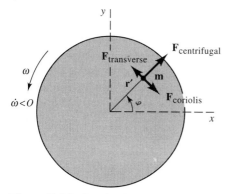

Figure 5.13   Inertial forces acting on a mass m moving radially outward on a platform rotating with angular velocity $\boldsymbol{\omega}$ and angular acceleration $\dot{\boldsymbol{\omega}} < 0$. The x-y axes are fixed. The direction of $\boldsymbol{\omega}$ is out of the paper.

EXAMPLE 5.7

In Example 5.6, find how far the bug can crawl before it starts to slip, given the coefficient of static friction μ_s between the bug and the spoke.

Solution:
Since the force of friction **F** has a maximum value of $\mu_s mg$, slipping will start when

$$|\mathbf{F}| = \mu_s mg$$

or

$$[(2m\omega v')^2 + (m\omega^2 x')^2]^{1/2} = \mu_s mg$$

Upon solving for x', we find

$$x' = \frac{[\mu_s^2 g^2 - 4\omega^2(v')^2]^{1/2}}{\omega^2}$$

for the distance the bug can crawl before slipping.   ■

EXAMPLE 5.8

A smooth rod of length l rotates in a plane with a constant angular velocity ω about an axis fixed at the end of the rod and perpendicular to the plane of rotation. A bead of mass m is initially positioned at the stationary end of the rod and given a slight push such that its initial speed directed down the rod is $\epsilon = \omega l$. See Figure 5.14. Calculate how long it takes for the bead to reach the other end of the rod.

Solution:
The best way to solve this problem is to examine it from the perspective of an (x', y') frame of reference rotating with the rod. If we let the x'-axis lie along the rod, then the problem is one-dimensional along that direction. The only real force acting on the bead is **F**, the reaction force that the rod exerts upon the bead. It points perpendicular to the rod, along the y'-direction as shown in Figure 5.14. **F** has no

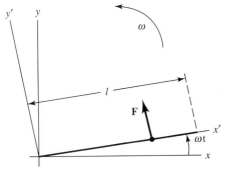

Figure 5.14 Bead sliding along smooth rod rotating at constant angular velocity ω about an axis fixed at one end.

x'-component since there is no friction. Thus, applying Equation 5.14 to the bead in this rotating frame, we obtain

$$F\mathbf{j}' - 2m\omega\mathbf{k}' \times \dot{x}'\mathbf{i}' - m\omega\mathbf{k}' \times (\omega\mathbf{k}' \times x'\mathbf{i}') = m\ddot{x}'\mathbf{i}'$$
$$F\mathbf{j} - 2m\omega\dot{x}'\mathbf{j}' + m\omega^2 x'\mathbf{i}' = m\ddot{x}'\mathbf{i}'$$

The first inertial force in the above equation is the Coriolis force. It appears in the expression because of the bead's velocity $\dot{x}'\mathbf{i}'$ along the x'-axis in the rotating frame. Note that it balances out the reaction force $\mathbf{F}$ that the rod exerts on the bead. The second inertial force is the centrifugal force, $m\omega^2 x'$. From the bead's perspective, this force shoves it down the end of the rod. These ideas are embodied in the two scalar equivalents of the above vector equation

$$F = 2m\omega\dot{x}' \qquad m\omega^2 x' = m\ddot{x}'$$

Solving the second equation above yields $x'(t)$, the position of the bead along the rod as a function of time

$$x'(t) = Ae^{\omega t} + Be^{-\omega t}$$
$$\dot{x}'(t) = \omega Ae^{\omega t} - \omega Be^{-\omega t}$$

The boundary conditions, $x'(t = 0) = 0$ and $\dot{x}(t = 0) = \epsilon$, allows us to determine the constants A and B

$$x'(0) = 0 = A + B \qquad \dot{x}'(0) = \epsilon = \omega(A - B)$$

$$A = -B = \frac{\epsilon}{2\omega}$$

which lead to the explicit solution

$$x'(t) = \frac{\epsilon}{2\omega}(e^{\omega t} - e^{-\omega t})$$

$$= \frac{\epsilon}{\omega} \sinh \omega t$$

The bead flies off the end of the rod at time T where

$$x'(T) = \frac{\epsilon}{\omega} \sinh \omega T = l$$

$$T = \frac{1}{\omega} \sinh^{-1}\left(\frac{\omega l}{\epsilon}\right)$$

Since the initial speed of the bead is $\epsilon = \omega l$, the above equation becomes

$$T = \frac{1}{\omega} \sinh^{-1}(l) = \frac{0.88}{\omega}$$

      ■

5.4   EFFECTS OF THE EARTH'S ROTATION

Let us apply the theory developed in the foregoing sections to a coordinate system that is moving with the earth. Since the angular speed of the earth's rotation is 2π radians per day, or about 7.27×10^{-5} radians per sec, we might expect the effects of such rotation to be relatively small. Nevertheless, it is the spin of the earth that produces the equatorial bulge; the equatorial radius is some 13 miles greater than the polar radius.

Static Effects. The Plumb Line

Let us consider the case of a plumb bob that is normally used to define the direction of the local "vertical" on the surface of the earth. We will discover that the plumb bob hangs perpendicular to the local surface (discounting bumps and surface irregularities). But due to the earth's rotation, it does not point toward the center of the earth unless it is suspended somewhere along the equator or just above one of the poles. Let us describe the motion of the plumb bob in a local frame of reference whose origin is at the position of the bob. Our frame of reference is attached to the surface of the earth. It is undergoing both translation as well as rotation. The translation of the frame takes place along a circle whose radius is $\rho = r_e \cos \lambda$ where r_e is the radius of the earth and λ is the geocentric latitude of the plumb bob (see Figure 5.15).

Its rate of rotation is $\boldsymbol{\omega}$, the same as that of the earth about its axis. Let us now examine the terms that make up Equation 5.14. The acceleration of the bob $\mathbf{a}'$ is zero; the bob is at rest in the local frame of reference. The centrifugal force on the bob relative to our local frame is zero since $\mathbf{r}'$ is zero; the origin of the local coordinate system is centered on the bob. The transverse force is zero since $\dot{\boldsymbol{\omega}} = 0$; the rotation of the earth is constant. The Coriolis force is zero since $\mathbf{v}'$, the velocity of the plumb bob, is zero; the plumb bob is at rest in the local frame. The only surviving terms in Equation 5.14 are the real forces $\mathbf{F}$ and the inertial term $-m\mathbf{A}_0$, which arises because the local frame of reference is accelerating. Thus,

$$\mathbf{F} - m\mathbf{A}_0 = 0$$

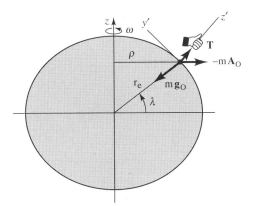

Figure 5.15 Gravitational force $m\mathbf{g}_0$, inertial force $-m\mathbf{A}_0$, and tension $\mathbf{T}$ acting on a plumb bob hanging just above the surface of the earth at latitude λ.

The rotation of the earth causes the acceleration of the local frame. In fact, the situation under investigation here is entirely analogous to that of Example 5.2—the linear accelerometer. There, the pendulum bob did not hang vertically because it experienced an inertial force directed opposite to the acceleration of the railroad car. The case here is almost completely identical. The bob does not hang on a line pointing toward the center of the earth because the inertial force $-m\mathbf{A}_0$ throws it outward, away from the earth's axis of rotation. This force, like the one of Example 5.2, is also directed opposite to the acceleration of the local frame of reference. It arises from the centripetal acceleration of the local frame toward the earth's axis. The magnitude of this force is $m\omega^2 r_e \cos\lambda$. It is a maximum when $\lambda = 0$ at the earth's equator and a minimum at either pole when $\lambda = \pm 90°$. It is instructive to compare the value of the acceleration portion of this term, $A_0 = \omega^2 r_e \cos\lambda$, to g, the acceleration due to gravity. At the equator, it is $3.4 \cdot 10^{-3} g$ or less than 1% of g.

$\mathbf{F}$ is the vector sum of all real, physical forces acting on the plumb bob. All forces, including the inertial force $-m\mathbf{A}_0$, are shown in the vector diagram of Figure 5.16a. The tension $\mathbf{T}$ in the string balances out the real gravitational force $m\mathbf{g}_0$ and the inertial force $-m\mathbf{A}_0$. In other words

$$(\mathbf{T} + m\mathbf{g}_0) - m\mathbf{A}_0 = 0$$

Now, when we hang a plumb bob, we normally think that the tension $\mathbf{T}$ balances out the local force of gravity, which we call $m\mathbf{g}$. We can see from the above equation and Figure 5.16b that $m\mathbf{g}$ is actually the vector sum of the real gravitational force $m\mathbf{g}_0$ and the inertial force $-m\mathbf{A}_0$. Thus,

$$m\mathbf{g}_0 - m\mathbf{g} - m\mathbf{A}_0 = 0 \qquad \therefore \mathbf{g} = \mathbf{g}_0 - \mathbf{A}_0 \qquad (5.15)$$

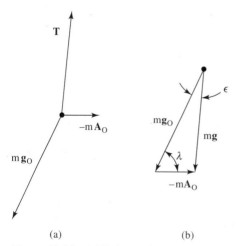

(a)   (b)

Figure 5.16   (a) Forces acting on a plumb
bob at latitude λ.
(b) Forces defining the weight
of the plumb bob, $m\mathbf{g}$.

As can be seen from Figure 5.16b, the local acceleration **g** due to gravity contains a term A_0 due to the rotation of the earth. The force $m\mathbf{g}_0$ is the true force of gravity and is directed toward the center of the earth. The inertial reaction $-m\mathbf{A}_0$, directed away from the earth's axis, causes the direction of the plumb line to deviate by a small angle ϵ away from the direction toward the earth's center. The plumb line direction defines the local direction of the vector **g.** The shape of the earth is also defined by the direction of **g.** Hence, the plumb line is always perpendicular to the earth's surface, which is not shaped like a true sphere but is flattened at the poles and bulged outward at the equator as depicted in Figure 5.15.

We can easily calculate the value of the angle ϵ. It is a function of the geocentric latitude of the plumb bob. Applying the law of sines to Figure 5.16b, we have

$$\frac{\sin \epsilon}{m\omega^2 r_e \cos \lambda} = \frac{\sin \lambda}{mg}$$

or, since ϵ is small

$$\sin \epsilon \approx \epsilon = \frac{\omega^2 r_e}{g} \cos \lambda \sin \lambda = \frac{\omega^2 r_e}{2g} \sin 2\lambda$$

Thus, ϵ vanishes at the equator ($\lambda = 0$) and the poles ($\lambda = \pm 90°$) as we have already surmised. The maximum deviation of the direction of the plumb line from the center of the earth occurs at $\lambda = 45°$ where

$$\epsilon_{max} = \frac{\omega^2 r_e}{2g} \approx 1.7 \times 10^{-3} \text{ radian} \approx \frac{1°}{10}$$

In this analysis, we have assumed that the real gravitational force $m\mathbf{g}_0$ is constant and directed toward the center of the earth. This is not valid, since the earth is not a true sphere. Its cross section is approximately elliptical as we have indicated in Figure 5.15. Hence, $\mathbf{g}_0$ varies with latitude. Moreover, local mineral deposits, mountains, and so on, affect the value of $\mathbf{g}_0$. Clearly, calculating the shape of the earth (essentially, the angle ϵ as a function of λ) is difficult. A more accurate solution can only be obtained numerically. The corrections to the above analysis are small.

Dynamic Effects. Motion of a Projectile

The equation of motion Equation 5.14 can be written

$$m\ddot{\mathbf{r}}' = \mathbf{F} + m\mathbf{g}_0 - m\mathbf{A}_0 - 2m\boldsymbol{\omega} \times \dot{\mathbf{r}}' - m\boldsymbol{\omega} \times (\boldsymbol{\omega} \times \mathbf{r}')$$

where **F** represents any applied forces other than gravity. But, from the static case considered above, the combination $m\mathbf{g}_0 - m\mathbf{A}_0$ is called $m\mathbf{g}$; hence, we can write the equation of motion as

$$m\ddot{\mathbf{r}}' = \mathbf{F} + m\mathbf{g} - 2m\boldsymbol{\omega} \times \dot{\mathbf{r}}' - m\boldsymbol{\omega} \times (\boldsymbol{\omega} \times \mathbf{r}')$$

Let us consider the motion of a projectile. If we neglect air resistance, then $\mathbf{F} = 0$. Furthermore, the term $-m\boldsymbol{\omega} \times (\boldsymbol{\omega} \times \mathbf{r}')$ is very small compared to the other terms, so we shall neglect it. The equation of motion then reduces to

$$m\ddot{\mathbf{r}}' = m\mathbf{g} - 2m\boldsymbol{\omega} \times \dot{\mathbf{r}}' \tag{5.16}$$

in which the last term is the Coriolis force.

To solve the above equation we shall choose the directions of the coordinate axes $O'x'y'z'$ such that the z'-axis is vertical (in the direction of the plumb line), the x'-axis is to the east, and the y'-axis points north (Figure 5.17). With this choice of axes, we have

$$\mathbf{g} = -\mathbf{k}'g$$

The components of $\boldsymbol{\omega}$ in the primed system are

$$\omega_{x'} = 0 \qquad \omega_{y'} = \omega \cos \lambda \qquad \omega_{z'} = \omega \sin \lambda$$

The cross product is therefore given by

$$\boldsymbol{\omega} \times \dot{\mathbf{r}}' = \begin{vmatrix} \mathbf{i}' & \mathbf{j}' & \mathbf{k}' \\ \omega_{x'} & \omega_{y'} & \omega_{z'} \\ \dot{x}' & \dot{y}' & \dot{z}' \end{vmatrix}$$

$$= \mathbf{i}'(\omega\dot{z}' \cos \lambda - \omega\dot{y}' \sin \lambda) + \mathbf{j}'(\omega\dot{x}' \sin \lambda) \tag{5.16a}$$
$$+ \mathbf{k}'(-\omega\dot{x}' \cos \lambda)$$

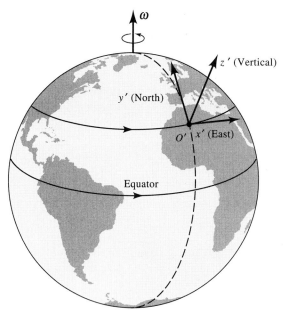

Figure 5.17   Coordinate axes for analyzing projectile motion.

Upon using the above expressions for $\boldsymbol{\omega} \times \dot{\mathbf{r}}'$ in Equation 5.16 and canceling the m's and equating components, we find

$$\ddot{x}' = -2\omega(\dot{z}' \cos \lambda - \dot{y}' \sin \lambda) \tag{5.17}$$

$$\ddot{y}' = -2\omega(\dot{x}' \sin \lambda) \tag{5.18}$$

$$\ddot{z}' = -g + 2\omega\dot{x}' \cos \lambda \tag{5.19}$$

for the component differential equations of motion. These equations are not of the separated type, but we can integrate once with respect to t to obtain

$$\dot{x}' = -2\omega(z' \cos \lambda - y' \sin \lambda) + \dot{x}_0' \tag{5.20}$$

$$\dot{y}' = -2\omega x' \sin \lambda + \dot{y}_0' \tag{5.21}$$

$$\dot{z}' = -gt + 2\omega\dot{x}' \cos \lambda + \dot{z}_0' \tag{5.22}$$

The constants of integration $\dot{x}_0'$, $\dot{y}_0'$, and $\dot{z}_0'$ are the initial components of the velocity. The values of $\dot{y}'$ and $\dot{z}'$ from Equations 5.21 and 5.22 may be substituted into Equation 5.17. The result is

$$\ddot{x}' = 2\omega gt \cos \lambda - 2\omega(\dot{z}_0' \cos \lambda - \dot{y}_0' \sin \lambda) \tag{5.23}$$

where terms involving ω^2 have been neglected. We now integrate again to get

$$\dot{x}' = \omega gt^2 \cos \lambda - 2\omega t(\dot{z}_0' \cos \lambda - \dot{y}_0' \sin \lambda) + \dot{x}_0'$$

and finally, by a third integration, we find x' as a function of t:

$$x'(t) = \frac{1}{3}\omega gt^3 \cos \lambda - \omega t^2(\dot{z}_0' \cos \lambda - \dot{y}_0' \sin \lambda) + \dot{x}_0' t + x_0' \tag{5.24}$$

The above expression for x' may be inserted into Equations 5.21 and 5.22. The resulting equations, when integrated, yield

$$y'(t) = \dot{y}_0' t - \omega \dot{x}_0' t^2 \sin \lambda + y_0' \tag{5.25}$$

$$z'(t) = -\frac{1}{2}gt^2 + \dot{z}_0' t + \omega \dot{x}_0' t^2 \cos \lambda + z_0' \tag{5.26}$$

where, again, terms of order ω^2 have been ignored.

In Equations 5.24, 5.25, and 5.26, the terms involving ω express the effect of the earth's rotation on the motion of a projectile in a coordinate system fixed to the earth.

EXAMPLE 5.9 *Falling Body*

Suppose a body is dropped from rest at a height h above the ground. Then at time $t = 0$ we have $\dot{x}_0' = \dot{y}_0' = \dot{z}_0' = 0$, and we shall set $x_0' = y_0' = 0$, $z_0' = h$ for the initial position. Equations 5.24, 5.25, and 5.26 then reduce to

$$x'(t) = \frac{1}{3}\omega g t^3 \cos \lambda$$

$$y'(t) = 0$$

$$z'(t) = -\frac{1}{2}g t^2 + h$$

Thus, as it falls, the body drifts to the east. When it hits the ground ($z' = 0$), we see that $t^2 = 2h/g$, and the eastward drift is given by the corresponding value of $x'(t)$, namely,

$$x'_h = \frac{1}{3}\omega \left(\frac{8h^3}{g}\right)^{1/2} \cos \lambda$$

For a height of, say, 100 meters at a latitude of 45°, the drift is

$$\frac{1}{3}(7.27 \times 10^{-5}\ \text{s}^{-1})(8 \times 100^3\ \text{m}^3/9.8\ \text{ms}^{-2})^{1/2} \cos 45°$$

$$= 1.55 \times 10^{-2}\ \text{m} = 1.55\ \text{cm}$$

Since the earth turns to the east, common sense would seem to say that the body should drift westward. Can the reader think of an explanation? ∎

EXAMPLE 5.10   *Deflection of a Rifle Bullet*

Consider a projectile that is fired with high initial speed v_0 in a nearly horizontal direction, and suppose this direction is east. Then $\dot{x}'_0 = v_0$ and $\dot{y}'_0 = \dot{z}'_0 = 0$. If we take the origin to be the point from which the projectile is fired, then $x'_0 = y'_0 = z'_0 = 0$ at time $t = 0$. Equation 5.25 then gives

$$y'(t) = -\omega v_0 t^2 \sin \lambda$$

which says that the projectile veers to the south or to the right in the northern hemisphere ($\lambda > 0$), and to the left in the southern hemisphere ($\lambda < 0$). If H is the horizontal range of the projectile, then we know that $H \simeq v_0 t_1$ where t_1 is the time of flight. The transverse deflection is then found by setting $t = t_1 = H/v_0$ in the above expression for $y'(t)$. The result is

$$\Delta \simeq \frac{\omega H^2}{v_0} |\sin \lambda|$$

for the magnitude of the deflection. It can be shown that this is the same for *any* direction in which the projectile is initially aimed, provided the trajectory is flat. This follows from the fact that the magnitude of the horizontal component of the Coriolis force on a body traveling parallel to the ground is independent of the direction of motion. (See Problem 5.9.) Since the deflection is proportional to the square of the horizontal range it becomes of considerable importance in long-range gunnery. ∎

Motion of a Projectile In a Rotating Cylinder

We would like to present one final example concerning the dynamics of projectiles in rotating frames of reference. The example is rather involved and is the first one that we shall encounter in which part of the solution makes use of applied numerical techniques. The details of the numerical portion of the solution will be presented in Appendix I. Those details, though interesting and useful as a problem-solving methodology, are non-essential insofar as understanding the example is concerned. We hope that its inclusion will give the student a better appreciation for the connection between the geometry of straight-line, force-free trajectories seen in an inertial frame of reference and the resulting curved geometry seen in a noninertial rotating frame of reference. Inertial forces appear in the noninertial frame that produce precisely the curved trajectory that may be calculated purely from geometrical considerations alone. This is obviously a necessity if the validity of Newton's laws of motion is to be preserved. Such a realization, though completely obvious with hindsight, should not be trivialized. It was ultimately just this sort of realization that led Einstein to formulate his general theory of relativity.

EXAMPLE 5.11

In several popular science fiction novels[2] spacecraft capable of supporting entire populations have been envisioned as large, rotating toroids or cylinders. Consider a cylinder of radius $R = 1000$ km and, for our purposes here, infinite length. Let it rotate about its axis with an angular velocity of $\omega = 0.18°/s$. It completes one revolution every 2000 s. This rotation rate leads to an apparent centrifugal acceleration for objects on the interior surface of $\omega^2 R$ equal to $1g$. Imagine several warring factions living on the interior of the cylinder. Let them fire projectiles at each other.

(a) Show that when projectiles are fired at low speeds ($v \ll \omega R$) and low "altitudes" at nearby points (say, $\Delta r' \leq R/10$), the equations of motion governing the resulting trajectories are identical to those of a similarly limited projectile on the surface of the earth.

(b) Find the general equations of motion for a projectile of unlimited speed and range using cylindrical coordinates rotating with the cylinder.

(c) Find the trajectory h versus ϕ' of a projectile fired vertically upward with a velocity $v' = \omega R$ in this noninertial frame of reference. $h = R - r'$ is the altitude of the projectile and ϕ' is its angular position in azimuth relative to the launch point. Calculate the angle Φ where it lands relative to the launch point. Also, calculate the maximum height H reached by the projectile.

(d) Finally, calculate h versus ϕ' solely from the geometrical basis that an inertial observer would employ to predict what the noninertial observer would see. Show that this result agrees with that of part (c), calculated from the perspective of the noninertial observer. In particular, show that Φ and H agree.

---

[2] For example, *Rendezvous with Rama* and *Rama II* by Arthur C. Clarke (Bantam Books) or *Titan* by John Varley (Berkeley Books).

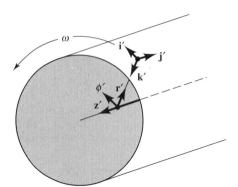

Figure 5.18   (a) Coordinates denoted by
unit vectors $\mathbf{i'}, \mathbf{j'}, \mathbf{k'}$ on the interior surface of
a rotating cylinder. (b) Unit vectors $\mathbf{r'}, \boldsymbol{\phi'}, \mathbf{z'}$
denoting cylindrical coordinates. Each set is
embedded in and rotates with the cylinder.

Solution:

(a) Since first we consider short, low-lying trajectories, we choose Cartesian coordinates (x', y', z') denoted by the unit vectors $\mathbf{i'}, \mathbf{j'}, \mathbf{k'}$ attached to and rotating with the cylinder shown in Figure 5.18.

The coordinate system is centered on the launch point. Since there is no real force acting upon the projectile after it is launched, the mass m common to all remaining terms in Equation 5.14 can be stripped and the equation then written in terms of accelerations only

$$-\mathbf{A}_0 - 2\boldsymbol{\omega} \times \mathbf{v'} - \boldsymbol{\omega} \times (\boldsymbol{\omega} \times \mathbf{r'}) = \mathbf{a'} \qquad (5.26a)$$

The transverse acceleration is zero, since the cylinder rotates at a constant rate. The first term on the left is the acceleration of the coordinate system origin. It is given by

$$\mathbf{A}_0 = \omega^2 R \mathbf{k'}$$

The second term is the Coriolis acceleration and is given by

$$\mathbf{a}_{Cor} = 2\boldsymbol{\omega} \times \mathbf{v'} = 2(-\mathbf{j'}\omega) \times (\mathbf{i'}\dot{x'} + \mathbf{j'}\dot{y'} + \mathbf{k'}\dot{z'})$$
$$= 2\omega\dot{x'}\mathbf{k'} - 2\omega\dot{z'}\mathbf{i'}$$

The third term is the centrifugal acceleration given by

$$\mathbf{a}_{centrif} = -\mathbf{j'}\omega \times ((-\mathbf{j'}\omega) \times \mathbf{r'})$$
$$= \mathbf{j'}\omega \times ((\mathbf{j'}\omega) \times (\mathbf{i'}x' + \mathbf{j'}y' + \mathbf{k'}z))$$
$$= \mathbf{j'}\ \omega \times (-\mathbf{k'}\omega x' + \mathbf{i'}\omega z')$$
$$= \mathbf{i'}\omega^2 x' - \mathbf{k'}\omega^2 z'$$

Upon gathering all appropriate terms, the x', y', and z' components of the resultant acceleration become

$$\ddot{x}' = 2\omega\dot{z}' + \omega^2 x'$$
$$\ddot{y}' = 0$$
$$\ddot{z}' = -2\omega\dot{x}' + \omega^2 z' - \omega^2 R$$

If projectiles are limited in both speed and range such that

$$|\dot{x}'| \sim |\dot{z}'| \ll \omega R \qquad |x'| \sim |z'| \ll R$$

and recalling that the rotation rate of the cylinder has been adjusted to $\omega^2 R = g$, the above acceleration components reduce to

$$\ddot{x}' \approx 0 \qquad \ddot{y}' = 0 \qquad \ddot{z}' \approx -g$$

which are equivalent to the equations of motion for a projectile of limited speed and range on the surface of the earth.

(b) In this case there is no limit placed on projectile velocity or range. We describe the motion using cylindrical coordinates (r', ϕ', z') attached to and rotating with the cylinder as indicated in Figure 5.18. r' denotes the radial position of the projectile measured from the central axis of the cylinder; ϕ' denotes its azimuthal position and is measured from the radius vector directed outward to the launch point; z' represents its position along the cylinder ($z' = 0$ corresponds to the z'-position of the launch point). The overall position, velocity, and acceleration of the projectile in cylindrical coordinates are given by Equations 1.50–1.52. We can use these relations to evaluate all the acceleration terms in Equation 5.26a. The term A_0 is zero, since the rotating coordinate system is centered on the axis of rotation. The Coriolis acceleration is

$$2\boldsymbol{\omega} \times \mathbf{v}' = 2\omega\mathbf{k} \times (\dot{r}'\mathbf{r}' + r'\dot{\phi}'\boldsymbol{\phi}' + \dot{z}'\mathbf{k}')$$
$$= 2\omega\dot{r}'(\mathbf{k}' \times \mathbf{r}') + 2\omega r'\dot{\phi}'(\mathbf{k}' \times \boldsymbol{\phi}')$$
$$= 2\omega\dot{r}'\boldsymbol{\phi}' - 2\omega r'\dot{\phi}'\mathbf{r}'$$

The centrifugal acceleration is

$$\boldsymbol{\omega} \times (\boldsymbol{\omega} \times \mathbf{r}') = \omega^2\mathbf{k}' \times (\mathbf{k}' \times (r'\mathbf{r}' + z'\mathbf{k}'))$$
$$= \omega^2\mathbf{k}' \times r'\boldsymbol{\phi}'$$
$$= -\omega^2 r'\mathbf{r}'$$

We can now rewrite Equation 5.26a in terms of components by gathering together all the above corresponding elements and equating them to those in Equation 1.52

$$\ddot{r}' - r'\dot{\phi}'^2 = 2\omega r'\dot{\phi}' + \omega^2 r'$$
$$2\dot{r}'\dot{\phi}' + r'\ddot{\phi}' = -2\omega\dot{r}'$$
$$\ddot{z}' = 0$$

In what follows we will neglect the z'-equation of motion since it contains no non-zero acceleration terms and simply gives rise to a "drift" along the axis of the cylinder of any trajectory seen in the $r'\phi'$ plane. Finally, we rewrite the radial and azimuthal equations in such a way that we can more readily see the dependency of the acceleration upon velocities and positions

$$\ddot{r}' = 2\omega r'\dot{\phi}' + (\omega^2 + \dot{\phi}'^2)r' \tag{5.26b}$$

$$\ddot{\phi}' = -\frac{2\dot{r}'}{r'}(\omega + \dot{\phi}')$$

(c) Before solving these equations of motion for a projectile fired vertically upward (from the viewpoint of a cylinder dweller), we will investigate the situation from the point of view of an inertial observer located outside the rotating cylinder. The rotational speed of the cylinder is ωR. If the projectile is fired vertically upward with a speed ωR from the point of view of the noninertial observer, the inertial observer sees the projectile launched with a speed $v = \sqrt{2}\omega R$ at 45° with respect to the vertical. Furthermore, according to this observer, there are no real forces acting on the projectile. Travel appears to be in a straight line. Its flight path is a chord of a quadrant. This situation is depicted in Figure 5.19.

As can be seen in Figure 5.19, by the time the projectile reaches a point in its trajectory denoted by the vector **r'**, the cylinder has rotated such that the launch point a has moved to the position labeled b. Therefore, the inertial observer concludes that the noninertial observer will think that the projectile has moved through the angle ϕ' and attained an altitude of $R - r'$. When the projectile lands, the noninertial observer would find that the projectile had moved through a total angle of $\Phi = \pi/2 - \omega T$ where T is the total time of flight. But $T = L/v = \sqrt{2}R/(\sqrt{2}\omega R) = 1/\omega$, or $\omega T = 1$ radian. Hence, the apparent deflection angle should be $\Phi = \pi/2 - 1$ radians or about 32.7°. The maximum height reached by

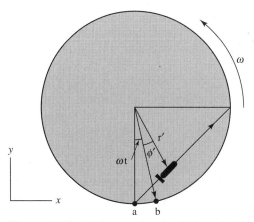

Figure 5.19 Trajectory of projectile launched inside rotating cylinder at 45° with respect to the "vertical" from point of view of an inertial observer.

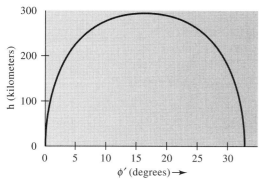

Figure 5.20 Trajectory of projectile fired vertically upward (toward the central axis) from the interior surface of a large cylinder rotating with an angular velocity ω, such that $\omega^2 R = g$.

the projectile occurs midway through its trajectory when $\omega t + \phi' = \pi/4$ radians.
At this point $r' = R/\sqrt{2}$ or $H = R - R/\sqrt{2} = 290$ km. At least, this is what the
inertial observer believes the noninertial observer would see. Let us see what the
noninertial observer does see according to Newton's laws of motion.

We have solved the above differential equations of motion (Equation 5.26b)
numerically (See Appendix I) and the result is shown in Figure 5.20.

It can be seen that the projectile is indeed launched vertically upward according
to the rotating observer. But the existence of the centrifugal and Coriolis inertial
forces cause the projectile to accelerate back toward the surface and toward the east,
in the direction of the angular rotation of the cylinder. Note that the rotating observer concludes that the vertically launched projectile has been pushed sideways
by the Coriolis force such that it lands 32.7° to the east of the launch point. The
centrifugal force has limited its altitude to a maximum value of 290 km. Each value
is in complete agreement with the conclusion of the noninertial observer. Clearly,
an intelligent military, aware of the dynamical equations of motion governing projectile trajectories on this cylindrical world, could launch all their missiles vertically
upward and hit any point around the cylinder by merely adjusting launch velocities.
(Positions located up or down the cylindrical axis could be hit by tilting the launcher
in that direction and firing the projectile at the required initial and thereafter constant axial velocity $\dot{z}_0$.)

(d) The inertial observer calculates the trajectory seen by the rotating observer
in the following way: first, look at Figure 5.21. It is a blow-up of the geometry
illustrated in Figure 5.19.

ϕ is the azimuthal angle of the projectile as measured in the fixed, inertial
frame. The azimuthal angle in the noninertial frame is $\phi' = \phi - \omega t$ (See Figure
5.19). As can be seen from the geometry of Figure 5.21, we can calculate the functional dependency of ϕ upon time

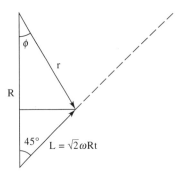

Figure 5.21 Geometry used to calculate trajectory seen by rotating observer according to inertial observer.

$$\tan \phi(t) = \frac{L(t) \sin 45°}{R - L(t) \cos 45°} = \frac{L(t)}{\sqrt{2}R - L(t)}$$

$$= \frac{\sqrt{2}\omega Rt}{\sqrt{2}R - \sqrt{2}\omega Rt} = \frac{\omega t}{1 - \omega t}$$

The projectile appears to be deflected toward the east by the angle ϕ' as a function of time given by

$$\phi'(t) = \phi(t) - \omega t = \tan^{-1}\left(\frac{\omega t}{1 - \omega t}\right) - \omega t$$

The dependency of r' upon time is given by

$$r'^2(t) = (L(t) \sin 45°)^2 + (R - L(t) \cos 45°)^2$$
$$= L(t)^2 + R^2 - \sqrt{2}L(t)R$$
$$= 2(\omega Rt)^2 + R^2 - \sqrt{2}(\sqrt{2}\omega Rt)R$$
$$= R^2(1 - 2\omega t(1 - \omega t))$$
$$\therefore r'(t) = R(1 - 2\omega t(1 - \omega t))^{1/2}$$

These final two parametric equations $r'(t)$ and $\phi'(t)$ describe a trajectory that the inertial observer predicts the noninertial observer should see. If we let time evolve and then plot $h = R - r'$ versus ϕ' we obtain exactly the same trajectory shown in Figure 5.20. That trajectory was calculated by the noninertial observer who used Newton's dynamical equations of motion in the rotating frame of reference. Thus, we see the equivalence between the curved geometry of straight lines seen from the perspective of an accelerated frame of reference and the existence of inertial forces that produce that geometry in the accelerated frame. ∎

5.5 THE FOUCAULT PENDULUM

In this section we shall study the effect of the earth's rotation on the motion of a pendulum that is free to swing in any direction, the so called *spherical pendulum.* As shown in Figure 5.22, the applied force acting on the pendulum bob is the vector sum of the weight $m\mathbf{g}$ and the tension $\mathbf{S}$ in the cord. The differential equation of motion is then

$$m\ddot{\mathbf{r}}' = m\mathbf{g} + \mathbf{S} - 2m\boldsymbol{\omega} \times \mathbf{r}' \tag{5.27}$$

Here we have neglected the term $-m\boldsymbol{\omega} \times (\boldsymbol{\omega} \times \mathbf{r}')$. It is vanishingly small in this context. Previously, we worked out the components of the cross product $\boldsymbol{\omega} \times \mathbf{r}'$ (See Equation 5.16a). Now the x' and y' components of the tension can be found simply by noting that the direction cosines of the vector $\mathbf{S}$ are $-x'/l$, $-y'/l$, and $-(l - z')/l$, respectively. Consequently $S_x = -x'S/l$, $S_y = -y'S/l$, and the corresponding components of the differential equation of motion (5.27) are

$$m\ddot{x}' = \frac{-x'}{l}S - 2m\omega(\dot{z}' \cos \lambda - \dot{y}' \sin \lambda) \tag{5.27a}$$

$$m\ddot{y}' = \frac{-y'}{l}S - 2m\omega\dot{x}' \sin \lambda \tag{5.27b}$$

We are interested in the case where the amplitude of oscillation of the pendulum is small so that the magnitude of the tension S is very nearly constant and equal to mg. Also, we shall neglect $\dot{z}'$ compared to $\dot{y}'$ in Equation 5.27a. The $x'y'$ motion is then governed by the following differential equations:

$$\ddot{x}' = -\frac{g}{l}x' + 2\omega'\dot{y}' \tag{5.28a}$$

$$\ddot{y}' = -\frac{g}{l}y' - 2\omega'\dot{x}' \tag{5.28b}$$

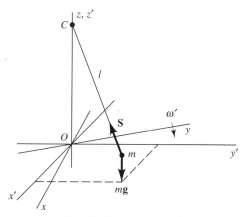

Figure 5.22 The Foucault pendulum.

in which we have introduced the quantity $\omega' = \omega \sin \lambda = \omega_{z'}$, which is the local vertical component of the earth's angular velocity.

Again we are confronted with a set of differential equations of motion that are not in separated form. A heuristic method of solving the equations is to transform to a new coordinate system $Oxyz$ that rotates relative to the primed system in such a way as to cancel the vertical component of the earth's rotation, namely, with angular rate $-\omega'$ about the vertical axis as shown in Figure 5.17. Thus, the unprimed system has no rotation about the vertical axis. The equations of transformation are

$$x' = x \cos \omega' t + y \sin \omega' t$$
$$y' = -x \sin \omega' t + y \cos \omega' t$$

Upon substituting the expressions for the primed quantities and their derivatives from the above equations into Equations 5.28a and 5.28b, the following result is obtained, after collecting terms and dropping terms involving ω'^2,

$$\left(\ddot{x} + \frac{g}{l}x \right) \cos \omega' t + \left(\ddot{y} + \frac{g}{l}y \right) \sin \omega' t = 0 \tag{5.29}$$

and an identical equation, except that the sine and cosine are reversed. Clearly, the above equation is satisfied if the coefficients of the sine and cosine terms both vanish, namely,

$$\ddot{x} + \frac{g}{l}x = 0$$

$$\ddot{y} + \frac{g}{l}y = 0$$

These are the differential equations of the two-dimensional harmonic oscillator discussed previously in Sec. 4.4. Thus, the path, projected on the xy plane, is an ellipse with *fixed* orientation in the unprimed system. In the primed system the path is an ellipse that undergoes a steady precession with angular speed $\omega' = \omega \sin \lambda$.

In addition to the above type of precession, there is another *natural* precession of the spherical pendulum, which is ordinarily much larger than the rotational precession under discussion. However, if the pendulum is carefully started by drawing it aside with a thread and letting it start from rest by burning the thread, the natural precession is rendered negligibly small.[3]

The rotational precession is clockwise in the northern hemisphere and counterclockwise in the southern. The period is $2\pi/\omega' = 2\pi/(\omega \sin \lambda) = 24/\sin \lambda$ hr. Thus, at a latitude of $45°$, the period is $(24/0.707)$ hr $= 33.94$ hr. The result was first demonstrated by the French physicist Jean Foucault in Paris in the year 1851. The Foucault pendulum has come to be a traditional display in major planetariums throughout the world.

---

[3] The natural precession will be discussed briefly in Chapter 10.

PROBLEMS

5.1 A 120 lb person stands on a bathroom spring scale while riding in an elevator. If the elevator has (a) upward and (b) downward acceleration of $g/4$, what is the weight indicated on the scale in each case?

5.2 An ultracentrifuge has a rotational speed of 500 rps. (a) Find the centrifugal force on a 1 µg particle in the sample chamber if the particle is 5 cm from the rotational axis. (b) Express the result as the ratio of the centrifugal force to the weight of the particle.

5.3 A plumb line is held steady while being carried along in a moving train. If the mass of the plumb bob is m, find the tension in the cord and the deflection from the local vertical if the train is accelerating forward with constant acceleration $g/10$. (Neglect any effects of the earth's rotation.)

5.4 If, in Problem 5.3, the plumb line is not held steady but oscillates as a simple pendulum, find the period of oscillation for small amplitude.

5.5 A hauling truck is traveling on a level road. The driver suddenly applies the brakes causing the truck to decelerate by an amount $g/2$. This causes a box in the rear of the truck to slide forward. If the coefficient of sliding friction between the box and the truckbed is 1/3, find the acceleration of the box relative to (a) the truck and (b) the road.

5.6 A cockroach crawls with constant speed in a circular path of radius b on a phonograph turntable rotating with constant angular speed ω. The circular path is concentric with the center of the turntable. If the mass of the insect is m and the coefficient of static friction with the surface of the turntable is μ_s, how fast, relative to the turntable, can the cockroach crawl before it starts to slip if it goes (a) in the direction of rotation and (b) opposite to the direction of rotation.

5.7 In the problem of the bicycle wheel rounding a curve, Example 5.5, what is the acceleration relative to the ground of the point at the very front of the wheel?

5.8 On the salt flats at Bonneville, Utah (latitude = 41°N) the British auto racer John Cobb in 1947 became the first man to travel at a speed of 400 mph on land. If he was headed due north at this speed, find the ratio of the magnitude of the Coriolis force on the racing car to the weight of the car. What is the direction of the Coriolis force?

5.9 A particle moves in a horizontal plane on the surface of the earth. Show that the magnitude of the horizontal component of the Coriolis force is independent of the direction of the motion of the particle.

5.10 If a pebble were dropped down an elevator shaft of the Empire State Building ($h = 1250$ ft, latitude = 41°N), find the deflection of the pebble due to Coriolis force.

5.11 In Yankee Stadium, New York, a baseball is driven a distance of 200 ft in a fairly flat trajectory. Is the amount of deflection due to Coriolis force alone of much importance? (Let the angle of elevation be 15°.)

5.12 Show that the third derivative with respect to time of the position vector (jerk) of a particle moving in a rotating coordinate system in terms of appropriate derivatives in the rotating system is given by

$$\dddot{\mathbf{r}} = \dddot{\mathbf{r}}' + 3\ddot{\omega} \times \dot{\mathbf{r}}' + 3\omega \times \ddot{\mathbf{r}}' + \ddot{\omega} \times \mathbf{r}' + 3\omega \times (\omega \times \dot{\mathbf{r}}')$$
$$+ \dot{\omega} \times (\omega \times \mathbf{r}') + 2\omega \times (\dot{\omega} \times \mathbf{r}') - \omega^2(\omega \times \mathbf{r}')$$

5.13 A bullet is fired straight up with initial speed v_0'. Assuming g is constant and neglecting air resistance, show that the bullet will hit the ground west of the initial point of upward motion by an amount $4\omega v_0^3 \cos \lambda / 3g^2$ where λ is the latitude and ω is the earth's angular velocity.

5.14 The force on a charged particle in an electric field $\mathbf{E}$ and a magnetic field $\mathbf{B}$ is given by

$$\mathbf{F} = q(\mathbf{E} + \mathbf{v} \times \mathbf{B})$$

in an inertial system where q is the charge and $\mathbf{v}$ is the velocity of the particle in the inertial system. Show that the differential equation of motion referred to a rotating coordinate system with angular velocity $\boldsymbol{\omega} = -(q/2m)\mathbf{B}$ is, for small ω,

$$m\ddot{\mathbf{r}}' = q\mathbf{E}$$

that is, the term involving $\mathbf{B}$ is eliminated. This result is known as *Larmor's theorem.*

5.15 Complete the steps leading to Equation 5.29 for the differential equation of motion of the Foucault pendulum.

5.16 The latitude of Mexico City is approximately 19°N. What is the period of precession of a Foucault pendulum there?

5.17 Work Example 5.5 using a coordinate system that is fixed to the bicycle wheel and rotates with it, as in Example 5.4.

COMPUTER/CALCULATOR APPLICATIONS

5.1 (a) Solve parts (c) and (d) of Example 5.11. Plot the trajectory h versus ϕ' as seen from each observer's (inertial and noninertial) point of view, as explained in the text. Your graphs should be identical to those in Figure 5.18. (b) Repeat (a) when the missile is fired with an initial velocity of $\mathbf{v}' = (2\omega R/\pi)\hat{\mathbf{r}}' - \omega R\hat{\boldsymbol{\phi}}'$ [see Figure 5.18]. In this case, what is the maximum altitude H attained by the missile and the angle Φ where it lands relative to the launch point?

Sir Isaac Newton. (North Wind Picture Archives)

6

GRAVITATION AND CENTRAL FORCES

"We have explained the phenomena of the heavens and of our sea by the power of gravity, but have not yet assigned the cause of this power. . . . I have not been able to discover the cause of those properties of gravity from phenomena, and I frame no hypotheses;—"

(Sir Isaac Newton, The Principia, 1687; Florian Cajori's translation, Berkeley, Univ. of Calif. Press, 1966)

"Gravity must be a scholastic occult quality or the effect of a miracle."

(Gottfried Wilhelm Leibniz; See Let Newton Be!, by J. Fauvel, R. Flood, M. Shorthand, and R. Wilson, Oxford Univ. Press, 1988)

6.1 INTRODUCTION

Throughout the year the ancient peoples observed the five visible planets slowly move through the fixed constellations of the zodiac in a fairly regular fashion. But occasionally, at times that occurred with astonishing predictability, they mysteriously halted their slow forward progression, suddenly reversing direction for as long as a few weeks before again resuming their steady march through the sky. This apparent quirk of planetary behavior is called *retrograde motion.* Unmasking its origin would consume the intellectual energy of ancient astronomers for centuries to come. Indeed, horribly complicated concoctions from minds shackled by philosophical dogma and fuzzy notions of physics, such as the cycles and epicycles of Ptolemy (125 A.D.) and others of like-minded mentality, would serve as models of physical reality for more than 2000 years. Ultimately, Nicolaus Copernicus (1473–1543) would demonstrate that retrograde motion was nothing other than a simple consequence of the relative motion between earth and the other planets each moving in a heliocentric orbit. Nonetheless, even Copernicus could not purge himself of the ludricous Ptolemaic epicycles, constrained by the dogma of uniform circular motion and the requirement of obtaining agreement between the observed and predicted irregularities of planetary motion.

It was not until Johannes Kepler (1571–1630) turned loose his potent intellect on the problem of solving the orbit of Mars, an endeavor that was to occupy him intensely for 20 years, that for the first time in history, scientists would glimpse the precise mathematical nature of the heavenly motions. Kepler painstakingly constructed a concise set of three mathematical formulas that accurately described the orbits of the planets around the sun. These three laws of planetary motion would soon be seen by Newton as nothing other than simple consequences of the interplay of a law of universal gravitation with three fundamental laws of mechanics that Newton had developed mostly from Galileo's investigations of the motions of terrestrial objects. Thus, Newton was to incorporate the physical workings of all the heavenly bodies within a framework of natural law that resided on earth. The world would never be seen in quite the same way again.

Newton's Law of Universal Gravitation

Newton formally announced the law of universal gravitation in the *Principia,* published in 1687. He actually worked out much of the theory at his family home in Woolsthorpe, England, as early as 1665–1666, during a six months hiatus from Cambridge University, which was closed while a plague ravaged most of London.

The law can be stated as follows: *Every particle in the universe attracts every other particle with a force whose magnitude is proportional to the product of the masses of the two particles and inversely proportional to the square of the distance between them. The direction of the force lies along the straight line connecting the two particles.*

We can express the law vectorially by the equation

$$\mathbf{F}_{ij} = G \frac{m_i m_j}{r_{ij}^2} \left(\frac{\mathbf{r}_{ij}}{r_{ij}} \right) \tag{6.1}$$

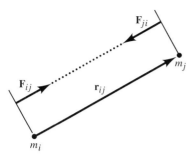

Figure 6.1 Action and reaction in
Newton's law of gravity.

where $\mathbf{F}_{ij}$ is the force on particle i of mass m_i exerted by particle j of mass m_j. The
vector $\mathbf{r}_{ij}$ is the directed line segment running from particle i to particle j, as shown in
Figure 6.1. The law of action and reaction requires that $\mathbf{F}_{ij} = -\mathbf{F}_{ji}$. The constant of
proportionality G is known as the *universal constant of gravitation*. Its value is determined in the laboratory by carefully measuring the force between two bodies of known
mass. The internationally accepted value at present is, in SI units,

$$G = (6.672 \pm 0.004) \times 10^{-11} \text{ Nm}^2\text{kg}^{-2}$$

All our present knowledge of the masses of astronomical bodies, including the earth, is
based on the value of this fundamental constant.[1]

This law is an example of a general class of forces termed *central;* that is, forces
whose lines of action either emanate from or terminate on a single point or center. Furthermore, if the magnitude of the force, as is the case with gravitation, is independent of
any direction, the force is isotropic. The behavior of such a force may be visualized in
the following way: Imagine being confined to a hypothetical, spherical surface centered
about a massive particle that serves as a source of gravity. When walking around that
surface, one would discover that the force of attraction would always be directed toward
the center, and the magnitude of this force would be independent of position on the
spherical surface. Nothing about this force could be used to determine position on the
sphere.

The main purpose of this chapter is to study the motion of a particle subject to a
central, isotropic force with particular emphasis on the force of gravity. In carrying out
this study, we will follow Newton's original line of inquiry, which led to the formulation
of his universal law of gravitation. In so doing, we hope to engender an appreciation for
the tremendous depth of Newton's intellectual achievement.

Gravitation: An Inverse-Square Law?

While home at Woolsthorpe in 1665, Newton took up the studies that were to occupy
him for the rest of his life: mathematics, mechanics, optics, and gravitation. Perhaps the

---

[1] G is the least accurately known of all the basic physical constants. This stems from the fact that the gravitational force between two bodies of laboratory size is extremely small. For a review of the current situation
regarding the determination of G, see an article by J. Maddox, *Nature*, **30**, 723 (1984).

most classic image we have of Newton depicts him sitting under an apple tree and being struck by a falling apple. This visual image is meant to convey the notion of Newton pondering the nature of gravity, most probably wondering whether or not the force that caused the apple to fall could be the same one that held the moon in its orbit about the earth.

Galileo, who very nearly postulated the law of inertia in its Newtonian form, inexplicably failed to apply it correctly to the motion of heavenly objects. He missed the most fundamental point of circular motion, namely, that objects moving in circles are accelerated inward and therefore require a resultant force in that direction. By Newton's time, a number of natural philosophers had come to the conclusion that some sort of force was required, not to accelerate a planet or satellite inward toward its parent body, but to "maintain it in its orbit." In 1665 the Italian astronomer Giovanni Borelli had presented a theory of the motion of the Galilean moons of Jupiter in which he stated that the centrifugal force of a moon's orbital motion was exactly in equilibrium with the attractive force of Jupiter.[2]

Newton was the first to realize that the earth's moon was not "balanced in its orbit" but was undergoing a centripetal acceleration toward the earth that had to be caused by a centripetal force. Newton surmised that this force was the same one that attracted all earthbound objects toward its surface. This had to be the case, since the kinematical behavior of the moon was no different than that of any object falling toward the earth. The falling moon never hits the earth because the moon has such a large tangential velocity that, as it falls a given distance, it moves far enough sideways that the earth's surface has curved away by that same distance. No one at the time even remotely suspected that the centripetal acceleration of the moon and the gravitational acceleration of an apple falling on the surface of the earth had a common origin.

Newton demonstrated that if a falling apple could also be given a large enough horizontal velocity, its motion would be identical to that of the orbiting moon (the apple's orbit would just be closer to the earth), thus making the argument for a common origin of an attractive gravitational force even more convincing. Newton further reasoned that the centripetal acceleration of an apple put in orbit about the earth just above its surface would be identical to its gravitational free-fall acceleration. (Imagine an apple shot horizontally out of a powerful cannon. Let there be no air resistance. If the initial horizontal speed of the projected apple was adjusted just right, the apple would never hit the earth because the earth's surface would fall away at the same rate that the apple would fall toward it, just as is the case for the orbiting moon. In other words, the apple would be in orbit and its centripetal acceleration would exactly equal the g of an

---

[2]Recall from Chapter 5 that centrifugal force is an inertial force exerted upon an object in a rotating frame of reference. In the context here, it arises from the centripetal acceleration of a Galilean moon traveling in essentially a circular orbit around Jupiter. To most pre-Newtonian thinkers, the centrifugal force acting on planets or satellites was a real one. Many of their arguments centered upon the nature of the force required to "balance out" the centrifugal force. They completely missed the point that from the perspective of an inertial observer, the satellite was undergoing centripetal acceleration inward. They were thus arguing from the perspective of a noninertial observer, although none of them had such a precise understanding regarding the distinction.

apple falling from rest.) Thus, with this single brilliant mental leap, Newton was about to uncover the first and one of the most beautiful of all unifying principles in physics, the law of universal gravitation.

The critical question for Newton was figuring out just how this attractive force depended on distance away from its center. Newton knew that the strength of the earth's attractive force was proportional to the acceleration of falling objects at whatever distance from the earth they happened to be. The moon's acceleration toward the earth is $a = v^2/r$ where v is the speed of the moon and r is the radius of its circular orbit. (This is equal to the local value of g.) Newton deduced, with the aid of Kepler's third law (the square of the orbital period τ^2 is proportional to the cube of the distance from the center of the orbit r^3), that this acceleration should vary as $1/r^2$. For example, if the moon were 4 times farther away from the earth than it actually is, then by Kepler's third law its period of revolution would be 8 times longer, and its orbital speed 2 times slower; consequently, its centripetal acceleration would be 16 times less than it is—or weaker as the inverse-square of the distance.

Newton thus hypothesized that the local value of g for all falling objects and, hence, the attractive force of gravity should vary accordingly. To confirm this hypothesis Newton had to calculate the centripetal acceleration of the moon, compare it to the acceleration g of a falling apple, and see if the ratio was equal to that of the inverse-square of their respective distances from the center of the earth. The moon's distance is 60 earth radii. The force of the earth's gravity must, therefore, weaken by a factor of 3600. The rate of fall of an apple must be 3600 times larger than that of the moon or, put another way, the distance an apple falls in 1 s should equal the distance the moon falls toward the earth in 1 min., the distance of fall being proportional to time squared. Unfortunately, Newton made a mistake in carrying out this calculation. He assumed that an angle of 1° subtended an arc length of 60 mi on the surface of the earth. He got this from a sailor's manual, the only book at hand. (This distance is, in fact, 60 nautical miles, or 69 English miles.) However, setting this equal to 60 English miles of 5280 ft each, he computed the moon's distance of fall in 1 s to be 0.0036 ft, or 13 ft in 1 min. Through Galileo's experiments with falling bodies, repeated later with more accuracy, an apple (or any other body) had been measured to fall about 15 ft in 1 s on earth. The values are very close, differing by about 1 part in 8, but such a difference was great enough that Newton abandoned his brilliant idea! Later, he was to use the correct values, get it exactly right, and thus demonstrate an inverse-square law for the law of gravity.

Proportional to Mass?

Newton also concluded that the force of gravity acting upon any object must be proportional to its mass (as opposed to, say, mass squared or something else). This conclusion is derivable from his second law of motion and Galileo's finding that the rate of fall of all objects is independent of their weight and composition. For example, let the force of gravity of the earth acting on some object of inertial mass m be proportional to that mass. Then, according to Newton's second law of motion, $F_{grav} = k \cdot m/r^2 = m \cdot a = m \cdot g$. Thus, $g = k/r^2$. The masses cancel out in this dynamical equation, and the acceleration g depends only on some constant k (which, in some way, must depend on the mass of the earth but, obviously, is the same for all bodies attracted to the earth) and the distance r to the center of the earth. So all bodies fall with the same acceleration regardless of their mass or composition. The gravitational force must be directly proportional to the inertial mass, or this precise cancellation would not occur. Then all falling bodies would exhibit mass-dependent accelerations, contrary to all experiments designed to test such a hypothesis. In fact the equivalence of inertial and gravitational mass of all objects is one of the cornerstones of Einstein's general theory of relativity. For Newton this equivalence remained a mystery to his death.

Product of Masses, Universality?

Newton also realized that if the force of gravity were to obey his third law of motion and if the force of gravity was proportional to the mass of the object being attracted, then it must also be proportional to the mass of the attracting object. Such a requirement leads us inevitably to the conclusion that the law of gravity must therefore be "universal"; that is, every object in the universe must attract (albeit very weakly, in most cases) every other object in the universe. Let us see how this comes about. Imagine two masses m_1 and m_2 separated by a distance r. The forces of attraction on 1 by 2 and on 2 by 1 are $F_{12} = k_2 m_1 / r^2$ and $F_{21} = k_1 m_2 / r^2$ where k_1 and k_2 are "constants" that, as we will be forced to conclude, must depend on the mass of the attracting object. According to Newton's third law, these forces have to be equal in magnitude (and opposite in direction). Hence, $k_2 m_1 = k_1 m_2$ or $k_2 / k_1 = m_2 / m_1$. In order to ensure the equality of this ratio, the strength of the attraction of gravity must be proportional to the mass of the attractive body, that is, $k_i = G m_i$. Thus, the force of gravity between two particles is a central, isotropic law of force possessing a wonderful symmetry: particle 1 attracts particle 2 and particle 2 attracts particle 1 with a magnitude and direction, obeying Newton's third law, proportional to the product of each of their masses and varying inversely as the square of their distance of separation. This conclusion was the work of true genius!

6.2   GRAVITATIONAL FORCE BETWEEN A UNIFORM SPHERE AND A PARTICLE

Newton did not publish the *Principia* until 1687. There was one particular problem that bothered him and made him reluctant to publish. We quickly glossed over that problem in our above discussion. Newton derived the inverse-square law by assuming that the relevant distance of separation between two objects, such as the earth and the moon, is the distance between their respective geometrical centers. This does not seem to be unreasonable for spherical objects like the sun and the planets, or the earth and the moon, whose distances of separation are large compared with their radii. But what about the earth and the apple? If you or I were the apple, looking around at all the stuff in the earth attracting us, we would see lines of gravitational force tugging on us from directions all over the place. There is stuff to the east and stuff to the west whose directions of pull differ by 180°. Who is to say that when we properly add up all the force vectors, due to all this attractive stuff, that we get a resultant vector that points to the center of the earth and whose strength depends on the mass of the earth and inversely on the distance to its center squared, as though all the earth's mass were completely concentrated at its center?

Yet, this is the way it works out. It is a tricky problem in calculus that requires a vector sum of infinitessimal contributions over an infinite number of mass elements that lead to a finite result. At that time no one knew calculus because Newton had just invented it, probably to solve this very problem! He was understandably reluctant to publish such a proof, couched in a framework of nonexistent mathematics. However, since everyone knows calculus in this age of enlightenment, we will go ahead and use it to solve the problem, proving that, for any uniform spherical body or any spherically symmetric distribution of matter, the gravitational force exerted by it on any external particle can be calculated by simply assuming that the entire mass of the distribution acts as though concentrated at its geometric center. Only an inverse-square force law works this way.

Consider first a thin uniform shell of mass M and radius R. Let r be the distance from the center O to a test particle P of mass m (Fig. 6.2). It is assumed that $r > R$. We shall divide the shell into circular rings of width $R\,\Delta\theta$ where, as shown in the figure, the angle POQ is denoted by θ, Q being a point on the ring. The circumference of our representative ring element is, therefore, $2\pi R \sin\theta$, and its mass ΔM is given by

$$\Delta M \simeq \rho 2\pi R^2 \sin\theta\,\Delta\theta$$

where ρ is the mass per unit area of the shell.

Now the gravitational force exerted on P by a small subelement Q of the ring (which we shall regard as a particle) is in the direction PQ. Let us resolve this force $\Delta\mathbf{F}_q$ into two components, one component along PO, of magnitude $\Delta F_q \cos\phi$, the other perpendicular to PO, of magnitude $\Delta F_q \sin\phi$. Here ϕ is the angle OPQ, as shown in Figure 6.2. From symmetry we can easily see that the vector sum of all of the perpendicular components exerted on P by the whole ring vanishes. The force $\Delta\mathbf{F}$ exerted by

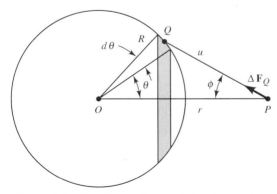

Figure 6.2 Coordinates for calculating the gravitational field of a spherical shell.

the entire ring is, therefore, in the direction *PO,* and its magnitude ΔF is obtained by summing the components $\Delta F_q \cos \phi$. The result is clearly

$$\Delta F = G\frac{m\Delta M}{u^2}\cos \phi = G\frac{m2\pi\rho R^2 \sin \theta \cos \phi}{u^2}\Delta\theta$$

where u is the distance *PQ* (the distance from the particle *P* to the ring) as shown. The magnitude of the force exerted on *P* by the whole shell is then obtained by taking the limit of $\Delta\theta$ and integrating

$$F = Gm2\pi\rho R^2\int_0^\pi \frac{\sin \theta \cos \phi \; d\theta}{u^2}$$

The integral is most easily evaluated by expressing the integrand in terms of u. From the triangle *OPQ* we have, from the law of cosines,

$$r^2 + R^2 - 2rR \cos \theta = u^2$$

Differentiating, we have, since both R and r are constant,

$$rR \sin \theta \; d\theta = u \; du$$

Also, in the same triangle *OPQ,* we can write

$$\cos \phi = \frac{u^2 + r^2 - R^2}{2ru}$$

Upon performing the substitutions given by the above two equations, we obtain

$$\begin{aligned}
F &= Gm2\pi\rho R^2\int_{\theta=0}^{\theta=\pi} \frac{u^2 + r^2 - R^2}{2Rr^2u^2} \; du \\
&= \frac{GmM}{4Rr^2}\int_{r-R}^{r+R}\left(1 + \frac{r^2 - R^2}{u^2}\right)du \\
&= \frac{GmM}{r^2}
\end{aligned}$$

where $M = 4\pi\rho R^2$ is the mass of the shell. We can then write vectorially

$$\mathbf{F} = -G\frac{Mm}{r^2}\mathbf{e}_r \qquad\qquad (6.2)$$

where $\mathbf{e}_r$ is the unit radial vector from the origin *O.* The above result means that a uniform spherical shell of matter attracts an external particle as if the whole mass of the shell were concentrated at its center. This will be true for every concentric spherical portion of a solid uniform sphere. *A uniform spherical body, therefore, attracts an external particle as if the entire mass of the sphere were located at the center.* The same is true also for a nonuniform sphere provided the density depends only on the radial distance *r.*

    It can be shown that the gravitational force on a particle located *inside* a uniform spherical shell is zero. The proof is left as an exercise.

6.3 KEPLER'S LAWS OF PLANETARY MOTION

Kepler's laws of planetary motion were a landmark in the history of physics. They played a crucial role in Newton's development of the law of gravitation. Kepler deduced these laws from a detailed analysis of planetary motions, primarily the motion of Mars, the closest outer planet and one whose orbit is, unlike that of Venus, highly elliptical. Mars had been most accurately observed, and its positions on the celestial sphere dutifully recorded by Kepler's irascible but brilliant patron, Tycho de Brahe (1546–1601). Kepler even used some sightings that had been made by the early Greek astronomer Hipparchus (190–125 B.C.). Kepler's three laws are:

I. Law of Ellipses (1609)
The orbit of each planet is an ellipse, with the sun located at one of its foci.

II. Law of Equal Areas (1609)
A line drawn between the sun and the planet sweeps out equal areas in equal times as the planet orbits the sun.

III. Harmonic Law (1618)
The square of the sidereal period of a planet (the time it takes a planet to complete one revolution about the sun relative to the stars) is directly proportional to the cube of the semimajor axis of the planet's orbit.

The derivation of these laws from Newton's theories of gravitation and mechanics was one of the most stupendous achievements in the annals of science. A number of Newton's colleagues who were prominent members of the British Royal Society were convinced that there was a force of gravitation exerted on the planets by the sun, that its strength must diminish by the square of the distance from it, and that this fact could be used to explain Kepler's laws. (Kepler's second law is, however, a statement that the angular momentum of a planet in orbit is conserved, a consequence only of the central nature of the gravitational force, not its inverse-square feature.) The trouble was, as noted by Edmond Halley over lunch with Robert Hooke and Christopher Wren in January of 1684, that no one could make the connection mathematically. Part of the problem was that no one, except the silent Newton, could show that the gravitational forces of spherical bodies could be treated as though they emanated from and terminated on their geometric centers. Hooke brashly stated that he could prove the fact that the planets traveled in elliptical orbits but had not told anyone how to do it so that they might, in attempting a solution themselves, appreciate the magnitude of the problem. Wren offered a prize of 40 shillings—in those days the price of an expensive book—to the one who could produce such a proof within two months. Neither Hooke, nor anyone else, won the prize!

In August of 1684, while visiting Cambridge, Halley stopped in to see Newton and asked him what would be the shape of the planets' orbits if they were subject to an inverse-square attractive force by the sun? Newton replied, without hesitation, "An ellipse!" Halley wanted to know how Newton knew this, and Newton said that he had calculated it years ago. Halley was stunned. They looked through thousands of Newton's papers but could not find the calculation. Newton told Halley that he would redo it and send it to him.

Newton had actually done the calculation five years earlier, in 1679, stimulated in part by Robert Hooke, the aforementioned claimant to the inverse-square law, who had written Newton with questions about the trajectory of objects falling toward a gravitationally attractive body. Unfortunately, there was a mistake in the calculation of Newton's written reply to Hooke. Hooke, with glee, pointed out the mistake, and the angry Newton, concentrating on the problem with renewed vigor, apparently straightened things out. However, these subsequent calculations also contained a mistake, and perhaps this is why Newton failed to find them when queried by Halley. At any rate, Newton furiously attacked the problem again and within three months sent Halley a paper in which he correctly derived all of Kepler's laws from an inverse-square law of gravitation and the laws of mechanics. Thus was the *Principia* born. In the sections that follow we, too, will derive Kepler's laws from Newton's fundamental principles.

6.4   KEPLER'S SECOND LAW. EQUAL AREAS: CONSERVATION OF ANGULAR MOMENTUM

Kepler's second law is nothing other than the statement that the angular momentum of a planet about the sun is a conserved quantity. To show this, we first define angular momentum and then show that its conservation is a general consequence of the central nature of the gravitational force.

The *angular momentum* of a particle located a vector distance $\mathbf{r}$ from a given origin and moving with momentum $\mathbf{p}$ is defined to be the quantity $\mathbf{L} = \mathbf{r} \times \mathbf{p}$. The time derivative of this quantity is

$$\frac{d\mathbf{L}}{dt} = \frac{d(\mathbf{r} \times \mathbf{p})}{dt} = \mathbf{v} \times \mathbf{p} + \mathbf{r} \times \frac{d\mathbf{p}}{dt}$$

but

$$\mathbf{v} \times \mathbf{p} = \mathbf{v} \times m\mathbf{v} = m\mathbf{v} \times \mathbf{v} = 0$$

Thus,

$$\mathbf{r} \times \mathbf{F} = \mathbf{r} \times \frac{d\mathbf{p}}{dt} = \frac{d\mathbf{L}}{dt} \tag{6.3}$$

where we have used Newton's second law, $\mathbf{F} = d\mathbf{p}/dt$.

The cross product $\mathbf{N} = \mathbf{r} \times \mathbf{F}$ is the moment of force, or torque, on the particle about the origin of the coordinate system. If $\mathbf{r}$ and $\mathbf{F}$ are collinear, this cross product vanishes and so does $\dot{\mathbf{L}}$. The angular momentum $\mathbf{L}$, in such cases, is a constant of the motion. This is quite obviously the case for a particle (or a planet) subject to a *central force* $\mathbf{F}$, that is, one that either emanates or terminates from a single point and whose line of action lies along the radius vector $\mathbf{r}$.

Furthermore, since the vectors $\mathbf{r}$ and $\mathbf{v}$ define an "instantaneous" plane within which the particle moves, and since the angular momentum vector $\mathbf{L}$ is normal to this plane and is constant in both magnitude and direction, the orientation of this plane is fixed in space. Thus, the problem of motion of a particle in a central field is really a two-dimensional problem and can be treated that way without any loss of generality.

Magnitude of the Angular Momentum

In order to determine the magnitude of the angular momentum, it is convenient to resolve the velocity vector **v** into radial and transverse components in polar coordinates. Thus, we can write

$$\mathbf{v} = \dot{r}\mathbf{e}_r + r\dot{\theta}\mathbf{e}_\theta$$

in which $\mathbf{e}_r$ is the unit radial vector and $\mathbf{e}_\theta$ is the unit transverse vector. The magnitude of the angular momentum is then given by

$$L = |\mathbf{r} \times m\mathbf{v}| = |r\mathbf{e}_r \times m(\dot{r}\mathbf{e}_r + r\dot{\theta}\mathbf{e}_\theta)|$$

Since $|\mathbf{e}_r \times \mathbf{e}_r| = 0$ and $|\mathbf{e}_r \times \mathbf{e}_\theta| = 1$, we find

$$L = |mr^2\dot{\theta}| = \text{constant} \tag{6.4}$$

for a particle moving in a central field of force.

In order to prove Kepler's second law, we must first calculate the rate at which the position vector **r** of a planet, subject to the sun's central gravitational force, sweeps out area and then demonstrate the proportionality of this "areal velocity" to its angular momentum. Figure 6.3 shows a planet at two successive positions in its eliptical orbit about the sun. The area dA swept out by the radius vector **r** during a time interval dt is half the area of the parallelogram formed by the vectors **r** and **dr** in Figure 6.3 or

$$dA = \frac{1}{2}|\mathbf{r} \times d\mathbf{r}| = \frac{1}{2}|\mathbf{r} \times \mathbf{v}dt| = \frac{L}{2m}dt \tag{6.4a}$$

$$\therefore \dot{A} = \frac{L}{2m} = \text{constant} \tag{6.4b}$$

Thus, the areal velocity $\dot{A}$ is directly proportional to the angular momentum of the planet about the sun, so it too is a constant of the motion, precisely as stated by Mr. Kepler.

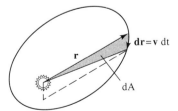

Figure 6.3 Area dA swept out by the radius vector **r** in a time dt as a planet orbits the sun.

6.5 KEPLER'S FIRST LAW: THE LAW OF ELLIPSES

In order to prove Kepler's first law, we will develop a general differential equation for the orbit of a particle in any central, isotropic field of force. Then we solve the orbital equation for the specific case of an inverse-square law of force.

First we express Newton's differential equations of motion using two-dimensional polar coordinates instead of three, remembering from our previous discussion that no loss of generality is incurred since the motion is confined to a plane. The equation of motion in polar coordinates is

$$m\ddot{\mathbf{r}} = f(r)\ \mathbf{e}_r$$

where $f(r)$ is the central, isotropic force that acts on the particle of mass m. Note that it is a function only of the scalar distance r to the force center (hence, it is isotropic) and its direction is along the radius vector (hence, it is central). As shown in Chapter 1, the radial component of $\ddot{\mathbf{r}}$ is $\ddot{r} - r\dot{\theta}^2$ and the transverse component is $2\dot{r}\dot{\theta} + r\ddot{\theta}$. Thus, the component differential equations of motion are

$$m(\ddot{r} - r\dot{\theta}^2) = f(r) \tag{6.5}$$

$$m(2\dot{r}\dot{\theta} + r\ddot{\theta}) = 0 \tag{6.6}$$

From the latter equation it follows that

$$\frac{d}{dt}(r^2\dot{\theta}) = 0$$

or

$$r^2\dot{\theta} = \text{constant} = l \tag{6.7}$$

From Equation 6.4 we see that

$$|l| = \frac{L}{M} = |\mathbf{r} \times \mathbf{v}| \tag{6.8}$$

Thus, l is the angular momentum per unit mass. Its constancy is simply a restatement of a fact that we already know, namely, that the angular momentum of a particle is constant when it is moving under the action of a central force.

Given a certain radial force function $f(r)$, we could, in theory, solve the pair of differential equations (Equations 6.5 and 6.6) to obtain r and θ as functions of t. It is often the case that one is interested only in the path in space (the *orbit*) without regard to the time t. To find the equation of the orbit, we shall use the variable u defined by

$$r = \frac{1}{u} \tag{6.9}$$

Then

$$\dot{r} = -\frac{1}{u^2}\dot{u} = -\frac{1}{u^2}\dot{\theta}\frac{du}{d\theta} = -l\frac{du}{d\theta} \tag{6.10}$$

The last step follows from the fact that

$$\dot{\theta} = lu^2 \tag{6.11}$$

according to Equations 6.7 and 6.9.

Differentiating a second time, we have

$$\ddot{r} = -l\frac{d}{dt}\frac{du}{d\theta} = -l\dot{\theta}\frac{d^2u}{d\theta^2} = -l^2u^2\frac{d^2u}{d\theta^2} \tag{6.12}$$

From these values of r, $\dot{\theta}$, and $\ddot{r}$, we readily find that Equation 6.5 transforms to

$$\frac{d^2u}{d\theta^2} + u = -\frac{1}{ml^2u^2}f(u^{-1}) \tag{6.13}$$

Equation 6.13 is the *differential equation of the orbit* of a particle moving under a central force. The solution gives u (hence r) as a function of θ. Conversely, if one is given the polar equation of the orbit, namely, $r = r(\theta) = u^{-1}$, then the force function can be found by differentiating to get $d^2u/d\theta^2$ and inserting this into the differential equation.

EXAMPLE 6.1

A particle in a central field moves in the spiral orbit

$$r = c\theta^2$$

Determine the force function.

Solution:
We have

$$u = \frac{1}{c\theta^2}$$

and

$$\frac{du}{d\theta} = \frac{-2}{c}\theta^{-3} \qquad \frac{d^2u}{d\theta^2} = \frac{6}{c}\theta^{-4} = 6cu^2$$

Then from Equation 6.13

$$6cu^2 + u = -\frac{1}{ml^2u^2}f(u^{-1})$$

Hence,

$$f(u^{-1}) = -ml^2(6cu^4 + u^3)$$

and

$$f(r) = -ml^2\left(\frac{6c}{r^4} + \frac{2}{r^3}\right)$$

Thus, the force is a combination of an inverse cube and inverse-fourth power law.

∎

EXAMPLE 6.2

In Example 6.1 determine how the angle θ varies with time.

Solution:

Here we use the fact that $l = r^2\dot\theta$ is constant. Thus,

$$\dot\theta = lu^2 = l\frac{1}{c^2\theta^4}$$

or

$$\theta^4 d\theta = \frac{l}{c^2} dt$$

and so, by integrating, we find

$$\frac{\theta^5}{5} = lc^{-2}t$$

where the constant of integration is taken to be zero, so that $\theta = 0$ at $t = 0$. Then we can write

$$\theta = \alpha t^{1/5}$$

where $\alpha = $ constant $= (5lc^{-2})^{1/5}$.

∎

Inverse-Square Law

We can now solve the Equation 6.13 for the orbit of a particle subject to the force of gravity. In this case

$$f(r) = -\frac{k}{r^2}$$

where the constant $k = GMm$. In this chapter we will always assume that $M \gg m$ and remains fixed in space. The small mass m is the one whose orbit we will calculate. (A modification in our treatment will be required when $M \approx m$, or at least not much greater than m. We will present such a treatment in Chapter 7.) The equation of the orbit (Equation 6.13) then becomes

$$\frac{d^2u}{d\theta^2} + u = \frac{k}{ml^2} \tag{6.14}$$

The general solution is clearly

$$u = A\cos(\theta - \theta_0) + \frac{k}{ml^2}$$

or

$$r = \frac{1}{A \cos(\theta - \theta_0) + k/ml^2} \tag{6.15}$$

The constants of integration A and θ_0 are determined from the initial conditions. The value of θ_0 merely determines the orientation of the orbit, so we can, without loss of generality in discussing the form of the orbit, choose $\theta_0 = 0$. Then

$$r = \frac{1}{A \cos \theta + k/ml^2} \tag{6.16}$$

This is the polar equation of the orbit. It is the equation of a conic section (ellipse, parabola, or hyperbola) with the origin at a focus. The equation can be written in the standard form (see Appendix C):

$$r = r_0 \frac{1 + e}{1 + e \cos \theta} \tag{6.17}$$

where

$$e = \frac{Aml^2}{k} \tag{6.18}$$

and

$$r_0 = \frac{ml^2}{k(1 + e)} \tag{6.19}$$

The constant e is called the *eccentricity*. The different cases, illustrated in Figure 6.4 for constant r_0, are as follows:

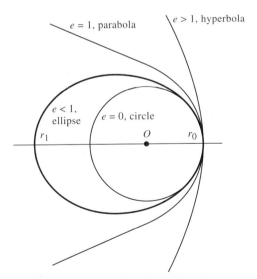

Figure 6.4 The family of central conics.

$e < 1$ *ellipse*
$e = 0$ *circle (special case of an ellipse)*
$e = 1$ *parabola*
$e > 1$ *hyperbola*

From Equation 6.17, r_0 is the value of r for $\theta = 0$. The value of r for elliptic orbits at $\theta = \pi$ is given by

$$r_1 = r_0 \frac{1 + e}{1 - e} \tag{6.20}$$

In reference to the elliptic orbits of the planets around the sun, the distance r_0 is called the *perihelion* distance (closest to the sun), and the distance r_1 is called the *aphelion* distance (farthest from the sun). The corresponding distances for the orbit of the moon around the earth—and for the orbits of the earth's artificial satellites—are called the *perigee* and *apogee* distances, respectively.

The orbital eccentricities of the planets are quite small. (See Table 6.1.) For example, in the case of the earth's orbit $e = 0.017$, $r_0 = 91,000,000$ mi, and $r_1 = 95,000,000$ mi. On the other hand, the comets generally have large orbital eccentricities (highly elongated orbits). Halley's comet, for instance, has an orbital eccentricity of 0.967 with a perihelion distance of only 55,000,000 mi, while at aphelion it is beyond the orbit of Neptune. Many comets (the nonrecurring type) have parabolic or hyperbolic orbits.

The energy of the object is the primary factor that determines whether or not its orbit is an open (parabola, hyperbola) or closed (circle, ellipse) conic section. "High" energy objects follow open, unbound orbits, while "low" energy objects follow closed, bound ones. We will treat this subject in greater detail in Section 6.10. Using the language of a noninertial observer, perfectly circular orbits correspond to a situation where the gravitational and centrifugal forces of a planet are exactly balanced. It should surprise you to see that the orbits of the planets are nearly circular.

It is very difficult to envision just how such a situation might arise from initial conditions. If a planet is hurtling around the sun a little too fast, the centrifugal force will slightly outweigh the gravitational force, and the planet will move away from the sun a little bit. In doing so, it slows down until the force of gravity begins to overwhelm the centrifugal force. The planet then falls back in a little bit closer to the sun, picking up speed along the way. The centrifugal force builds up to a point where it again outweighs the force of gravity and the process repeats itself. Thus, elliptical orbits can be seen as the result of a continuing tug of war between the slightly unbalanced gravitational and centrifugal forces that inevitably occurs whenever the tangential velocity of the planet is not adjusted just so. These forces must grow and shrink in such a way that the stability of the orbit is ensured. The criterion for stability will be discussed in Section 6.13.

One way in which these two forces could be perfectly balanced all the way around the orbit would be if the planet started off just right; that is, very special initial conditions would have to have been set up in the beginning, so to speak. It is difficult to imagine how any natural process could have established such nearly perfect prerequisites. Hence, if planets are bound to the sun at all, one would think that they would be most likely to travel in elliptical orbits just like Kepler said, unless something happened during the course of solar system evolution that brought the planets into circular orbits. We leave it to the student to think about just what sort of thing might do this.

Orbital Parameters from the Conditions at Closest Approach

From Equation 6.19 we find the eccentricity can be expressed as

$$e = \frac{ml^2}{kr_0} - 1 \tag{6.21}$$

Let v_0 be the speed of the particle at $\theta = 0$. Then, from the definition of the constant l, we have

$$l = r^2\dot{\theta} = r_0^2\dot{\theta}_0 = r_0v_0$$

The eccentricity is then given by

$$e = \frac{mr_0v_0^2}{k} - 1 \tag{6.22}$$

For a circular orbit ($e = 0$) we have then $k = mr_0v_0^2$ or

$$\frac{k}{r_0^2} = \frac{mv_0^2}{r_0} \tag{6.23}$$

Now let us denote the quantity k/mr_0 by v_c^2, so that if $v_0 = v_c$, the orbit is a circle. The expression for the eccentricity (Equation 6.22) can then be written, for $v_0 \geq v_c$, as

$$e = (v_0/v_c)^2 - 1 \tag{6.24}$$

and the equation of the orbit can be written

$$r = r_0 \frac{(v_0/v_c)^2}{1 + [(v_0/v_c)^2 - 1] \cos\theta} \tag{6.25}$$

The value of r_1 is given by setting $\theta = \pi$; thus,

$$r_1 = r_0 \frac{(v_0/v_c)^2}{2 - (v_0/v_c)^2}$$

Note: Equation 6.24 is also valid when $v_0 < v_c$, in which case $r_1 < r_0$ and the eccentricity is given by $e = 1 - (v_0/v_c)^2$, and $\theta_0 = \pi$ in Equation 6.15.

EXAMPLE 6.3

In the earth's gravitational field the force constant $k = GM_e m$ in which M_e is the mass of the earth and m is the mass of the body. Thus, for circular orbits of earth satellites, we have

$$v_c^2 = \frac{k}{mr_0} = \frac{GM_e}{r_0}$$

Now, as pointed out in Example 2.3, the product GM_e can be found simply by noting that the force of gravity at the earth's surface is $mg = GM_e m/R_e^2$ or $GM_e = gR_e^2$ where R_e is the earth's radius. Hence, the speed for a circular orbit around the earth is

$$v_c = \left(\frac{gR_e^2}{r_0}\right)^{1/2}$$

In particular, for satellites in circular orbits very near the earth's surface $r_0 \simeq R_e$, so the speed for such orbits is simply

$$v_c = (gR_e)^{1/2} = (9.8 \text{ ms}^{-2} \times 6.4 \times 10^6 \text{ m})^{1/2} = 7920 \text{ m/s}$$

or about 8 km/s.   ∎

EXAMPLE 6.4

A rocket satellite is going around the earth in a circular orbit of radius r_0. A sudden blast of the rocket motor increases the speed by 15%. Find the equation of the new orbit, and compute the apogee distance.

Solution:
Let v_c be the speed in the circular orbit, and let v_0 be the new initial speed; that is,

$$v_0/v_c = 1.15$$

Equation 6.25 for the new orbit then reads

$$r = r_0 \frac{1.3225}{1 + 0.3225 \cos \theta}$$

and the apogee distance is

$$r_1 = r_0 \frac{1.3225}{2 - 1.3225} = 1.95 r_0$$

The orbits are shown in Figure 6.5.   ∎

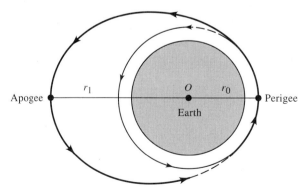

Figure 6.5   Space rocket changing from a circular orbit to an elliptical orbit.

6.6 KEPLER'S THIRD LAW. THE HARMONIC LAW

Why is Kepler's third law, relating orbital period to distance from the sun, called the harmonic law? Kepler's work, more than that of any of the other great scientists who were involved in the pursuit of unlocking the mysteries of planetary motion, is a wonderful illustration of the profound effect that intense hunger for knowledge and personal belief have on the growth of science. Kepler held the conviction that the world, that had treated him so harshly at times, nonetheless, was fundamentally a beautiful place. Kepler believed in the Pythagorean doctrine of celestial harmony. The world was a tumultuous place, and the planets were discordant only because man had not yet learned how to hear the true harmony of the worlds. In his work *Harmonice Mundi* (The Harmony of the World), Kepler, like the Pythagoreans almost 2000 years earlier, tried to connect the planetary motions with all fields of abstraction and harmony: geometrical figures, numbers, and musical harmonies. In this attempt he failed. But in the midst of all this work, indicative of his yearnings and strivings, we find his final precious jewel, always cited as Kepler's third law, the harmonic law. It was this law that gave us the sheets to the music of the spheres.

We will show how the third law can be derived from Newton's laws of motion and the inverse-square law of gravity. It is easy for circular orbits which we will examine first. We will show that it is true for elliptical orbits as well. In the case of a circular orbit, from Newton's second law, we have

$$\frac{GMm}{r^2} = \frac{mv^2}{r}$$

but $v = 2\pi r/\tau$. (As an aside, we note that the orbital period can be expressed as

$$\tau = \frac{2\pi r}{v} = \frac{2m(\pi r^2)}{mvr}$$

$$= \frac{2mA}{L} = \frac{2A}{l}$$

This result is true for elliptical orbits as well, and we will use it when proving Kepler's third law for the general case.)

Inserting the above expression for the orbital velocity into Newton's second law reduces to

$$\frac{GM}{r^2} = \frac{\frac{4\pi^2 r^2}{\tau^2}}{r}$$

$$\therefore \tau^2 = \frac{4\pi^2}{GM} r^3$$

which is Kepler's third law.

The more general proof for elliptical orbits proceeds as follows: Starting with Equation 6.4, Kepler's second law

$$\dot{A} = \frac{L}{2m}$$

we can relate the area of the orbit to its period and angular momentum per unit mass, $l = L/m$, by integrating the areal velocity over the entire orbital period

$$\int_0^\tau \dot{A}dt = A = \frac{l}{2}\tau$$

$$\tau = \frac{2A}{l}$$

and we arrive at the result stated above for the case of a circular orbit. The area of an ellipse is πab where a and b are the semimajor and semiminor axes, respectively. a and b are related to the orbital eccentricity by (see Appendix C)

$$\frac{b}{a} = \sqrt{(1 - e^2)}$$

where e is the eccentricity of the ellipse. Thus, we can express the period as

$$\tau = \frac{2\pi a^2}{l}\sqrt{(1 - e^2)}$$

We can find the major axis $2a$ by using Equations 6.19 and 6.20

$$2a = r_0 + r_1 = \frac{ml^2}{k}\left(\frac{1}{1 + e} + \frac{1}{1 - e}\right) = \frac{2ml^2}{k(1 - e^2)}$$

Now, upon squaring the expression for the period, we obtain

$$\tau^2 = \frac{4\pi^2 a^4}{l^2}(1 - e^2)$$

but from the previous expression, we have

$$(1 - e^2) = \frac{ml^2}{ka}$$

and upon inserting this expression into the one for the square of the period, we get

$$\tau^2 = \frac{4\pi^2 m}{k}a^3$$

and since $k = GMm$ we finally arrive at Kepler's third law

$$\tau^2 = \frac{4\pi^2}{GM}a^3 \tag{6.26}$$

In the case of an elliptical orbit, we obtain the same relationship that we derived for a circular orbit: The square of a planet's orbital period is proportional to the cube of its distance from the sun. Only, in this case, the relevant distance is the length of the semimajor axis a.

The constant $4\pi^2/GM$ is the same for all objects in orbit about the sun, regardless

TABLE 6.1 Planetary Data.

Planet	Period τ(yr)	Square τ²(yr²)	Semimajor Axis a (AU)	Cube a³(AU³)	Eccentricity e
Mercury	0.241	**0.0581**	0.387	**0.0580**	0.206
Venus	0.615	**0.378**	0.723	**0.378**	0.007
Earth	1.000	**1.000**	1.000	**1.000**	0.017
Mars	1.881	**3.538**	1.524	**3.540**	0.093
Jupiterd	11.86	**140.7**	5.203	**140.8**	0.048
Saturn	29.46	**867.9**	9.539	**868.0**	0.056
Uranus	84.01	**7058.**	19.18	**7056.**	0.047
Neptune	164.8	**27160.**	30.06	**27160.**	0.009
Pluto	247.7	**61360.**	39.44	**61350**	0.249

of their mass.[3] If distances are measured in astronomical units (1 AU = 1.50 · 10⁸ km) and periods are expressed in earth years, then $4\pi^2/GM = 1$. Kepler's third law then takes the very simple form $\tau^2 = a^3$. Listed in Table 6.1 are the periods and their squares, the semimajor axes and their cubes, along with the eccentricities of all the planets. It is left to the student to see how well Kepler's third law works.

Note: Most of the planets have nearly circular orbits with the exception of Pluto, Mercury, and Mars.

EXAMPLE 6.5

Find the period of a comet whose semimajor axis is 4 AU.

Solution:
With τ measured in years and a in AU, we have

$$\tau = 4^{3/2} \text{ yr} = 8 \text{ yr}$$

There are about 20 comets in the solar system with periods like this one whose aphelia lie close to Jupiter's orbit. They are known as Jupiter's family of comets. They do not include Halley's comet. ∎

Universality of Gravitation

A tremendous triumph of Newtonian physics ushered in the 19th century—Urbain Jean Leverrier's (1811–1877) discovery of Neptune. It signified a turning point in the history of science, when the newly emerging methodology, long embroiled in a struggle with

---

[3] Actually the sun and a planet orbit their common center of mass (if only a two-body problem is being considered). A more accurate treatment of the orbital motion will be carried out in Section 7.3 where it will be shown that a more correct value for this "constant" is given by $4\pi^2/G(M + m)$ where m is the mass of the orbiting planet. Clearly, the correction is very small.

biblical ideas, began to dominate world concepts. The episode started when Alexis Bouvard, a farmer's boy from the Alps, came to Paris to study science and there perceived irregularities in the motion of Uranus that could not be accounted for by the attraction of the other known planets. In the years to follow, as the irregularities in the motion of Uranus mounted up, the opinion became fairly widespread among astronomers that there had to be an unknown planet disturbing the motion of Uranus.

In 1842–1843 John Couch Adams, a gifted student at Cambridge, began work on this problem, and by September of 1845 he presented Sir George Airy, the Astronomer Royal, and James Challis, the director of the Cambridge observatory, the likely coordinates of the offending, but then unknown, planet. It seemed impossible to them that a mere student, armed only with paper and pencil could take observations of Uranus, invoke the known laws of physics, and predict the existence and precise location of an undiscovered planet. Besides, Airy had grave doubts concerning the validity of the inverse-square law of gravity. In fact, he believed that it fell off faster than inverse-square at great distances. Thus, somewhat understandably, Airy was reluctant to place much credence in Adams' work, and so the two great men chose to ignore him, sealing forever their fate as the astronomers who failed to discover Neptune.

It was about this time that Leverrier began work on the problem. By 1846 he had calculated the orbit of the unknown planet and made a precise prediction of its position on the celestial sphere. Airy and Challis saw that Leverrier's result miraculously agreed with the prediction of Adams. Challis immediately initiated a search for the unknown planet in the suspicious sector of the sky, but owing to Cambridge's lack of detailed star maps in that area, the search was laboriously painstaking and the data reduction problem prodigious. Had Challis proceeded with vigor and tenacity, he most assuredly would have found Neptune for it was there on his photographic plates. Unfortunately, he dragged his feet. However, by this time, an impatient Leverrier had written Johann Galle (1812–1910), astronomer at the Berlin Observatory, asking him to use their large refractor to examine the stars in the suspect area in order to see if one showed a disc, a sure signature of a planet. A short time before the arrival of Leverrier's letter containing this request, the Berlin Observatory had received a detailed star map of this sector of the sky from the Berlin Academy. Upon receipt of Leverrier's letter, September 23, 1846, the map was compared with an image of the sky taken that night, and the planet was identified as a foreign star of eighth magnitude, not seen on the star map. It was named Neptune. Newtonian physics had triumphed in a way never seen before—laws of physics had been used to make a verifiable prediction to the world at large, an unexpected demonstration of the power of science.

Since that time, celestial objects observed at increasingly remote distances continue to exhibit behavior consistent with the laws of Newtonian physics (ignoring those special cases involving large gravitational fields or very extreme distances, each requiring treatment with general relativity). The behavior of binary star systems within our galaxy serves as a classic example. Such stars are bound together gravitationally and their orbital dynamics are well described by Newtonian mechanics. We shall discuss them in the next chapter. So strongly do we believe in the universality of gravitation and the laws of physics, that apparent violations by celestial objects, as in the case of Uranus and the subsequent discovery of Neptune, are usually greeted by searches for unseen distur-

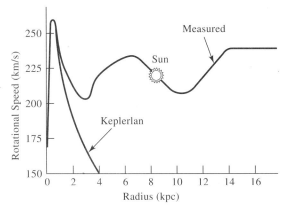

Figure 6.6 Galactic rotation curve. The Sun's speed is about 220 km/s and its distance from the galactic center is about 8.5 kpc ($\approx$28,000 light years).

bances. Rarely do we instead demand the overthrow of the laws of physics. (Although, astonishingly enough, two famous examples discussed subsequently in this chapter had precisely this effect and helped revolutionize physics.)

The more likely scenario, ferreting out the unseen disturbance, is currently in progress in many areas of contemporary astronomical research, as illustrated by the search for dark matter in the universe. One of the reasons we think that an enormous amount of unseen matter fills the universe (perhaps ten times as much as is visible) can be gleaned from the dynamics of spiral galaxies—thin disc-shaped, rotating aggregates of as many as 100 billion stars. At first sight the rotation curve for spiral galaxies (a plot of the rotational velocity of the stars as a function of their radial distance from the galactic center) seems to violate Kepler's laws. An example of such a curve is shown in Figure 6.6. Most of the luminous matter of a spiral galaxy is contained in its central nucleus whose extent is on the order of several thousand light years in radius. The rest of the luminous matter is in the spiral arms that extend in radius out to a distance of about 50,000 light years. The whole thing slowly rotates about its center of gravity, exactly as one would expect for a self-gravitating conglomerate of stars, gas, dust, and so on. The surprising thing about the rotation curve in Figure 6.6 is that it is apparently non-Keplerian.

We can illustrate what we mean by this with a simple example. Assume that the entire galactic mass is concentrated within a nucleus of radius R and that stars fill the nucleus at uniform density. This is an oversimplification, but a calculation based on such a model should serve as a guide for what we might expect a rotation curve to look like. The rotational velocity of stars at some radius $r < R$ within the nucleus is determined only by the amount of mass M within the radius r. Stars external to r have no effect. Since the density of stars within the total nuclear radius R is constant, we calculate M as

$$M = \frac{4}{3}\pi\rho r^3$$

where

$$\rho = \frac{M_{gal}}{\frac{4}{3}\pi R^3}$$

and from Newton's second law

$$\frac{GMm}{r^2} = \frac{mv^2}{r}$$

for the gravitational force exerted on a star of mass m at a distance r from the center of the nucleus by the mass M interior to that distance r. Solving for v, we get

$$v = \sqrt{GM_{gal}/R^3}\ r$$

or, the rotational velocity of stars at $r < R$ is proportional to r. For stars in the spiral arms at distances $r > R$, we obtain

$$\frac{GM_{gal}m}{r^2} = \frac{mv^2}{r}$$

thus,

$$v = \sqrt{\frac{GM_{gal}}{r}}$$

or, the rotational velocity of stars at $r > R$ is proportional to $1/\sqrt{r}$. This is what we mean by Keplerian rotation. It is the way the velocities of planets depend on their distance from the sun. We show such a curve in Figure 6.6, where we have assumed that the entire mass of the galaxy is uniformly distributed in a sphere whose radius is 1 kpc (1 kpc = 3.26 light-year).

Let us examine the measured rotation curve. Initially, it climbs rapidly from zero at the galactic center to about 250 km/s at 1 kpc, more or less as expected, but the astonishing thing is that the curve does not fall off in the expected Keplerian manner. It stays more or less flat all the way out to the edges of the spiral arms (the zero on the vertical axis has been suppressed so the curve is flatter than it appears to be). The conclusion is inescapable. As we move away from the galactic center, we must "pick up" more and more matter within any given radius, which causes even the most remote objects in the galaxy to orbit at velocities that exceed those expected for a highly centralized matter distribution. Since most of the luminosity of a galaxy comes from its nucleus, we conclude that dark, unseen matter must permeate spiral galaxies all the way out to their very edges and beyond. (In fact it should be a simple matter to deduce the radial distribution of dark matter required to generate this flat rotation curve.) Of course, Newton's laws could be wrong, but we think not. It looks like a case of Neptune revisited.

6.7 POTENTIAL ENERGY IN A GRAVITATIONAL FIELD. GRAVITATIONAL POTENTIAL

In Section 2.3 we proved that the inverse-square law of force leads to an inverse first power law for the potential energy function. In this section we shall derive this same relationship in a more physical way.

Let us consider the work W required to move a test particle of mass m along some prescribed path in the gravitational field of another particle of mass M.

We shall place the particle of mass M at the origin of our coordinate system, as shown in Figure 6.7(a). Since the force $\mathbf{F}$ on the test particle is given by $\mathbf{F} = -(GMm/r^2)\mathbf{e}_r$, then to overcome this force an external force $-\mathbf{F}$ must be applied. The work dW done in moving the test particle through a distance $d\mathbf{r}$ is thus given by

$$dW = -\mathbf{F} \cdot d\mathbf{r} = \frac{GMm}{r^2}\,\mathbf{e}_r \cdot d\mathbf{r} \qquad (6.27)$$

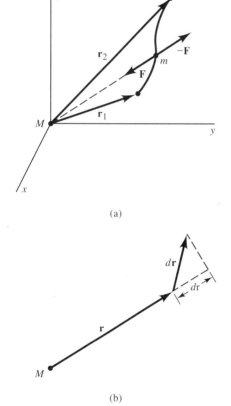

(a)

(b)

Figure 6.7   Diagram for finding the work required to move a test particle in a gravitational field.

Now we can resolve $d\mathbf{r}$ into two components: $\mathbf{e}_r\, dr$ parallel to $\mathbf{e}_r$ (the radial component) and the other at right angles to $\mathbf{e}_r$ [Figure 6.7(b)]. Clearly,

$$\mathbf{e}_r \cdot d\mathbf{r} = dr$$

and so W is given by

$$W = GMm \int_{r_1}^{r_2} \frac{dr}{r^2} = -GMm\left(\frac{1}{r_2} - \frac{1}{r_1}\right) \tag{6.28}$$

where r_1 and r_2 are the radial distances of the particle at the beginning and end of the path. Thus, the work is independent of the particular path taken; it depends only upon the endpoints. This verifies a fact we already knew, namely that the inverse-square law is conservative.

Thus, we can define the potential energy of a test particle of mass m at a given point in the gravitational field of another particle of mass M as the work done in moving the test particle from some arbitrary reference position r_1 to the position r_2. We take the reference position to be $r_1 = \infty$. This assignment is usually a convenient one, since the gravitational force between two particles vanishes when they are separated by ∞. Thus, putting $r_1 = \infty$ and $r_2 = r$ in Equation 6.28, we have

$$V(r) = GMm \int_{\infty}^{r} \frac{dr}{r^2} = -\frac{GMm}{r} \tag{6.29}$$

Like the gravitational force, the gravitational potential energy of two particles separated by ∞ also vanishes.

Both the gravitational force and potential energy between two particles involve the concept of action at a distance. Newton himself was never able to explain or describe the mechanism by which such a force worked. We will not attempt to either. But we would like to introduce the concept of field in such a way that forces and potential energies can be thought of as being generated not by actions at a distance but by local actions of matter with an existing field. To do this, we introduce the quantity Φ, called the *gravitational potential*

$$\Phi = \lim_{m \to 0}\left(\frac{V}{m}\right)$$

In essence Φ is the gravitational potential energy per unit mass that a very small test particle would have in the presence of other surrounding masses. We take the limit as $m \to 0$ to ensure that the presence of the test particle does not affect the distribution of the other matter and change the thing we are trying to define. Clearly, the potential should depend only upon the magnitude of the other masses and their positions in space, not those of the particle we are using to test for the presence of gravitation. We can think of the potential as a scalar function of spatial coordinates, $\Phi(x, y, z)$, or a field, set up by all the other surrounding masses. We test for its presence by placing the test mass m at any point (x, y, z). The potential energy of that test particle is then given by

$$V(x, y, z) = m\Phi(x, y, z)$$

We can think of this potential energy as being generated by the local interaction of the mass m and the field Φ that is present at the point (x, y, z).

The gravitational potential at a distance r from a particle of mass M is

$$\Phi = -\frac{GM}{r} \tag{6.30}$$

If we have a number of particles $M_1, M_2, \ldots, M_i, \ldots$ located at positions $\mathbf{r}_1, \mathbf{r}_2, \ldots,$ $\mathbf{r}_i, \ldots$, then the gravitational potential at the point $\mathbf{r}(x, y, z)$ is the sum of the gravitational potentials of all the particles, that is,

$$\Phi(x, y, z) = \sum \Phi_i = -G \sum \frac{M_i}{u_i} \tag{6.31}$$

in which u_i is the distance of the field point $\mathbf{r}(x, y, z)$ from the position $\mathbf{r}_i(x_i, y_i, z_i)$ of the ith particle

$$u_i = |\mathbf{r} - \mathbf{r}_i|$$

We define a vector field $\mathbf{G}$, called the *gravitational field intensity*, in a way that is completely analogous to the above definition of the gravitational potential scalar field

$$\mathbf{G} = \lim_{m \to 0} \left(\frac{\mathbf{F}}{m} \right)$$

Thus, the gravitational field intensity is the gravitational force per unit mass acting upon a test particle of mass m positioned at the point (x, y, z). Clearly, if the test particle experiences a gravitational force given by

$$\mathbf{F} = m\mathbf{G}$$

then we know that there are other masses nearby responsible for the presence of the local field intensity $\mathbf{G}$.

The relationship between field intensity and the potential is the same as that between the force $\mathbf{F}$ and the potential energy V, namely

$$\mathbf{G} = -\nabla\Phi \tag{6.32}$$
$$\mathbf{F} = -\nabla V$$

The gravitational field intensity can be calculated by first finding the potential function from Equation 6.31 and then calculating the gradient. This method is usually simpler than the method of calculating the field directly from the inverse-square law. The reason is that the potential energy is a scalar sum whereas the field is given by a vector sum. The situation is analogous to the theory of electrostatic fields. In fact one can apply any of the corresponding results from electrostatics to find gravitational fields and potentials with the proviso, of course, that there are no negative masses.

EXAMPLE 6.6   *Potential of a Uniform Spherical Shell*

As an example, let us find the potential function for a uniform spherical shell.

Solution:
By using the same notation as that of Figure 6.2, we have

$$\Phi = -G \int \frac{dM}{u} = -G \int \frac{2\pi\rho R^2 \sin\theta \; d\theta}{u}$$

From the same relation between u and θ that we used earlier, we find that the above equation may be simplified to read

$$\Phi = -G \frac{2\pi\rho R^2}{rR} \int_{r-R}^{r+R} du = -\frac{GM}{r} \tag{6.33}$$

where M is the mass of the shell. This is the same potential function as that of a single particle of mass M located at O. Hence, the gravitational field outside the shell is the same as if the entire mass were concentrated at the center. It is left as a problem to show that, with an appropriate change of the integral and its limits, the potential inside the shell is constant and, hence, that the field there is zero.   ■

EXAMPLE 6.7   *Potential and Field of a Thin Ring*

We now wish to find the potential function and the gravitational field intensity in the plane of a thin circular ring.

Solution:
Let the ring be of radius R and mass M. Then, for an exterior point lying in the plane of the ring, Figure 6.8, we have

$$\Phi = -G \int \frac{dM}{u} = -G \int_{0}^{2\pi} \frac{\mu R \; d\theta}{u}$$

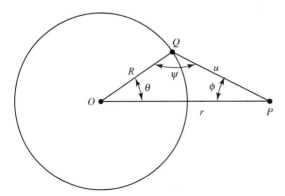

Figure 6.8   Coordinates for calculating the gravitational field of a ring.

where μ is the linear mass density of the ring. In order to evaluate the integral, we first express u as a function of θ using the law of cosines

$$u^2 = R^2 + r^2 - 2Rr \cos \theta$$

The integral becomes

$$\Phi = -2R\mu G \int_0^\pi \frac{d\theta}{(r^2 + R^2 - 2Rr \cos \theta)^{1/2}}$$

$$= -\frac{2R\mu G}{r} \int_0^\pi \frac{d\theta}{\left(1 + \dfrac{R^2}{r^2} - 2\dfrac{R}{r} \cos \theta\right)^{1/2}}$$

First, let us use the so-called far field approximation $r > R$ and expand the integrand in a power series of $x\ (=R/r)$, making certain to keep all terms of order x^2.

$$\Phi = -2x\mu G \int_0^\pi \left[\left(1 - \frac{1}{2}x^2 + x \cos \theta\right) + \frac{3}{8}\left(x^2 - 2x \cos \theta\right)^2 + \cdots\right] d\theta$$

$$= -2x\mu G \int_0^\pi \left(1 - \frac{1}{2}x^2 + x \cos \theta + \frac{3}{2}x^2 \cos^2 \theta\right.$$

$$\left. - \frac{3}{2}x^3 \cos \theta + \frac{3}{8}x^4 + \cdots\right) d\theta$$

Now, dropping all terms of order x^3 or higher and noting that the term containing $\cos \theta$ has zero integral over a half cycle, we obtain

$$\Phi = -2x\mu G \left(\pi + \pi\frac{x^2}{4} + \cdots\right.$$

$$= \frac{-2\pi R\mu G}{r}\left(1 + \frac{R^2}{4r^2} + \cdots\right.$$

$$= -\frac{GM}{r}\left(1 + \frac{R^2}{4r^2} + \cdots\right.$$

The field intensity at a distance r from the center of the ring is in the radial direction (since Φ is not a function of θ) and is given by

$$\mathbf{G} = -\frac{\partial \Phi}{\partial r}\mathbf{e}_r = -\frac{GM}{r^2}\left(1 + \frac{3}{4}\left(\frac{R}{r}\right)^2 + \cdots\right)\mathbf{e}_r$$

The field is not given by an inverse-square law. If $r \gg R$, the term in parentheses approaches unity, and the field intensity approaches the inverse-square field of a single particle of mass M. This is true for a finite-sized body of any shape; that is, for distances large compared to the linear dimensions of the body, the field intensity approaches that of a single particle of mass M.

The potential for a point near the center of the ring can be found by invoking the near field, or $r < R$, approximation. The solution proceeds more or less as above, but in this case we expand the above integrand in powers of r/R to obtain

$$\Phi = -\frac{GM}{R}\left(1 + \frac{r^2}{4R^2} + \cdots\right)$$

G can again be found by differentiation

$$G = \left(\frac{GM}{2R^3}\,r\right)e_r + \cdots$$

Thus, it can be seen that a ring of matter exerts an approximately linear *repulsive* force, directed away from the center, on a particle located somewhere near the center of the ring. It is easy to see that this must be so. Imagine that you are at the center of such a ring of radius R with a field of view both in front of you and behind you which subtends some definite angle. If you move slowly a distance r away from the center, the matter you see attracting you in the forward direction diminishes by a factor of r, whereas the matter you see attracting you from behind grows by r. But since the force of gravity from any material element falls as $1/r^2$, the force exerted on you by the forward mass and the backward mass varies as $1/r$, and thus the force difference between the two of them is $[1/(R - r) - 1/(R + r)]$ or proportional to r for $r < R$. The gravitational force of the ring repels objects from its center. ∎

6.8   POTENTIAL ENERGY IN A GENERAL CENTRAL FIELD

We have previously shown that a central field of the inverse-square type is conservative. Let us now consider the question as to whether or not any (isotropic) central field of force is conservative. A general isotropic central field can be expressed in the following way:

$$\mathbf{F} = f(r)\,\mathbf{e}_r \tag{6.34}$$

in which $\mathbf{e}_r$ is the unit radial vector. To apply the test for conservativeness, we calculate the curl of $\mathbf{F}$. It is convenient here to employ spherical coordinates for which the curl is given in Appendix F. We find

$$\nabla \times \mathbf{F} = \frac{1}{r^2 \sin\theta}\begin{vmatrix} \mathbf{e}_r & \mathbf{e}_\theta r & \mathbf{e}_\phi r \sin\theta \\ \dfrac{\partial}{\partial r} & \dfrac{\partial}{\partial\theta} & \dfrac{\partial}{\partial\phi} \\ F_r & rF_\theta & rF_\phi \sin\theta \end{vmatrix}$$

For our central force $F_r = f(r)$, $F_\theta = 0$, and $F_\phi = 0$. The curl then reduces to

$$\nabla \times \mathbf{F} = \frac{\mathbf{e}_\theta}{r \sin\theta}\frac{\partial f}{\partial\phi} - \frac{\mathbf{e}_\phi}{r}\frac{\partial f}{\partial\theta} = 0$$

The two partial derivatives both vanish since $f(r)$ does not depend on the angular coordinates ϕ and θ. Thus, the curl vanishes and so the general central field defined by Equation 6.34 is conservative. We recall that the same test was applied to the inverse-square field in Example 4.5.

We can now define a potential energy function

$$V(r) = -\int_{r_{ref}}^{r} \mathbf{F} \cdot d\mathbf{r} = -\int_{r_{ref}}^{r} f(r) dr \tag{6.35}$$

where the lower limit r_{ref} is the reference value of r at which the potential energy is *defined* to be zero. For inverse-power type forces, r_{ref} is often taken to be at infinity. This allows us to calculate the potential energy function, given the force function. Conversely, if we know the potential energy function, we have

$$f(r) = -\frac{dV(r)}{dr} \tag{6.36}$$

giving the force function for a central field.

6.9 ENERGY EQUATION OF AN ORBIT IN A CENTRAL FIELD

The square of the speed is given in polar coordinates by

$$v^2 = \dot{r}^2 + r^2\dot{\theta}^2$$

Since a central force is conservative, the total energy $T + V$ is constant and is given by

$$\frac{1}{2} m(\dot{r}^2 + r^2\dot{\theta}^2) + V(r) = E = \text{constant} \tag{6.37}$$

We can also write the above equation in terms of the variable $u = 1/r$. From Equations 6.10 and 6.11 we obtain

$$\frac{1}{2} ml^2 \left[\left(\frac{du}{d\theta} \right)^2 + u^2 \right] + V(u^{-1}) = E \tag{6.38}$$

In the above equation the only variables occurring are u and θ. We shall call this equation, *the energy equation of the orbit.*

EXAMPLE 6.8

In Example 6.1 we had for the spiral orbit $r = c\theta^2$

$$\frac{du}{d\theta} = \frac{-2}{c} \theta^{-3} = -2c^{1/2}u^{3/2}$$

so the energy equation of the orbit is

$$\frac{1}{2} ml^2(4cu^3 + u^2) + V = E$$

Thus,

$$V(r) = E - \frac{1}{2} ml^2 \left(\frac{4c}{r^3} + \frac{1}{r^2} \right)$$

This readily gives the force function of Example 6.1, since $f(r) = -dV/dr$. ∎

6.10   ORBITAL ENERGIES IN AN INVERSE-SQUARE FIELD

Since the potential energy function $V(r)$ for an inverse-square force field is given by

$$V(r) = -\frac{k}{r} = -ku$$

the energy equation of the orbit (Equation 6.37) then reads

$$\frac{1}{2} ml^2 \left[\left(\frac{du}{d\theta} \right)^2 + u^2 \right] - ku = E$$

or, upon separating variables,

$$d\theta = \left(\frac{2E}{ml^2} + \frac{2ku}{ml^2} - u^2 \right)^{-1/2} du$$

Upon integrating, we find

$$\theta = \sin^{-1} \left[\frac{ml^2 u - k}{(k^2 + 2Eml^2)^{1/2}} \right] + \theta_0$$

where θ_0 is a constant of integration. If we let $\theta_0 = -\pi/2$ and solve for u, we obtain

$$u = \frac{k}{ml^2} [1 + (1 + 2Eml^2 k^{-2})^{1/2} \cos \theta]$$

or

$$r = \frac{ml^2 k^{-1}}{1 + (1 + 2Eml^2 k^{-2})^{1/2} \cos \theta} \tag{6.39}$$

This is the polar equation of the orbit. If we compare it with Equations 6.31 and 6.32, we see that the eccentricity is given by

$$e = (1 + 2Eml^2 k^{-2})^{1/2} \tag{6.40}$$

Equation 6.40 allows us to classify the orbits according to the total energy E as follows:

$$
\begin{array}{lll}
E < 0 & e < 1 & \textit{closed orbits (ellipse or circle)} \\
E = 0 & e = 1 & \textit{parabolic orbit} \\
E > 0 & e > 1 & \textit{hyperbolic orbit}
\end{array}
$$

Since $E = T + V$ and is constant, the closed orbits are those for which $T < |V|$, and the open orbits are those for which $T \geq |V|$.

In the sun's gravitational field the force constant $k = GMm$ where M is the mass of the sun and m is the mass of the body. The total energy is then

$$\frac{mv^2}{2} - \frac{GMm}{r} = E = \text{constant}$$

so the orbit is an ellipse, a parabola, or a hyperbola depending on whether v^2 is less than, equal to, or greater than the quantity $2GM/r$, respectively.

EXAMPLE 6.9

A comet is observed to have a speed v_{com} when it is a distance r_{com} from the sun, and its direction of motion makes an angle ϕ with the radius vector from the sun, Figure 6.9. Find the eccentricity of the comet's orbit.

Solution:
To use the formula for the eccentricity (Equation 6.40), we need the square of the angular momentum constant l. It is given by

$$l^2 = |\mathbf{r} \times \mathbf{v}|^2 = (r_{com} v_{com} \sin \phi)^2$$

The eccentricity, therefore, has the value

$$e = \left[1 + \left(v_{com}^2 - \frac{2GM}{r_{com}} \right) \left(\frac{r_{com} v_{com} \sin \phi}{GM} \right)^2 \right]^{1/2}$$

Note that the mass m of the comet cancels out. Now the product GM can be ex-

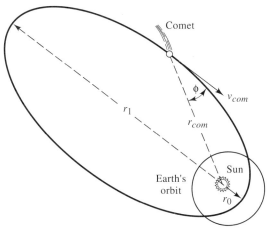

Figure 6.9 Orbit of a comet.

pressed in terms of the earth's speed v_e and orbital radius a_e (assuming a circular orbit), namely

$$GM = a_e v_e^2$$

The above expression for the eccentricity then becomes

$$e = \left[1 + \left(\mathsf{V}^2 - \frac{2}{\mathsf{R}} \right) (\mathsf{R}\mathsf{V} \sin \phi)^2 \right]^{1/2}$$

where we have introduced the *dimensionless ratios*

$$\mathsf{V} = \frac{v_{com}}{v_e} \qquad \mathsf{R} = \frac{r_{com}}{a_e}$$

which simplify the computation of e.

As a numerical example, let v_{com} be one-half the earth's speed, let r_{com} be four times the earth–sun distance, and $\phi = 30°$. Then $\mathsf{V} = 0.5$ and $\mathsf{R} = 4$, so the eccentricity is

$$e = [1 + (0.25 - 0.5)(4 \times 0.5 \times 0.5)^2]^{1/2} = (0.75)^{1/2} = 0.866$$

For an ellipse the quantity $(1 - e^2)^{-1/2}$ is equal to the ratio of the major (long) axis to the minor (short) axis. For the orbit of the comet in this example this ratio is $(1 - 0.75)^{-1/2} = 2$, or $2:1$, as shown in Figure 6.9. ∎

6.11   LIMITS OF THE RADIAL MOTION. EFFECTIVE POTENTIAL

We have seen that the angular momentum of a particle moving in any isotropic central field is a constant of the motion, as expressed by Equations 6.7 and 6.8 defining l. This fact allows us to write the general energy equation (Equation 6.37) in the following form:

$$\frac{m}{2} \left(\dot{r}^2 + \frac{l^2}{r^2} \right) + V(r) = E$$

or

$$\frac{m}{2} \dot{r}^2 + U(r) = E \tag{6.41}$$

in which

$$U(r) = \frac{ml^2}{2r^2} + V(r) \tag{6.42}$$

The function $U(r)$ defined above is called the *effective potential*. The term $ml^2/2r^2$ is called the *centrifugal potential*. Looking at Equation 6.41 we see that, as far as the radial motion is concerned, the particle behaves in exactly the same way as a particle of mass m moving in one-dimensional motion under a potential energy function $U(r)$. As in

Section 3.3 where we discussed harmonic motion, the limits of the radial motion (turning points) are given by setting $\dot{r} = 0$ in Equation 6.41. These limits are, therefore, the roots of the equation

$$U(r) - E = 0 \qquad (6.43)$$

or

$$\frac{ml^2}{2r^2} + V(r) - E = 0 \qquad (6.43a)$$

Furthermore, the *allowed* values of r are those for which $U(r) \leq E$, since $\dot{r}^2$ is necessarily positive or zero.

Thus, it is possible to determine the range of the radial motion without knowing anything about the orbit. A plot of $U(r)$ is shown in Figure 6.10. Also shown are the radial limits r_0 and r_1 for a particular value of the total energy E. The graph is drawn for the inverse-square law, namely,

$$U(r) = \frac{ml^2}{2r^2} - \frac{k}{r} \qquad (6.44)$$

In this case Equation 6.43, upon rearranging terms, becomes

$$-2Er^2 - 2kr + ml^2 = 0$$

which is a quadratic equation in r. The two roots are

$$r_{1,0} = \frac{k \pm (k^2 + 2Eml^2)^{1/2}}{-2E} \qquad (6.45)$$

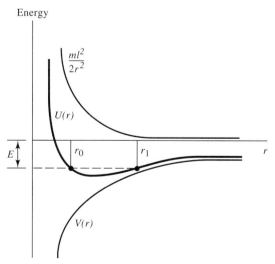

Figure 6.10 Illustrating the effective potential and limits of the radial motion for the inverse-square law of force.

giving the maximum (upper sign) and minimum (lower sign) values of the radial distance r under the inverse-square law of force. Since the energy E is a negative quantity for all bound orbits, the two roots are both positive, as they should be.

Now we have already shown that closed orbits under the inverse-square law are ellipses for which the major axis $2a$ is the sum $r_1 + r_0$. Thus, by adding the two roots above, we have

$$2a = r_1 + r_0 = \frac{k}{-E} \qquad (6.46)$$

This result shows that the value of a, the semimajor axis, is determined entirely by the force constant k and the total energy E.

EXAMPLE 6.10

Find the semimajor axis of the orbit of the comet of Example 6.9.

Solution:
Equation 6.46 gives directly

$$a = \frac{k}{-2E} = \frac{GMm}{-2\left(\dfrac{mv_{com}^2}{2} - \dfrac{GMm}{r_{com}}\right)}$$

where m is the mass of the comet. Clearly, m again cancels out. Also, as stated above, $GM = a_e v_e^2$. So the final result is the simple expression

$$a = \frac{a_e}{\dfrac{2}{R} - V^2}$$

where R and V are as defined in the previous example.

For the previous numerical values, $R = 4$ and $V = 0.5$, we find $a = a_e/[0.5 - (0.5)^2] = 4a_e$. ∎

Examples 6.9 and 6.10 bring out an important fact, namely, that the orbital parameters are independent of the mass of a body. Given the same initial position, speed, and direction of motion, a grain of sand, a coasting spaceship, or a comet would all have identical orbits, provided that no other bodies came near enough to have an effect on the motion of the body. (We also assume, of course, that the mass of the body in question is small compared to the sun's mass.)

6.12   MOTION IN AN INVERSE-SQUARE REPULSIVE FIELD. SCATTERING OF ALPHA PARTICLES

It is ironic that one of the crowning achievements of Newtonian mechanics contained its own seeds of destruction. In 1911 Ernest Rutherford (1871–1937), attempting to solve the problem of the scattering of alpha particles by thin metal foils, went for help back to the very source of classical mechanics, the *Principia* of Sir Isaac Newton. Paradoxically, in the process of finding a solution to the problem based upon classical mechanics, the idea of the nuclear atom was born, an idea that would forever remain incomprehensible within the confines of the classical paradigm. A complete, self-consistent theory of the nuclear atom would emerge only when many of the notions of Newtonian mechanics were given up and replaced by the novel and astounding concepts of quantum mechanics. It is not that Newtonian mechanics was "wrong"; its concepts, which worked so well time and again when applied to the macroscopic world of falling balls and orbiting planets, simply broke down when applied to the microscopic world of atoms and nuclei. Indeed, the architects of the laws of quantum physics constructed them in such a way that the results of calculations based on the new laws agreed with those of Newtonian mechanics when applied to problems in the macroscopic world. The domain of Newtonian physics would be seen to be merely limited, rather than "wrong," and its practitioners from that time on would now have to be aware of these limits.

In the early 1900s the atom was thought to be sort of a distributed blob of positive charge within which were embedded the negatively charged electrons discovered in 1897 by J. J. Thomson (1856–1940). The model was first suggested by Lord Kelvin in 1902 but mathematically refined a year later by Thomson. Thomson developed the model with emphasis on the mechanical and electrical stability of the system. In his honor it became known as the *Thomson atom.*

In 1907 Rutherford accepted a position at the University of Manchester where he encountered Hans Geiger (1882–1945), a bright, young German experimental physicist, who was about to embark on an experimental program designed to test the validity of the Thomson atom. His idea was to direct a beam of the recently discovered alpha particles emitted from radioactive atoms toward thin metal foils. A detailed analysis of the way they scattered should provide information on the structure of the atom. With the help of Ernest Marsden, a young undergraduate, Geiger would carry out these investigations over several years. Things behaved more or less as expected, except there were many more large angle scatterings than could be accounted for by the Thomson model. In fact, some of the alpha particles scattered completely backward at angles of 180°. When Rutherford heard of this, he was dumbfounded. It was as though an on-rushing freight train had been hurled backward upon striking a chicken sitting in the middle of the track.

In searching for a model that would lead to such a large force being exerted upon a fast moving projectile, Rutherford envisioned a comet swinging around the sun and coming back out again, just like the alpha particles scattered at large angles. This suggested the idea of a hyperbolic orbit for a positively charged alpha attracted by a negatively charged nucleus. Of course, Rutherford realized that the only important thing in the dynamics of the problem was the inverse-square nature of the law, which, as we have seen, leads to conic sections as solutions for the orbit. Whether the force is attractive or repulsive is completely irrelevant. Rutherford then remembered a theorem about conics from geometry that related the eccentricity of the hyperbola to the angle between its asymptotes. Using this relation, along with conservation of angular momentum and energy, he obtained a complete solution to the alpha particle scattering problem, which agreed well within the data of Geiger and Marsden. Thus, the current model of the nuclear atom was born. We will solve the problem below.

Be aware, though, that an identical solution could be obtained for an attractive force. The Rutherford solution says nothing about the sign of the nuclear charge. The sign becomes obvious from other arguments.

Consider a particle of charge q and mass m (the incident high-speed particle) passing near a heavy particle of charge Q (the nucleus, assumed fixed). The incident particle is repelled with a force given by Coulomb's law:

$$f(r) = \frac{Qq}{r^2} \tag{6.47}$$

where the position of Q is taken to be the origin. (We shall use cgs electrostatic units for Q and q. Then r is in centimeters, and the force is in dynes.) The differential equation of the orbit then takes the form

$$\frac{d^2u}{d\theta^2} + u = -\frac{Qq}{ml^2}$$

and so the equation of the orbit is

$$u^{-1} = r = \frac{1}{A \cos(\theta - \theta_0) - Qq/ml^2}$$

We can also write the equation of the orbit in the form given by Equation 6.39, namely,

$$r = \frac{ml^2 Q^{-1} q^{-1}}{-1 + (1 + 2Eml^2 Q^{-2} q^{-2})^{1/2} \cos(\theta - \theta_0)} \tag{6.48}$$

since $k = -Qq$. The orbit is a hyperbola. This may be seen from the physical fact that the energy E is always greater than zero in a repulsive field of force. (In our case $E = \frac{1}{2}mv^2 + Qq/r$.) Hence, the eccentricity e, the coefficient of $\cos(\theta - \theta_0)$, is greater than unity, which means that the orbit must be hyperbolic.

The incident particle approaches along one asymptote and recedes along the other, as shown in Figure 6.11. We have chosen the direction of the polar axis such that the initial position of the particle is $\theta = 0$, $r = \infty$. It is clear from either of the two equations of the orbit that r assumes its minimum value when $\cos(\theta - \theta_0) = 1$, that is, when $\theta = \theta_0$. Since $r = \infty$ when $\theta = 0$, then r is also infinite when $\theta = 2\theta_0$. Hence, the angle between the two asymptotes of the hyperbolic path is $2\theta_0$, and the angle θ_s through which the incident particle is deflected is given by

$$\theta_s = \pi - 2\theta_0$$

Furthermore, in Equation 6.48 the denominator on the right vanishes at $\theta = 0$ and $\theta = 2\theta_0$. Thus,

$$-1 + (1 + 2Eml^2 Q^{-2} q^{-2})^{1/2} \cos \theta_0 = 0$$

from which we readily find

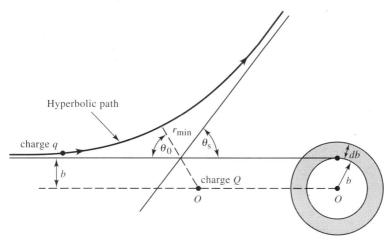

Figure 6.11 Hyperbolic path (orbit) of a charged particle moving in the inverse-square repulsive force field of another charged particle.

$$\tan \theta_0 = (2Em)^{1/2} l Q^{-1} q^{-1} = \cot \frac{\theta_s}{2} \qquad (6.49)$$

The last step follows from the angle relationship given above.

In applying the above equation to scattering problems, it is convenient to express the constant l in terms of another quantity b called the *impact parameter*. The impact parameter is the perpendicular distance from the origin (scattering center) to the initial line of motion of the particle, as shown in Figure 6.11. We have then

$$|l| = |\mathbf{r} \times \mathbf{v}| = b v_0$$

where v_0 is the initial speed of the particle. We know also that the energy E is constant and is equal to the initial kinetic energy $\frac{1}{2} m v_0^2$, because the initial potential energy is zero ($r = \infty$). Accordingly, we can write the scattering formula (Equation 6.49) in the form

$$\cot \frac{\theta_s}{2} = \frac{b m v_0^2}{Qq} = \frac{2bE}{Qq} \qquad (6.50)$$

giving the relationship between the scattering angle and the impact parameter.

In a typical scattering experiment a beam of particles is projected at a target, such as a thin foil. The nuclei of the target atoms are the scattering centers. The fraction of incident particles that are deflected through a given angle θ_s can be expressed in terms of a *differential scattering cross section* $\sigma(\theta_s)$ defined by the equation

$$\frac{dN}{N} = n \, \sigma(\theta_s) \, d\Omega$$

Here dN is the number of incident particles scattered through an angle between θ_s and $\theta_s + d\theta_s$, N is the total number of incident particles, n is the number of scattering centers per unit area of the target foil, and $d\Omega$ is the element of solid angle corresponding to the increment $d\theta_s$. Thus, $d\Omega = 2\pi \sin \theta_s d\theta_s$.

Now an incident particle approaching a scattering center will have an impact parameter lying between b and $b + db$ if the projection of its path lies in a ring of inner radius b and outer radius $b + db$, Figure 6.11. The area of this ring is $2\pi \, b \, db$. The total number of such particles must correspond to the number scattered through a given angle, that is

$$dN = Nn\sigma(\theta_s)2\pi \sin \theta_s d\theta_s = Nn2\pi b \, db$$

Thus,

$$\sigma(\theta_s) = \frac{b}{\sin \theta_s} \left| \frac{db}{d\theta_s} \right| \tag{6.51}$$

To find the scattering cross section for charged particles, we differentiate with respect to θ_s in Equation 6.50:

$$\frac{1}{2 \sin^2\left(\dfrac{\theta_s}{2}\right)} = \frac{2E}{Qq} \left| \frac{db}{d\theta_s} \right| \tag{6.52}$$

(The absolute value sign is inserted because the derivative is negative.) By eliminating b and $|db/d\theta_s|$ among Equations 6.50, 6.51, and 6.52 and using the identity $\sin \theta_s = 2 \sin(\theta_s/2) \cos(\theta_s/2)$, we find the following result:

$$\sigma(\theta_s) = \frac{Q^2 q^2}{16E^2} \frac{1}{\sin^4\left(\dfrac{\theta_s}{2}\right)} \tag{6.53}$$

This is the famous Rutherford scattering formula. It shows that the differential cross section varies as the inverse fourth power of $\sin(\theta_s/2)$. Its experimental verification in the first part of this century marked one of the early milestones of nuclear physics.

EXAMPLE 6.11

An alpha particle emitted by radium ($E = 5$ million eV $= 5 \times 10^6 \times 1.6 \times 10^{-12}$ erg) suffers a deflection of $90°$ upon passing near a gold nucleus. What is the value of the impact parameter? For alpha particles $q = 2e$, and for gold $Q = 79e$, where e is the elementary charge. (The charge carried by a single electron is $-e$.) In our units $e = 4.8 \times 10^{-10}$ esu. Thus, from Equation 6.50

$$b = \frac{Qq}{2E} \cot 45° = \frac{2 \times 79 \times (4.8)^2 \times 10^{-20} \text{ cm}}{2 \times 5 \times 1.6 \times 10^{-6}}$$

$$= 2.1 \times 10^{-12} \text{ cm} \qquad \blacksquare$$

EXAMPLE 6.12

Calculate the distance of closest approach of the alpha particle in the above problem.

Solution:
The distance of closest approach is given by the equation of the orbit (Equation 6.48) for $\theta = \theta_0$; thus,

$$r_{min} = \frac{ml^2 Q^{-1} q^{-1}}{-1 + (1 + 2Eml^2 Q^{-2} q^{-2})^{1/2}}$$

Upon using Equation 6.50 and a little algebra, the above equation can be written

$$r_{min} = \frac{b \cot (\theta_s/2)}{-1 + [1 + \cot^2 (\theta_s/2)]^{1/2}} = \frac{b \cos (\theta_s/2)}{1 - \sin (\theta_s/2)}$$

Thus, for $\theta_s = 90°$, we find $r_{min} = 2.41 \, b = 5.1 \times 10^{-12}$ cm.

Notice that the expressions for r_{min} become indeterminate when $l = b = 0$. In this case the particle is aimed directly at the nucleus. It approaches the nucleus along a straight line, and, being continually repelled by the coulomb force, its speed is reduced to zero when it reaches a certain point r_{min}, from which point it returns along the same straight line. The angle of deflection is 180°. The value of r_{min} in this case is found by using the fact that the energy E is constant. At the turning point the potential energy is Qq/r_{min}, and the kinetic energy is zero. Hence, $E = \frac{1}{2}mv_0^2 = Qq/r_{min}$, and

$$r_{min} = \frac{Qq}{E}$$

For radium alpha particles and gold nuclei we find $r_{min} \simeq 10^{-12}$ cm when the angle of deflection is 180°. The fact that such deflections are actually observed shows that the order of magnitude of the radius of the nucleus is at least as small as 10^{-12} cm. ■

6.13 NEARLY CIRCULAR ORBITS IN CENTRAL FIELDS. STABILITY

A circular orbit is possible under any attractive central force, but not all central forces result in *stable* circular orbits. We wish to investigate the following question: If a particle traveling in a circular orbit suffers a slight disturbance, will the ensuing orbit remain close to the original circular path? In order to answer the query, we refer to the radial differential equation of motion (Equation 6.5). Since $\dot{\theta} = l/r^2$, we can write the radial equation as follows:

$$m\ddot{r} = \frac{ml^2}{r^3} + f(r)$$

[This is the same as the differential equation for one-dimensional motion under the effective potential $U(r) = (ml^2/2r^2) + V(r)$, so that $m\ddot{r} = -dU(r)/dr = (ml^2/r^3) - dV(r)/dr.$]

Now for a circular orbit, r is constant, and $\ddot{r} = 0$. Thus, calling a the radius of the circular orbit, we have

$$-\frac{ml^2}{a^3} = f(a) \tag{6.54}$$

for the force at $r = a$. It will be convenient to express the radial motion in terms of the variable x defined by

$$x = r - a$$

The differential equation for radial motion then becomes

$$m\ddot{x} = ml^2(x + a)^{-3} + f(x + a)$$

Expanding the two terms involving $x + a$ as power series in x, we obtain

$$m\ddot{x} = ml^2a^{-3}\left(1 - 3\frac{x}{a} + \cdots\right) + [f(a) + f'(a)x + \cdots]$$

The above equation, by virtue of the relation shown in Equation 6.54, reduces to

$$m\ddot{x} + \left[\frac{-3}{a}f(a) - f'(a)\right]x = 0 \tag{6.55}$$

if we neglect terms involving x^2 and higher powers of x. Now, if the coefficient of x (the quantity in brackets) in Equation 6.55 is positive, then the equation is the same as that of the simple harmonic oscillator. In this case the particle, if perturbed, oscillates harmonically about the circle $r = a$, so the circular orbit is a stable one. On the other hand, if the coefficient of x is negative, the motion is nonoscillatory, and the result is that x eventually increases exponentially with time; the orbit is unstable. (If the coefficient of x is zero, then higher terms in the expansion must be included in order to determine the stability.) Hence, we can state that a circular orbit of radius a is stable if the force function $f(r)$ satisfies the inequality

$$f(a) + \frac{a}{3}f'(a) < 0 \tag{6.56}$$

For example, if the radial force function is a power law, namely,

$$f(r) = -cr^n$$

then the condition for stability reads

$$-ca^n - \frac{a}{3}cna^{n-1} < 0$$

which reduces to

$$n > -3$$

Thus, the inverse-square law ($n = -2$) gives stable circular orbits, as does the law of direct distance ($n = 1$). The latter case is that of the two-dimensional isotropic harmonic oscillator. For the inverse-fourth power ($n = -4$) circular orbits are unstable. It can be shown that circular orbits are also unstable for the inverse-cube law of force ($n = -3$). To show this it is necessary to include terms of higher power than one in the radial equation. (See Problem 6.22.)

6.14 APSIDES AND APSIDAL ANGLES FOR NEARLY CIRCULAR ORBITS

An *apsis,* or *apse,* is a point in an orbit at which the radius vector assumes an extreme value (maximum or minimum). The perihelion and aphelion points are the apsides of planetary orbits. The angle swept out by the radius vector between two consecutive apsides is called the *apsidal angle.* Thus, the apsidal angle is π for elliptic orbits under the inverse-square law of force.

In the case of motion in a nearby circular oribt, we have seen that r oscillates about the circle $r = a$ (if the orbit is stable). From Equation 6.55 it follows that the period τ_r of this oscillation is given by

$$\tau_r = 2\pi \left[\frac{m}{-\dfrac{3}{a} f(a) - f'(a)} \right]^{1/2}$$

The apsidal angle in this case is just the amount by which the polar angle θ increases during the time that r oscillates from a minimum value to the succeeding maximum value. This time is clearly $\frac{1}{2}\tau_r$. Now $\dot\theta = l/r^2$; therefore, $\dot\theta$ remains approximately constant, and we can write

$$\dot\theta \simeq \frac{l}{a^2} = \left[-\frac{f(a)}{ma} \right]^{1/2}$$

The last step above follows from Equation 6.54. Hence, the apsidal angle is given by

$$\psi = \frac{1}{2}\tau_r\dot\theta = \pi\left[3 + a\frac{f'(a)}{f(a)} \right]^{-1/2} \tag{6.57}$$

Thus, for the power law of force $f(r) = -cr^n$, we obtain

$$\psi = \pi(3 + n)^{-1/2}$$

The apsidal angle is independent of the size of the orbit in this case. The orbit is *re-entrant,* or repetitive, in the case of the inverse-square law ($n = -2$) for which $\psi = \pi$ and also in the case of the linear law ($n = 1$) for which $\psi = \pi/2$. If, however, say $n = 2$, then $\psi = \pi/\sqrt{5}$, which is an irrational multiple of π, and so the motion does not repeat itself.

If the law of force departs slightly from the inverse-square law, then the apsides will

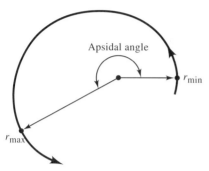

Figure 6.12   Illustrating the apsidal
angle.

either advance or regress steadily, depending on whether the apsidal angle is slightly
greater or slightly less than π. (See Figure 6.12.)

EXAMPLE 6.13

Let us assume that the gravitational force field acting on the planet Mercury takes
the form

$$f(r) = -\frac{k}{r^2} + \epsilon r$$

where ϵ is very small. The first term is the gravitational field due to the sun, while
the second term is a repulsive perturbation due to a surrounding ring of matter. We
assume this matter distribution as a simple model to represent the gravitational effects of all the other planets, primarily Jupiter. The perturbation is linear for points
near the sun and in the plane of the surrounding ring, as previously explained in
Example 6.2. The apsidal angle, from Equation 6.57, is

$$\Psi = \pi\left(3 + a\frac{2ka^{-3} + \epsilon}{-ka^{-2} + \epsilon a}\right)^{-1/2}$$

$$= \pi\left(\frac{1 - 4k^{-1}\epsilon a^3}{1 - k^{-1}\epsilon a^3}\right)^{-1/2} = \pi\left(1 - \frac{\epsilon}{k}a^3\right)^{1/2}\left(1 - 4\frac{\epsilon}{k}a^3\right)^{-1/2}$$

$$\approx \pi\left(1 - \frac{1}{2}\frac{\epsilon}{k}a^3\right)\left(1 + 2\frac{\epsilon}{k}a^3\right)$$

$$\approx \pi\left(1 + \frac{3}{2}\frac{\epsilon}{k}a^3\right)$$

In the last step above, we have used the binomial expansion theorem to expand the
terms in brackets in powers of ϵ/k and kept only the first-order term. The apsidal
angle advances if ϵ is positive and regresses if it is negative.   ■

By 1877, Urbain Leverrier, using perturbation methods, had succeeded in calculating the gravitational effects of all the known planets on each other's orbit. Depending on the planet, the apsidal angles were found to advance or regress in good agreement with theory with the sole exception of the planet Mercury. Observations of Mercury's solar transits since 1631 indicated an advance of the perihelion of its orbit by 565″ of arc per century. According to Leverrier, it should advance only 527″ per century, a discrepancy of 38″. Simon Newcomb (1835–1909), chief of the office for the American Nautical Almanac, improved Leverrier's calculations, and by the beginning of the 20th century, the accepted values for the advance of Mercury's perihelion per century were 575″ and 534″, respectively, or a discrepancy of 41″ ± 2″ of arc. Leverrier himself had decided that the discrepancy was real and that it could be accounted for by an as yet unseen planet with a diameter of about 1000 miles circling the sun within Mercury's orbit at a distance of about 0.2 AU. (You can easily extend the above example to show that an interior planet would lead to an advance in the perihelion of Mercury's orbit by the factor δ/ka^2.) Leverrier called the unseen planet Vulcan. No such planet was found.

Another possible explanation was put forward by Asaph Hall, the discoverer of the satellites of Mars in 1877. He proposed that the exponent in Newton's law of gravitation might not be exactly 2, that instead, it might be 2.0000001612 and that this would do the trick. Einstein was to comment that the discrepancy in Mercury's orbit "could be explained by means of classical mechanics only on the assumption of hypotheses which have little probability, and which were devised solely for this purpose." The discrepancy, of course, was nicely explained by Einstein himself in a paper presented to the Berlin Academy in 1915. The paper was based on Einstein's calculations of general relativity even before he had fully completed the theory. Thus, here we have the highly remarkable event of a discrepancy between observation and existing theory leading to the confirmation of an entirely new superceding theory.

If the sun were oblate (football-shaped) enough, its gravitational field would depart slightly from an inverse-square law, and the perihelion of Mercury's orbit would advance. Measurements to date have failed to validate this hypothesis as a possible explanation. Similar effects, however, have been observed in the case of artificial satellites in orbit about the earth. Not only does the perihelion of a satellite's orbit advance, but the plane of the orbit precesses if the satellite is not in the earth's equatorial plane. Detailed analysis of these orbits show that the earth is basically "pear-shaped and somewhat lumpy."

PROBLEMS

6.1 Find the gravitational attraction between two solid lead spheres of 1 kg mass each if the spheres are almost in contact. Express the answer as a fraction of the weight of either sphere. (The density of lead is 11.35 g/cm^3.)

6.2 Show that the gravitational force on a test particle inside a thin spherical shell is zero
    (a) By finding the force directly
    (b) By showing that the gravitational potential is constant

6.3 Assuming the earth to be a uniform solid sphere, show that if a straight hole were drilled from pole to pole, a particle dropped into the hole would execute simple harmonic motion. Show also that the period of this oscillation depends only on the density of the earth and is independent of the size. What is the period in hours? ($r_{earth} = 6.38 \times 10^6$ m.)

6.4 Show that the motion is simple harmonic with the same period as the previous problem for a particle sliding in a straight smooth tube passing obliquely through the earth. (Neglect any effects of rotation.)

6.5 Assuming a circular orbit, show that Kepler's third law follows directly from Newton's second law and his law of gravity: $GMm/r^2 = mv^2/r$.

6.6 (a) Show that the radius for a circular orbit of a synchronous (24-hr) earth satellite is about 7 earth radii.

(b) The distance to the moon is about 60 earth radii. From this calculate the length of the month (period of the moon's orbital revolution).

6.7 Show that the orbital period for an earth satellite in a circular orbit just above the earth's surface is the same as the period of oscillation of the particle dropped into a hole drilled through the earth (Problem 6.3).

6.8 If the solar system were embedded in a uniform dust cloud of density ρ, show that the law of force on a planet a distance r from the center of the sun would be given by

$$F(r) = -GMm/r^2 - (4/3)\pi\rho mGr$$

6.9 A particle moving in a central field describes the spiral orbit $r = r_0 e^{k\theta}$. Show that the force law is inverse-cube and that θ varies logarithmically with t.

6.10 A particle moves in an inverse-cube field of force. Show that, in addition to the exponential spiral orbit of Problem 6.9, there are two other possible types of orbit, and give their equations.

6.11 The orbit of a particle moving in a central field is a circle passing through the origin, namely $r = r_0 \cos \theta$. Show that the force law is inverse-fifth power.

6.12 A particle moves in a spiral orbit given by $r = a\theta$. If θ increases linearly with t, is the force a central field? If not, determine how θ would have to vary with t for a central force.

6.13 A rocket ship is initially going in a circular orbit close to the earth. It is desired to place the ship into a new orbit such that the apogee distance is equal to the radius of the moon's orbit around the earth.

(a) If a single rocket thrust is used to accomplish this, determine the ratio of the final and initial speeds. Assume that the radius of the original circular orbit is $\frac{1}{60}$ the distance to the moon.

(b) Find the new orbit if the speed ratio is 1% too great.

[This problem illustrates the extreme accuracy needed to achieve a circumlunar orbit.]

6.14 Compute the period of Halley's comet from the data given in Section 6.5. Find also the comet's speed at perihelion and aphelion.

6.15 A comet is first seen at a distance of d astronomical units from the sun and it is traveling with a speed of q times the earth's speed. Show that the orbit of the comet is hyperbolic, parabolic, or elliptic, depending on whether the quantity $q^2 d$ is greater than, equal to, or less than 2, respectively.

6.16 A particle moves in an elliptic orbit in an inverse-square force field. Prove that the product of the minimum and maximum speeds is equal to $(2\pi a/\tau)^2$ where a is the semimajor axis and τ is the periodic time.

6.17 Prove that the time average of the potential energy of a particle describing an elliptic orbit, in the inverse-square force field $f(r) = -k/r^2$, is $-k/a$ where a is the semimajor axis of the ellipse.

6.18 Find the apsidal angle for nearly circular orbits in a central field for which the law of force is

$$f(r) = -k\frac{e^{-br}}{r^2}$$

6.19 If the solar system were embedded in a uniform dust cloud (Problem 6.8), what would the apsidal angle of a planet be for motion in a nearly circular orbit? This was once suggested as a possible explanation for the advance of the perihelion of M.

6.20 Show that the stability condition for a circular orbit of radius a is equivalent to the condition that $d^2 U/dr^2 > 0$ for $r = a$ where $U(r)$ is the effective potential defined in Section 6.11.

6.21 Find the condition for which circular orbits are stable if the force function is of the form in Example 6.12, namely,

$$f(r) = -\frac{k}{r^2} - \frac{\epsilon}{r^4}$$

6.22 (a) Show that a circular orbit of radius r is stable in Problem 6.18 if r is less than b^{-1}.
(b) Show that circular orbits are unstable in an inverse-cube force field.

6.23 A comet is going in a parabolic orbit lying in the plane of the earth's orbit. Regarding the earth's orbit as circular of radius a, show that the points where the comet intersects the earth's orbit are given by

$$\cos\theta = -1 + \frac{2p}{a}$$

where p is the perihelion distance of the comet defined at $\theta = 0$.

6.24 Use the result of Problem 6.23 to show that the time interval that the comet remains inside the earth's orbit is the fraction

$$\frac{2^{1/2}}{3\pi}\left(\frac{2p}{a} + 1\right)\left(1 - \frac{p}{a}\right)^{1/2}$$

of a year and that the maximum value of this time interval is $2/3\pi$ year, or 77.5 days, corresponding to $p = a/2$. Compute the time interval for Halley's comet ($p = 0.6a$).

6.25 In advanced texts on potential theory, it is shown that the potential energy of a particle of mass m in the gravitational field of an oblate spheroid, like the earth, is approximately

$$V(r) = -\frac{k}{r}\left(1 + \frac{\epsilon}{r^2}\right)$$

where r refers to distances in the equatorial plane, $k = GMm$ as before, and $\epsilon = \frac{2}{5}R\Delta R$ in which R is the equatorial radius and ΔR is the difference between the equatorial and polar radii. From this, find the apsidal angle for a satellite moving in a nearly circular orbit in the equatorial plane of the earth where $R = 4000$ mi and $\Delta R = 13$ mi.

6.26 According to the special theory of relativity, a particle moving in a central field with potential energy $V(r)$ will describe the same orbit that a particle with a potential energy

$$V(r) - \frac{[E - V(r)]^2}{2m_0 c^2}$$

would describe according to nonrelativistic mechanics. Here E is the total energy, m_0 is the rest mass of the particle, and c is the speed of light. From this, find the apsidal angle for motion in an inverse-square force field, $V(r) = -k/r$.

6.27 A comet is observed to have a speed v_{com} when it is a distance r_{com} from the sun, and its direction of motion makes an angle ϕ with the radius vector from the sun. Show that the major axis of the elliptical orbit of the comet makes an angle θ with the initial radius vector of the comet given by

$$\theta = \cot^{-1}\left(\tan \phi - \frac{2}{V^2 R} \csc 2\phi\right)$$

where $V = v_{com}/v_{earth}$ and $R = r_{com}/a_e$ are dimensionless ratios as defined in Example 6.9. Apply the result to the numerical values of Example 6.9.

COMPUTER/CALCULATOR APPLICATIONS

6.1 In Example 6.7 we calculated the gravitational potential at a point P external to a ring of matter of mass M and radius R. P was in the same plane as the ring and a distance $r > R$ from its center. Assume now that the point P is at a distance $r < R$ from the center of the ring but still in the same plane.

(a) Show that the gravitational potential acting at the point r due to the ring of mass is given by

$$\Phi = -\frac{GM}{R}\left(1 + \frac{r^2}{4R^2} + \cdots\right)$$

Let r = radius of the Earth's orbit = 1.496×10^{11} m, R = radius of Jupiter's orbit = 7.784×10^{11} m, and M = mass of Jupiter = 1.90×10^{27} kg. Assume that the average gravitational potential produced by Jupiter on Earth is equivalent to that of a uniform ring of matter around the Sun whose mass is equal to that of Jupiter and whose radius from the Sun is equal to Jupiter's radius.

(b) Using this assumption and the values given above, calculate a numerical value for the average gravitational potential that Jupiter exerts on Earth.

(c) Assume that we can approximate this ring of mass by a sum of N strategically deployed discrete point masses M_i, such that $NM_i = M$. As a first approximation, let $N = 2$ and $M_i = M/2$. Deploy these two masses at radii = R and along a line directed between the center of the ring and the position of Earth at radius r. Calculate a numerical value for the potential at r due to these two masses. Repeat this calculation for the case $N = 4$, with the four masses being deployed at quadrants of the circle of radius R and two of them again lining up along the line connecting the center of the circle and Earth. Continue approximating the ring of matter in this fashion (successive multiplications of N by 2 and divisions of M_i by 2) and calculating the resultant potential at r. Stop the iteration when the calculated potential changes by no more than 1 part in 10^4 from the previous value. Compare your result with that obtained in part (b). How many individual masses were required to achieve this accuracy?

(d) Repeat (c) for values of r equal to 0, .2, .4, .6, .8 times r given above. Plot the absolute values of the difference $|\Phi(r) - \Phi(0)|$ versus r and show that this difference varies quadratically with r as predicted by the equation given in part (a).

7

DYNAMICS OF
SYSTEMS OF PARTICLES

*"Two equal bodies which are in direct impact with each other and have equal and
opposite velocities before impact, rebound with velocities that are, apart from the
sign, the same."*

*"The sum of the products of the magnitudes of each hard body, multiplied by the
square of the velocities, is always the same, before and after the collision."*

(Christiaan Huygens, memoir, De Motu Corporum ex mutuo impulsu Hypothesis,
composed in Paris, 5-Jan-1669, to Oldenburg, Secretary of the Royal Society)

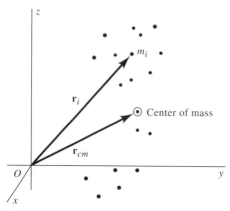

Figure 7.1   Center of mass of a system of particles.

7.1   INTRODUCTION. CENTER OF MASS AND LINEAR MOMENTUM OF A SYSTEM

We now expand our study of mechanics to systems of many particles (two or more). These particles may or may not move independently of one another. Special systems, called *rigid bodies,* in which the relative positions of all the particles are fixed will be taken up in the next two chapters. For the present, we shall develop some general theorems that apply to all systems. These will then be applied to some simple systems of free particles.

Our general system consists of n particles of masses $m_1, m_2, \ldots, m_n$ whose position vectors are, respectively, $\mathbf{r}_1, \mathbf{r}_2, \ldots, \mathbf{r}_n$. We define the *center of mass* of the system as the point whose position vector $\mathbf{r}_{cm}$ (Figure 7.1) is given by

$$\mathbf{r}_{cm} = \frac{m_1\mathbf{r}_1 + m_2\mathbf{r}_2 + \cdots + m_n\mathbf{r}_n}{m_1 + m_2 + \cdots + m_n} = \frac{\sum\limits_i \mathbf{m}_i\mathbf{r}_i}{m} \tag{7.1}$$

where $m = \Sigma\, m_i$ is the total mass of the system. The above definition is clearly equivalent to the three equations

$$x_{cm} = \frac{\sum\limits_i m_i x_i}{m} \qquad y_{cm} = \frac{\sum\limits_i m_i y_i}{m} \qquad z_{cm} = \frac{\sum\limits_i m_i z_i}{m}$$

We define the *linear momentum* $\mathbf{p}$ of the system as the vector sum of the linear momenta of the individual particles, namely,

$$\mathbf{p} = \sum_i \mathbf{p}_i = \sum_i m_i\mathbf{v}_i \tag{7.2}$$

Upon calculating $\dot{\mathbf{r}}_{cm} = \mathbf{v}_{cm}$ from Equation 7.1 and comparing with Equation 7.2, it readily follows that

$$\mathbf{p} = m\mathbf{v}_{cm} \tag{7.3}$$

that is, *the linear momentum of a system of particles is equal to the velocity of the center of mass multiplied by the total mass of the system.*

Suppose now that there are external forces $\mathbf{F}_1, \mathbf{F}_2, \ldots, \mathbf{F}_i, \ldots, \mathbf{F}_n$ acting on the respective particles. In addition, there may be internal forces of interaction between any two particles of the system. We shall denote these internal forces by $\mathbf{F}_{ij}$, meaning the force exerted on particle i by particle j, with the understanding that $\mathbf{F}_{ii} = 0$. The equation of motion of particle i is then

$$\mathbf{F}_i + \sum_{j=1}^{n} \mathbf{F}_{ij} = m_i \ddot{\mathbf{r}}_i = \dot{\mathbf{p}}_i \tag{7.4}$$

where $\mathbf{F}_i$ means the total external force acting on particle i. The second term in the above equation represents the vector sum of all the internal forces exerted on particle i by all other particles of the system. Adding Equation 7.4 for the n particles, we have

$$\sum_{i=1}^{n} \mathbf{F}_i + \sum_{i=1}^{n}\sum_{j=1}^{n} \mathbf{F}_{ij} = \sum_{i=1}^{n} \dot{\mathbf{p}}_i \tag{7.5}$$

In the double summation above, for every force $\mathbf{F}_{ij}$ there is also a force $\mathbf{F}_{ji}$, and these two forces are equal and opposite

$$\mathbf{F}_{ij} = -\mathbf{F}_{ji} \tag{7.6}$$

from the law of action and reaction, Newton's third law. Consequently, the internal forces cancel in pairs, and the double sum vanishes. We can, therefore, write Equation 7.5 in the following way:

$$\sum_i \mathbf{F}_i = \dot{\mathbf{p}} = m\mathbf{a}_{cm} \tag{7.7}$$

In words: *The acceleration of the center of mass of a system of particles is the same as that of a single particle having a mass equal to the total mass of the system and acted upon by the sum of the external forces.*

Consider, for example, a swarm of particles moving in a *uniform* gravitational field. Then, since $\mathbf{F}_i = m_i\mathbf{g}$ for each particle,

$$\sum_i \mathbf{F}_i = \Sigma m_i\mathbf{g} = m\mathbf{g}$$

The last step follows from the fact that $\mathbf{g}$ is constant. Hence,

$$\mathbf{a}_{cm} = \mathbf{g} \tag{7.8}$$

This is the same as the equation for a single particle or projectile. Thus, the center of mass of the shrapnel from an artillery shell that has burst in midair will follow the same parabolic path that the shell would have taken had it not burst (until any of the pieces strikes something).

In the special case in which there are no external forces acting on a system (or if $\Sigma\mathbf{F}_i = 0$), then $\mathbf{a}_{cm} = 0$ and $\mathbf{v}_{cm} =$ constant. Thus, the linear momentum of the system remains constant:

$$\sum_i \mathbf{p}_i = \mathbf{p} = m\mathbf{v}_{cm} = \text{constant} \tag{7.9}$$

This is the *principle of conservation of linear momentum.* In Newtonian mechanics the constancy of the linear momentum of an isolated system is directly related to, and is in fact a consequence of, the third law. But even in those cases in which the forces between particles do not directly obey the law of action and reaction, such as the magnetic forces between moving charges, the principle of conservation of linear momentum still holds when due account is taken of the total linear momentum of the particles and the electromagnetic field.[1]

EXAMPLE 7.1

At some point in its trajectory a ballistic missile (ICBM) of mass m breaks into three fragments of mass $m/3$ each. One of the fragments continues on with an initial velocity of one half the velocity $\mathbf{v}_0$ of the missile just before breakup. The other two pieces go off at right angles to each other with equal speeds. Find the initial speeds of the latter two fragments in terms of v_0.

Solution:
At the point of breakup, conservation of linear momentum is expressed as

$$m\mathbf{v}_{cm} = m\mathbf{v}_0 = \frac{m}{3}\mathbf{v}_1 + \frac{m}{3}\mathbf{v}_2 + \frac{m}{3}\mathbf{v}_3$$

The given conditions are: $\mathbf{v}_1 = \mathbf{v}_0/2$, $\mathbf{v}_2 \cdot \mathbf{v}_3 = 0$, and $v_2 = v_3$. From the first we get, upon cancellation of the m's, $3\mathbf{v}_0 = (\mathbf{v}_0/2) + \mathbf{v}_2 + \mathbf{v}_3$, or

$$\frac{5}{2}\mathbf{v}_0 = \mathbf{v}_2 + \mathbf{v}_3$$

Taking the dot product of each side with itself, we have

$$\frac{25}{4}v_0^2 = (\mathbf{v}_2 + \mathbf{v}_3) \cdot (\mathbf{v}_2 + \mathbf{v}_3) = v_2^2 + 2\mathbf{v}_2 \cdot \mathbf{v}_3 + v_3^2 = 2v_2^2$$

Therefore

$$v_2 = v_3 = \frac{5}{2\sqrt{2}}v_0 = 1.77v_0 \qquad\blacksquare$$

7.2   ANGULAR MOMENTUM AND KINETIC ENERGY OF A SYSTEM

We previously stated that the angular momentum of a single particle is defined as the cross product $\mathbf{r} \times m\mathbf{v}$. The angular momentum $\mathbf{L}$ of a system of particles is defined accordingly, as the vector sum of the individual angular momenta, namely,

---

[1] See, for example, P. M. Fishbane, S. Gasiorowicz, S. T. Thornton, *Physics for Scientists and Engineers.* Prentice-Hall, Englewood Cliffs, NJ, 1993.

$$L = \sum_{i=1}^{n} (\mathbf{r}_i \times m_i \mathbf{v}_i)$$

Let us calculate the time derivative of the angular momentum. Using the rule for differentiating the cross product, we find

$$\frac{d\mathbf{L}}{dt} = \sum_{i=1}^{n} (\mathbf{v}_i \times m_i \mathbf{v}_i) + \sum_{i=1}^{n} (\mathbf{r}_i \times m_i \mathbf{a}_i) \tag{7.10}$$

Now the first term on the right vanishes, because $\mathbf{v}_i \times \mathbf{v}_i = 0$ and, since $m_i \mathbf{a}_i$ is equal to the total force acting on particle i, we can write

$$\frac{d\mathbf{L}}{dt} = \sum_{i=1}^{n} \left[\mathbf{r}_i \times \left(\mathbf{F}_i + \sum_{j=1}^{n} \mathbf{F}_{ij} \right) \right] \tag{7.11}$$

$$= \sum_{i=1}^{n} \mathbf{r}_i \times \mathbf{F}_i + \sum_{i=1}^{n} \sum_{j=1}^{n} \mathbf{r}_i \times \mathbf{F}_{ij}$$

where, as in Section 7.1, $\mathbf{F}_i$ denotes the total external force on particle i, and $\mathbf{F}_{ij}$ denotes the (internal) force exerted on particle i by any other particle j. Now the double summation on the right consists of pairs of terms of the form

$$(\mathbf{r}_i \times \mathbf{F}_{ij}) + (\mathbf{r}_j \times \mathbf{F}_{ji}) \tag{7.12}$$

Denoting the vector displacement of particle j relative to particle i by $\mathbf{r}_{ij}$, we see from the triangle shown in Figure 7.2 that

$$\mathbf{r}_{ij} = \mathbf{r}_j - \mathbf{r}_i \tag{7.13}$$

Therefore, since $\mathbf{F}_{ji} = -\mathbf{F}_{ij}$, expression 7.12 reduces to

$$-\mathbf{r}_{ij} \times \mathbf{F}_{ij}$$

which clearly vanishes if the internal forces are central, that is, if they act along the lines connecting pairs of particles. Hence, the double sum in Equation 7.11 vanishes. Now

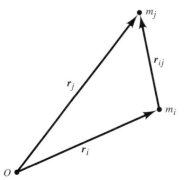

Figure 7.2 Definition of the vector $\mathbf{r}_{ij}$.

the cross product $\mathbf{r}_i \times \mathbf{F}_i$ is the moment of the external force $\mathbf{F}_i$. The sum $\Sigma \mathbf{r}_i \times \mathbf{F}_i$ is therefore the total moment of all the external forces acting on the system. If we denote the total external torque, or moment of force, by $\mathbf{N}$, Equation 7.11 takes the form

$$\frac{d\mathbf{L}}{dt} = \mathbf{N} \tag{7.14}$$

That is, *the time rate of change of the angular momentum of a system is equal to the total moment of all the external forces acting on the system.*

   If a system is isolated, then $\mathbf{N} = 0$, and the angular momentum remains constant in both magnitude and direction:

$$\mathbf{L} = \sum_i \mathbf{r}_i \times m_i \mathbf{v}_i = \text{constant vector} \tag{7.15}$$

This is a statement of the *principle of conservation of angular momentum.* It is a generalization for a single particle in a central field. Like the constancy of linear momentum discussed in the preceding section, the angular momentum of an isolated system is also constant in the case of a system of moving charges when the angular momentum of the electromagnetic field is considered.[2]

   It is sometimes convenient to express the angular momentum in terms of the motion of the center of mass. As shown in Figure 7.3, we can express each position vector $\mathbf{r}_i$ in the form

$$\mathbf{r}_i = \mathbf{r}_{cm} + \bar{\mathbf{r}}_i \tag{7.16}$$

where $\bar{\mathbf{r}}_i$ is the position of particle i relative to the center of mass. Taking the derivative with respect to t, we have

$$\mathbf{v}_i = \mathbf{v}_{cm} + \bar{\mathbf{v}}_i \tag{7.17}$$

---

[2] See footnote 1.

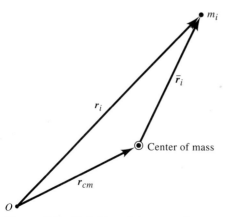

Figure 7.3   Definition of the vector $\bar{\mathbf{r}}_i$.

Here $\mathbf{v}_{cm}$ is the velocity of the center of mass and $\bar{\mathbf{v}}_i$ is the velocity of particle i relative to the center of mass. The expression for $\mathbf{L}$ can, therefore, be written

$$\mathbf{L} = \sum_i (\mathbf{r}_{cm} + \bar{\mathbf{r}}_i) \times m_i(\mathbf{v}_{cm} + \bar{\mathbf{v}}_i)$$

$$= \sum_i (\mathbf{r}_{cm} \times m_i\mathbf{v}_{cm}) + \sum_i (\mathbf{r}_{cm} \times m_i\bar{\mathbf{v}}_i)$$

$$+ \sum_i (\bar{\mathbf{r}}_i \times m_i\mathbf{v}_{cm}) + \sum_i (\bar{\mathbf{r}}_i \times m_i\bar{\mathbf{v}}_i)$$

$$= \mathbf{r}_{cm} \times \left(\sum_i m_i\right)\mathbf{v}_{cm} + \mathbf{r}_{cm} \times \sum_i m_i\bar{\mathbf{v}}_i$$

$$+ \left(\sum_i m_i\bar{\mathbf{r}}_i\right) \times \mathbf{v}_{cm} + \sum_i (\bar{\mathbf{r}}_i \times m_i\bar{\mathbf{v}}_i)$$

Now, from Equation 7.16, we have

$$\sum_i m_i\bar{\mathbf{r}}_i = \sum_i m_i(\mathbf{r}_i - \mathbf{r}_{cm}) = \sum_i m_i\mathbf{r}_i - m\mathbf{r}_{cm} = 0$$

Similarly, we obtain

$$\sum_i m_i\bar{\mathbf{v}}_i = \sum_i m_i\mathbf{v}_i - m\mathbf{v}_{cm} = 0$$

by differentiation with respect to t. (These two equations merely state that the position and velocity of the center of mass, relative to the center of mass, are both zero.) Consequently, the second and third summations in the expansion of $\mathbf{L}$ vanish, and we can write

$$\mathbf{L} = \mathbf{r}_{cm} \times m\mathbf{v}_{cm} + \sum_i \bar{\mathbf{r}}_i \times m_i\bar{\mathbf{v}}_i \tag{7.18}$$

expressing the angular momentum of a system in terms of an "orbital" part (motion of the center of mass) and a "spin" part (motion about the center of mass).

EXAMPLE 7.2

A long, thin rod of length l and mass m hangs from a pivot point about which it is free to swing in a vertical plane like a simple pendulum. Calculate the total angular momentum of the rod as a function of its instantaneous angular velocity ω. Show that the theorem represented by Equation 7.18 is true by comparing the angular momentum obtained using that theorem to that obtained by direct calculation.

Solution:
The rod is shown in Figure 7.4(a). First we calculate the angular momentum $\mathbf{L}_{cm}$ of the center of mass of the rod about the pivot point. Since the velocity $\mathbf{v}_{cm}$ of the center of mass is always perpendicular to the radius vector $\mathbf{r}$ denoting its location relative to the pivot point, the sine of the angle between those two vectors is unity. Thus, the magnitude of $\mathbf{L}_{cm}$ is given by

$$L_{cm} = \frac{l}{2}p_{cm} = m\frac{l}{2}v_{cm} = m\frac{l}{2}\left(\frac{l}{2}\omega\right) = \frac{l}{4}ml^2\omega$$

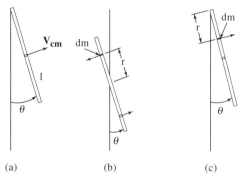

(a)   (b)   (c)

Figure 7.4   Rod of mass m and length l free to
swing in a vertical plane about a fixed pivot.

Figure 7.4(b) depicts the motion of the rod as seen from the perspective of its center
of mass. The angular momentum dL_{rel} of two small mass elements, each of size dm
symmetrically disposed about the center of mass of the rod, is given by

$$dL_{rel} = 2r\,dp = 2rv\,dm = 2r(r\omega)\lambda\,dr$$

where λ is the mass per unit length of the rod. The total relative angular momentum
is obtained by integrating this expression from $r = 0$ to $r = l/2$.

$$L_{rel} = 2\lambda\omega\int_0^{1/2} r^2 dr = \frac{1}{12}(\lambda l)l^2\omega = \left(\frac{1}{12}ml^2\right)\omega$$

We can see in the equation above that the angular momentum of the rod about its
center of mass is directly proportional to the angular velocity ω of the rod. The
constant of proportionality $ml^2/12$ is called the *moment of inertia* I_{cm} of the rod
about its center of mass. Moment of inertia plays a role in rotational motion similar
to that of inertial mass in translational motion as we shall see in the next chapter.

Finally, the total angular momentum of the rod is

$$L_{tot} = L_{cm} + L_{rel} = \frac{1}{3}ml^2\omega$$

Again, note that the total angular momentum of the rod is directly proportional to
the angular velocity of the rod. Here, though, the constant of proportionality is the
moment of inertia of the rod about the pivot point at the end of the rod. This moment
of inertia is larger than that about the center of mass. The reason is that more of the
mass of the rod is distributed farther away from its end than from its center, thus
making it more difficult to rotate a rod about an end.

Finally, The total angular momentum can also be obtained by integrating down the rod,
starting from the pivot point, to obtain the contribution from each mass element dm,
as shown in Figure 7.4(c).

$$dL_{tot} = rdp = r(v\,dm) = r(r\omega)\lambda\,dr$$

$$L_{tot} = \lambda\omega \int_0^1 r^2 dr = \frac{1}{3}ml^2\omega$$

And, indeed, the two methods yield the same result. ∎

Kinetic Energy of a System

The total kinetic energy T of a system of particles is given by the sum of the individual energies, namely,

$$T = \sum_i \frac{1}{2}m_i v_i^2 = \sum_i \frac{1}{2}m_i(\mathbf{v}_i \cdot \mathbf{v}_i) \tag{7.19}$$

As before, we can express the velocities relative to the mass center giving

$$T = \sum_i \frac{1}{2}m_i(\mathbf{v}_{cm} + \bar{\mathbf{v}}_i) \cdot (\mathbf{v}_{cm} + \bar{\mathbf{v}}_i)$$

$$= \sum_i \frac{1}{2}m_i v_{cm}^2 + \sum_i m_i(\mathbf{v}_{cm} \cdot \bar{\mathbf{v}}_i) + \sum_i \frac{1}{2}m_i \bar{v}_i^2$$

$$= \frac{1}{2}v_{cm}^2 \sum_i m_i + \mathbf{v}_{cm} \cdot \sum_i m_i \bar{\mathbf{v}}_i + \sum_i \frac{1}{2}m_i \bar{v}_i^2$$

Since the second summation $\sum_i m_i \bar{\mathbf{v}}_i$ vanishes, we can express the kinetic energy as follows:

$$T = \frac{1}{2}mv_{cm}^2 + \sum_i \frac{1}{2}m_i \bar{v}_i^2 \tag{7.20}$$

The first term is the kinetic energy of translation of the whole system, and the second is the kinetic energy of motion relative to the mass center.

The separation of angular momentum and kinetic energy into a center-of-mass part and a relative-to-center-of-mass part finds important applications in atomic and molecular physics and also in astrophysics. We shall find the above two theorems useful in the study of rigid bodies in the following chapters.

EXAMPLE 7.3

Calculate the total kinetic energy of the rod of Example 7.2. Use the theorem represented by Equation 7.20. As in example 7.2, show that the total energy obtained for the rod via this theorem is equivalent to that obtained by direct calculation.

Solution:
The translational kinetic energy of the center of mass of the rod is

$$T_{cm} = \frac{1}{2}m\mathbf{v}_{cm} \cdot \mathbf{v}_{cm} = \frac{1}{2}m\left(\frac{l}{2}\omega\right)^2 = \frac{1}{8}ml^2\omega^2$$

The kinetic energy of two equal mass elements dm symetrically disposed about the center of mass is

$$dT_{rel} = \frac{1}{2}(2dm)\,\mathbf{v}\cdot\mathbf{v} = \lambda dr(r\omega)^2 = \lambda\omega^2 r^2 dr$$

where λ, again, is the mass per unit length of the rod. The total energy relative to the center of mass can be obtained by integrating the above expression from $r = 0$ to $r = l/2$.

$$T_{rel} = \lambda\omega^2 \int_0^{l/2} r^2 dr = \frac{1}{24}\lambda\omega^2 l^3 = \frac{1}{2}\left(\frac{1}{12}ml^2\right)\omega^2 = \frac{1}{2}I_{cm}\omega^2$$

Note: As in Example 7.2, the moment of inertia term I_{cm} appears as the constant of proportionality to ω^2 in the above expression for the rotational kinetic energy of the rod about its center of mass. Again, the moment of inertia term that occurs in the expression for rotational kinetic energy can be seen to be completely analogous to the inertial mass term in an expression for the translational kinetic energy of a particle.

The total kinetic energy of the rod is then

$$T = T_{cm} + T_{rel} = \frac{1}{8}ml^2\omega^2 + \frac{1}{24}ml^2\omega^2 = \frac{1}{2}\left(\frac{1}{3}ml^2\right)\omega^2 = \frac{1}{2}I\omega^2$$

where we have expressed the final result in terms of the total moment of inertia of the rod about its endpoint, exactly as in Example 7.2.

We will leave it as an exercise for the reader to calculate the kinetic energy directly and show that it is equal to the value obtained above. The calculation proceeds in a fashion completely analogous to that in Example 7.2. ∎

7.3   MOTION OF TWO INTERACTING BODIES. THE REDUCED MASS

Let us consider the motion of a system consisting of two bodies, treated here as particles, that interact with one another by a central force. We shall assume the system is isolated, and hence the center of mass moves with constant velocity. For simplicity, we shall take the center of mass as the origin. We have then

$$m_1\bar{\mathbf{r}}_1 + m_2\bar{\mathbf{r}}_2 = 0 \tag{7.21}$$

where, as shown in Figure 7.5, the vectors $\bar{\mathbf{r}}_1$ and $\bar{\mathbf{r}}_2$ represent the positions of the particles m_1 and m_2, respectively, relative to the center of mass. Now, if $\mathbf{R}$ is the position vector of particle 1 relative to particle 2, then

$$\mathbf{R} = \bar{\mathbf{r}}_1 - \bar{\mathbf{r}}_2 = \bar{\mathbf{r}}_1\left(1 + \frac{m_1}{m_2}\right) \tag{7.22}$$

The last step follows from Equation 7.21.

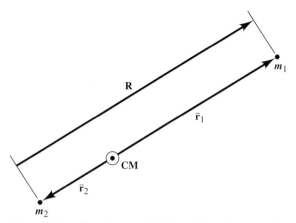

Figure 7.5 Showing the definition of the relative position vector for the two-body problem.

The differential equation of motion of particle 1 relative to the center of mass is

$$m_1\frac{d^2\bar{\mathbf{r}}_1}{dt^2} = \mathbf{F}_1 = f(R)\frac{\mathbf{R}}{R} \tag{7.23}$$

in which $|f(R)|$ is the magnitude of the mutual force between the two particles. By using Equation 7.22, we can write

$$\mu\frac{d^2\mathbf{R}}{dt^2} = f(R)\frac{\mathbf{R}}{R} \tag{7.24}$$

where

$$\mu = \frac{m_1m_2}{m_1 + m_2} \tag{7.25}$$

The quantity μ is called the *reduced mass*. The new equation of motion (Equation 7.24) gives the motion of particle 1 relative to particle 2, and an exactly similar equation gives the motion of particle 2 relative to particle 1. This equation is precisely the same as the ordinary equation of motion of a single particle of mass μ moving in a central field of force given by $f(R)$. Thus, the fact that both particles are moving relative to the center of mass is automatically accounted for by replacing m_1 by the reduced mass μ. If the bodies are of equal mass m, then $\mu = m/2$. On the other hand, if m_2 is very much greater than m_1, so that m_1/m_2 is very small, then μ is nearly equal to m_1.

For two bodies attracting one another by gravitation

$$f(R) = -\frac{Gm_1m_2}{R^2} \tag{7.26}$$

In this case the equation of motion is

$$\mu\ddot{\mathbf{R}} = -\frac{Gm_1m_2}{R^2}\mathbf{e}_R \tag{7.27}$$

or, equivalently,

$$m_1\ddot{\mathbf{R}} = -\frac{G(m_1 + m_2)m_1}{R^2}\mathbf{e}_R \tag{7.28}$$

where $\mathbf{e}_R = \mathbf{R}/R$ is a unit vector in the direction of $\mathbf{R}$.

In Section 6.11 we derived an equation giving the periodic time of orbital motion of a planet of mass m moving in the sun's gravitational field, namely, $\tau = 2\pi(GM)^{-1/2}a^{3/2}$ where M is the sun's mass and a is the semimajor axis of the elliptical orbit of the planet about the sun. In this derivation it was assumed that the sun was stationary with the origin of our coordinate system at the center of the sun. To account for the sun's motion about the common center of mass, the correct equation is Equation 7.28 in which $m = m_1$ and $M = m_2$. The force constant k, which was taken to be GMm in the earlier treatment, should, therefore, be replaced by $G(M + m)m$ so that the correct equation for the period is

$$\tau = 2\pi[G(M + m)]^{-1/2}a^{3/2} \tag{7.29}$$

or, for any two-body system held together by gravity, the orbital period is

$$\tau = 2\pi[G(m_1 + m_2)]^{-1/2}a^{3/2} \tag{7.29a}$$

If m_1 and m_2 are expressed in units of the sun's mass and a is in astronomical units (the mean distance from the earth to the sun), then the orbital period in years is given by

$$\tau = (m_1 + m_2)^{-1/2}a^{3/2} \tag{7.29b}$$

For most planets in our solar system, the added mass term in the above expression for the period makes very little difference—the earth's mass is only 1/330,000 the sun's mass. The most massive planet, Jupiter, has a mass of about 1/1000 the mass of the sun, so the effect of the reduced-mass formula is to change the earlier calculation in the ratio $(1.001)^{-1/2} = 0.9995$ for the period of Jupiter's revolution about the sun.

Binary Stars. White Dwarfs and Black Holes

It is known that about half of all the stars in the galaxy in the vicinity of the sun are binary or double; that is, they occur in pairs held together by their mutual gravitational attraction, with each member of the pair revolving about their common center of mass. From the above analysis we can infer that either member of a binary system revolves about the other in an elliptical orbit for which the orbiting period is given by Equations 7.29a and 7.29b, where a is the semimajor axis of the ellipse and m_1 and m_2 are the masses of the two stars. Values of a for known binary systems range from the very least (*contact binaries* in which the stars touch each other) to values so large that the period is measured in millions of years. A typical example is the brightest star in the night sky, Sirius, which consists of a very luminous star with a mass of about three solar masses and a very small dim star, called a *white dwarf*, which can only be seen in large telescopes. The mass of this small companion is about the same as the sun's mass, but its size is roughly that of a large planet, so its density is extremely large (30,000 times the density of water). The value of a for the Sirius system is approximately 20 AU (about the distance from the sun to the planet Uranus) and the period, as calculated from Equation 7.29b, should be

$$\tau = (3 + 1)^{-1/2}(20)^{3/2} \text{ yr} = 44.7 \text{ yr}$$

The observed period is about 49 yr.

A binary system that has received considerable attention in recent years is the one known as Cygnus X-1. One component of this system is a massive luminous star of about ten solar masses. The other component has about the same mass as the visible member, as determined by the latter's motion, but is invisible in even the largest telescopes. Many astronomers and astrophysicists now believe that this invisible component is a *black hole,* an object that is so massive and dense that nothing, not even light, can escape its gravitational field. Black holes are predicted mathematically by the general theory of relativity,[3] and convincing proof that they actually exist would constitute a milestone in the science of astrophysics.

EXAMPLE 7.4

A certain binary star system is observed to be both eclipsing and spectroscopic. This means that the system is being seen from earth with its orbital plane edge-on and that the orbital velocities v_1 and v_2 of the two stars comprising the system can be determined from Doppler shift measurements of observed spectral lines. You don't need to understand the details of this last statement. The important point is that we know the orbital velocities. They are, in appropriate units, $v_1 = 1.257$ AU/yr and $v_2 = 5.027$ AU/yr. The period of revolution of each star about its center of mass is $\tau = 5$ yr. (That can easily be ascertained from observed frequency of eclipses.) Calculate the mass (in solar mass units $m_\odot$) of each star. Assume circular orbits.

Solution:

The radius of the orbit of each star about their common center of mass can be calculated from its velocity and period

$$r_1 = \frac{1}{2\pi} v_1 \tau = 1 \text{ } AU \qquad r_2 = \frac{1}{2\pi} v_2 \tau = 4 \text{ } AU$$

Thus, the semimajor axis a of the orbit is

$$a = r_1 + r_2 = 5 \text{ AU}$$

---

[3] According to the theory, a spherical body of mass m becomes a black hole if it is compressed to a radius r_B, known as the *Schwarzschild radius,* where

$$r_B = \frac{2Gm}{c^2}$$

in which c is the speed of light. The earth would become a black hole if it were compressed to the size of a small marble. For the sun r_B is about 3 km, which is much smaller than the white dwarf companion of Sirius.

The sum of the masses can be obtained from Equation 7.29b

$$m_1 + m_2 = \frac{a^3}{\tau^2} = 5\, m_\odot$$

The ratio of the two masses can be determined by differentiating Equation 7.21

$$m_1 \mathbf{v}_1 + m_2 \mathbf{v}_2 = 0 \qquad \frac{m_2}{m_1} = \left| \frac{\mathbf{v}_1}{\mathbf{v}_2} \right| = \frac{1}{4}$$

Combining these last two expressions yields the values for each mass, $m_1 = 4\, m_\odot$ and $m_2 = 1\, m_\odot$. ∎

7.4 COLLISIONS

Whenever two bodies undergo a collision, the force that either exerts on the other during the contact is an internal force, if the bodies are regarded together as a single system. The total linear momentum is unchanged. We can, therefore, write

$$\mathbf{p}_1 + \mathbf{p}_2 = \mathbf{p}_1' + \mathbf{p}_2' \tag{7.30}$$

or, equivalently,

$$m_1 \mathbf{v}_1 + m_2 \mathbf{v}_2 = m_1 \mathbf{v}_1' + m_2 \mathbf{v}_2' \tag{7.30a}$$

The subscripts 1 and 2 refer to the two bodies, and the primes indicate the respective momenta and velocities after the collision. The above equations are quite general. They apply to any two bodies regardless of their shapes, rigidity, and so on.

With regard to the energy balance, we can write

$$\frac{p_1^2}{2m_1} + \frac{p_2^2}{2m_2} = \frac{p_1'^2}{2m_1} + \frac{p_2'^2}{2m_2} + Q \tag{7.31}$$

or

$$\frac{1}{2} m_1 v_1^2 + \frac{1}{2} m_2 v_2^2 = \frac{1}{2} m_1 v_1'^2 + \frac{1}{2} m_2 v_2'^2 + Q \tag{7.31a}$$

Here the quantity Q is introduced to indicate the net loss or gain in kinetic energy that occurs as a result of the collision.

In the case of an *elastic* collision, there is no change in the total kinetic energy, so that $Q = 0$. If there is an energy loss, then Q is positive. This is called an *exoergic* collision. It may happen that there is an energy gain. This would occur, for example, if an explosive was present on one of the bodies at the point of contact. In this case Q is negative, and the collision is called *endoergic*.

The study of collisions is of particular importance in atomic and nuclear physics. Here the bodies involved may be atoms, nuclei, or various elementary particles, such as electrons, protons, and so on.

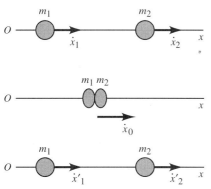

Figure 7.6 Head-on collision of two particles.

Direct Collisions

Let us consider the special case of a head-on collision of two bodies, or particles, in which the motion takes place entirely on a single straight line, the x-axis, as shown in Figure 7.6. In this case the momentum balance equation (Equation 7.30a) can be written

$$m_1 \dot{x}_1 + m_2 \dot{x}_2 = m_1 \dot{x}_1' + m_2 \dot{x}_2' \tag{7.32}$$

The direction along the line of motion is given by the signs of the $\dot{x}$'s.

In order to compute the values of the velocities after the collision, given the values before the collision, we can use the above momentum equation together with the energy balance equation (Equation 7.31a), if we know the value of Q. It is often convenient in this kind of problem to introduce another parameter ϵ called the *coefficient of restitution*. This quantity is defined as the ratio of the speed of separation v' to the speed of approach v. In our notation ϵ may be written as

$$\epsilon = \frac{|\dot{x}_2' - \dot{x}_1'|}{|\dot{x}_2 - \dot{x}_1|} = \frac{v'}{v} \tag{7.33}$$

The numerical value of ϵ depends primarily on the composition and physical makeup of the two bodies. It is easy to verify that in an elastic collision the value of $\epsilon = 1$. To do this, we set $Q = 0$ in Equation 7.31a and solve it together with Equation 7.32 for the final velocities. The steps are left as an exercise.

In the case of a *totally inelastic* collision, the two bodies stick together after colliding, so that $\epsilon = 0$. For most real bodies ϵ has a value somewhere between the two extremes of 0 and 1. For ivory billiard balls it is about 0.95. The value of the coefficient of restitution may also depend on the speed of approach. This is particularly evident in the case of a silicone compound known as Silly Putty. A ball of this material bounces when it strikes a hard surface at high speed, but at low speeds it acts like ordinary putty.

We can calculate the values of the final velocities from Equation 7.32 together with the definition of the coefficient of restitution (Equation 7.33). The result is

$$\dot{x}'_1 = \frac{(m_1 - \epsilon m_2)\dot{x}_1 + (m_2 + \epsilon m_2)\dot{x}_2}{m_1 + m_2}$$

$$\dot{x}'_2 = \frac{(m_1 + \epsilon m_1)\dot{x}_1 + (m_2 - \epsilon m_1)\dot{x}_2}{m_1 + m_2}$$

(7.34)

Taking the totally inelastic case by setting $\epsilon = 0$, we find, as we should, that $\dot{x}'_1 = \dot{x}'_2$; that is, there is no rebound. On the other hand, in the special case that the bodies are of equal mass $m_1 = m_2$, and are perfectly elastic $\epsilon = 1$, we obtain

$$\dot{x}'_1 = \dot{x}_2$$

$$\dot{x}'_2 = \dot{x}_1$$

The two bodies, therefore, just exchange their velocities as a result of the collision.

In the general case of a direct nonelastic collision, it is easily verified that the energy loss Q is related to the coefficient of restitution by the equation

$$Q = \frac{1}{2}\mu v^2(1 - \epsilon^2)$$

in which $\mu = m_1 m_2/(m_1 + m_2)$ is the reduced mass, and $v = |\dot{x}_2 - \dot{x}_1|$ is the relative speed before impact. The derivation is left as an exercise.

Impulse in Collisions

Forces of extremely short duration in time, such as those exerted by bodies undergoing collisions, are called *impulsive forces*. If we confine our attention to one body, or particle, the differential equation of motion is $d(m\mathbf{v})/dt = \mathbf{F}$, or in differential form $d(m\mathbf{v}) = \mathbf{F}\,dt$. Let us take the time integral over the interval $t = t_1$ to $t = t_2$. This is the time during which the force is considered to act. Then we have

$$\Delta(m\mathbf{v}) = \int_{t_1}^{t_2} \mathbf{F}\,dt$$

(7.35)

The time integral of the force is the impulse. It is customarily denoted by the symbol $\hat{\mathbf{P}}$. Equation 7.35 is, accordingly, expressed as

$$\Delta(m\mathbf{v}) = \hat{\mathbf{P}}$$

(7.35a)

We can think of an *ideal impulse* as produced by a force that tends to infinity but lasts for a time interval that approaches zero in such a way that the integral $\int\mathbf{F}\,dt$ remains finite. Such an ideal impulse would produce an instantaneous change in the momentum and velocity of a body without producing any displacement.

Relationship Between Impulse and Coefficient of Restitution in Direct Collisions

Let us apply the concept of impulse to the case of the direct collision of two bodies. We shall divide the impulse into two parts, namely, the impulse of compression $\hat{\mathbf{P}}_c$ and the impulse of restitution $\hat{\mathbf{P}}_r$. We are concerned only with direct collisions along the x-axis, as before. Therefore, for the compression we can write

$$m_1 \dot{x}_0 - m_1 \dot{x}_1 = \hat{P}_c \tag{7.36}$$

$$m_2 \dot{x}_0 - m_2 \dot{x}_2 = -\hat{P}_c$$

where $\dot{x}_0$ is the common velocity of both particles at the instant their relative speed is zero. Similarly, for the restitution, we have

$$m_1 \dot{x}_1' - m_1 \dot{x}_0 = \hat{P}_r \tag{7.37}$$

$$m_2 \dot{x}_2' - m_2 \dot{x}_0 = -\hat{P}_r$$

Upon eliminating $\dot{x}_0$ from Equations 7.36 and also from Equations 7.37, we obtain the following pair of equations:

$$m_1 m_2 (\dot{x}_2 - \dot{x}_1) = \hat{P}_c (m_1 + m_2) \tag{7.38}$$

$$m_1 m_2 (\dot{x}_1' - \dot{x}_2') = \hat{P}_r (m_1 + m_2)$$

Division of the second equation by the first yields the relation

$$\frac{|\dot{x}_2' - \dot{x}_1'|}{|\dot{x}_1 - \dot{x}_2|} = \frac{\hat{P}_r}{\hat{P}_c} \tag{7.39}$$

But the left-hand side is just the definition of the coefficient of restitution ϵ. Hence, we have

$$\epsilon = \frac{\hat{P}_r}{\hat{P}_c} \tag{7.40}$$

The coefficient of restitution is thus equal to the ratio of the impulse of restitution to the impulse of compression.

EXAMPLE 7.5 *Determining the Speed of a Bullet*

A gun is fired horizontally point-blank at a block of wood, which is initially at rest on a horizontal floor. The bullet becomes imbedded in the block, and the impact causes the system to slide a certain distance s before coming to rest. Given the mass of the bullet m, the mass of the block M, and the coefficient of sliding friction between the block and the floor μ_k, find the initial speed (muzzle velocity) of the bullet.

Solution:
First, from conservation of linear momentum, we can write

$$m \dot{x}_0 = (M + m) \dot{x}_0'$$

where $\dot{x}_0$ is the initial velocity of the bullet and $\dot{x}_0'$ is the velocity of the system (block + bullet) immediately after impact. (Note that the coefficient of restitution ϵ

is zero in this case.) Second, we know that the magnitude of the retarding frictional force is equal to $(M + m)\mu_k g = (M + m)a$ where $a = -\ddot{x}$ is the deceleration of the system after impact, so $a = \mu_k g$. Now, from Chapter 2 we recall that $s = v_0^2/2a$ for the case of uniform acceleration in one dimension. Thus, in our problem

$$s = \frac{\dot{x}_0'^2}{2\mu_k g} = \left(\frac{m\dot{x}_0}{M + m}\right)^2 \left(\frac{1}{2\mu_k g}\right)$$

Solving for $\dot{x}_0$ we obtain

$$\dot{x}_0 = \left(\frac{M + m}{m}\right)(2\mu_k g s)^{1/2}$$

for the initial velocity of the bullet in terms of the given quantities.

As a numerical example, let the mass of the block be 4 kg, and that of the bullet 10 g = 0.01 kg (about that of a .38 calibre slug). For the coefficient of friction (wood-on-wood) let us take $\mu_k = 0.4$. If the block slides a distance of 15 cm = 0.15 m, then we find

$$\dot{x}_0 = \frac{4.01}{0.01} (2 \times 0.4 \times 9.8 \text{ ms}^{-2} \times 0.15 \text{ m})^{1/2} = 435 \text{ m/s} \quad \blacksquare$$

7.5   OBLIQUE COLLISIONS AND SCATTERING. COMPARISON OF LABORATORY AND CENTER-OF-MASS COORDINATES

We now turn our attention to the more general case of collisions in which the motion is not confined to a single straight line. Here the vectorial form of the momentum equations must be employed. Let us study the special case of a particle of mass m_1 with initial velocity $\mathbf{v}_1$ (the incident particle) that strikes a particle of mass m_2 that is initially at rest (the target particle). This is a typical problem found in nuclear physics. The momentum equations in this case are

$$\mathbf{p}_1 = \mathbf{p}_1' + \mathbf{p}_2' \tag{7.41}$$

$$m_1\mathbf{v}_1 = m_1\mathbf{v}_1' + m_2\mathbf{v}_2' \tag{7.41a}$$

The energy balance condition is

$$\frac{p_1^2}{2m_1} = \frac{p_1'^2}{2m_1} + \frac{p_2'^2}{2m_2} + Q \tag{7.42}$$

or

$$\frac{1}{2} m_1 v_1^2 = \frac{1}{2} m_1 v_1'^2 + \frac{1}{2} m_2 v_2'^2 + Q \tag{7.42a}$$

Here, as before, the primes indicate the velocities and momenta after the collision, and Q represents the net energy that is lost or gained as a result of the impact. The quantity Q is of fundamental importance in atomic and nuclear physics, since it represents the energy released or absorbed in atomic and nuclear collisions. In many cases the target particle is broken up or changed by the collision. In such cases the particles that leave the collision are different from the particles that enter. This is easily taken into account by assigning different masses, say m_3 and m_4, to the particles leaving the collision. In any case, the law of conservation of linear momentum is always assumed to be valid.

Consider the particular case in which the masses of the incident and target particles are the same. Then the energy balance equation (Equation 7.42) can be written

$$p_1^2 = p_1'^2 + p_2'^2 + 2mQ \tag{7.43}$$

where $m = m_1 = m_2$. Now if we take the dot product of each side of the momentum equation (Equation 7.41) with itself, we get

$$p_1^2 = (\mathbf{p}_1' + \mathbf{p}_2') \cdot (\mathbf{p}_1' + \mathbf{p}_2') = p_1'^2 + p_2'^2 + 2\mathbf{p}_1' \cdot \mathbf{p}_2'$$

Comparing the above two equations, we see that

$$\mathbf{p}_1' \cdot \mathbf{p}_2' = mQ \tag{7.44}$$

For an elastic collision ($Q = 0$) we have, therefore,

$$\mathbf{p}_1' \cdot \mathbf{p}_2' = 0 \tag{7.44a}$$

so the two particles emerge from the collision at right angles to one another.

Center-of-Mass Coordinates

Theoretical calculations in nuclear physics are often done in terms of quantities referred to a coordinate system in which the center of mass of the colliding particles is at rest. On the other hand, the experimental observations on scattering of particles are carried out in terms of the laboratory coordinates. It is of interest, therefore, to consider briefly the problem of conversion from one coordinate system to the other.

The velocity vectors in the laboratory system and in the center-of-mass system are illustrated diagrammatically in Figure 7.7. In the figure ϕ_1 is the angle of deflection of

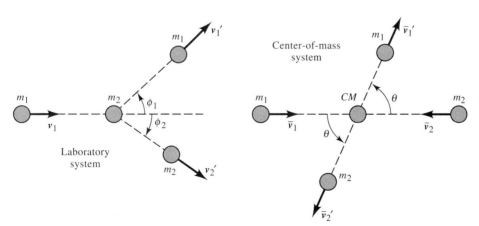

Figure 7.7 Comparison of laboratory and center-of-mass coordinates.

the incident particle after it strikes the target particle, and ϕ_2 is the angle that the line of motion of the target particle makes with the line of motion of the incident particle. Both ϕ_1 and ϕ_2 are measured in the laboratory system. In the center-of-mass system, since the center of mass must lie on the line joining the two particles at all times, both particles approach the center of mass, collide, and recede from the center of mass in opposite directions. The angle θ denotes the angle deflection of the incident particle in the center-of-mass system as indicated.

From the definition of the center of mass, the linear momentum in the center-of-mass system is zero both before and after the collision. Hence, we can write

$$\bar{\mathbf{p}}_1 + \bar{\mathbf{p}}_2 = 0 \tag{7.45}$$

$$\bar{\mathbf{p}}_1' + \bar{\mathbf{p}}_2' = 0$$

The bars are used to indicate that the quantity in question is referred to the center-of-mass system. The energy balance equation reads

$$\frac{\bar{p}_1^2}{2m_1} + \frac{\bar{p}_2^2}{2m_2} = \frac{\bar{p}_1'^2}{2m_1} + \frac{\bar{p}_2'^2}{2m_2} + Q \tag{7.46}$$

We can eliminate $\bar{p}_2$ and $\bar{p}_2'$ from the energy equation by using the momentum relations. The result, which is conveniently expressed in terms of the reduced mass, is

$$\frac{\bar{p}_1^2}{2\mu} = \frac{\bar{p}_1'^2}{2\mu} + Q \tag{7.47}$$

The momentum relations, Equations 7.45 expressed in terms of velocities, read

$$m_1 \bar{\mathbf{v}}_1 + m_2 \bar{\mathbf{v}}_2 = 0 \tag{7.48}$$

$$m_1 \bar{\mathbf{v}}_1' + m_2 \bar{\mathbf{v}}_2' = 0$$

The velocity of the center of mass is

$$\mathbf{v}_{cm} = \frac{m_1 \mathbf{v}_1}{m_1 + m_2} \tag{7.49}$$

Hence, we have

$$\bar{\mathbf{v}}_1 = \mathbf{v}_1 - \mathbf{v}_{cm} = \frac{m_2 \mathbf{v}_1}{m_1 + m_2} \tag{7.50}$$

The relationships among the velocity vectors $\mathbf{v}_{cm}$, $\mathbf{v}_1'$, and $\bar{\mathbf{v}}_1'$ are shown in Figure 7.8. From the figure, we see that

$$v_1' \sin \phi_1 = \bar{v}_1' \sin \theta \tag{7.51}$$

$$v_1' \cos \phi_1 = \bar{v}_1' \cos \theta + v_{cm}$$

Hence, by dividing, we find the equation connecting the scattering angles to be expressible in the form

$$\tan \phi_1 = \frac{\sin \theta}{\gamma + \cos \theta} \tag{7.52}$$

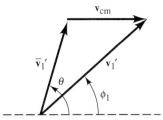

Figure 7.8 Velocity vectors in
the laboratory system and the
center-of-mass system.

in which γ is a numerical parameter whose value is given by

$$\gamma = \frac{v_{cm}}{\bar{v}'_1} = \frac{m_1 v_1}{\bar{v}'_1(m_1 + m_2)} \qquad (7.53)$$

The last step follows from Equation 7.49.

Now we can readily calculate the value of $\bar{v}'_1$ in terms of the initial energy of the incident particle from the energy equation (Equation 7.47). This gives us the necessary information to find γ and, thus, determine the relationship between the scattering angles. For example in the case of an elastic collision $Q = 0$, we find from the energy equation that $\bar{p}_1 = \bar{p}'_1$, or $\bar{v}_1 = \bar{v}'_1$. This result, together with Equation 7.50, yields the value

$$\gamma = \frac{m_1}{m_2} \qquad (7.54)$$

for an elastic collision.

Two special cases of such elastic collisions are instructive to consider. First, if the mass m_2 of the target particle is very much greater than the mass m_1 of the incident particle, then γ is very small. Hence, $\tan \phi_1 \approx \tan \theta$, or $\phi_1 \approx \theta$. That is, the scattering angles as seen in the laboratory and in the center-of-mass systems are nearly equal.

The second special case is that of equal masses of the incident and target particles $m_1 = m_2$. In this case $\gamma = 1$, and the scattering relation reduces to

$$\tan \phi_1 = \frac{\sin \theta}{1 + \cos \theta} = \tan \frac{\theta}{2} \qquad (7.55)$$

$$\phi_1 = \frac{\theta}{2}$$

That is, the angle of deflection in the laboratory system is just half that in the center-of-mass system. Furthermore, since the angle of deflection of the target particle is $\pi - \theta$ in the center-of-mass system, as shown in Figure 7.7, then the same angle in the laboratory system is $(\pi - \theta)/2$. Therefore, the two particles leave the point of impact at right angles to each other as seen in the laboratory system, in agreement with Equation 7.44a.

In the general case of nonelastic collisions, it is left as a problem to show that γ is expressible as

$$\gamma = \frac{m_1}{m_2}\left[1 - \frac{Q}{T}\left(1 + \frac{m_1}{m_2}\right)\right]^{-1/2} \tag{7.56}$$

in which T is the kinetic energy of the incident particle as measured in the laboratory system.

EXAMPLE 7.6

In a nuclear scattering experiment a beam of 4 MeV α-particles (helium nuclei) strikes a target consisting of helium gas, so that the incident and the target particles have equal mass. If a certain incident α-particle is scattered through an angle of 30° in the laboratory system, find its kinetic energy, and also the kinetic energy of recoil of the target particle, as a fraction of the initial kinetic energy T of the incident α-particle. (Assume that the target particle is at rest and that the collision is elastic.)

Solution:
For elastic collisions with particles of equal mass, we know from Equation 7.44a that $\phi_1 + \phi_2 = 90°$ (Figure 7.7). Hence, if we take components parallel to and perpendicular to the momentum of the incident particle, the momentum balance equation (Equation 7.41) becomes

$$p_1 = p_1' \cos \phi_1 + p_2' \sin \phi_1$$
$$0 = p_1' \sin \phi_1 - p_2' \cos \phi_1$$

in which $\phi_1 = 30°$. Solving the above pair of equations for the primed components, we find

$$p_1' = p_1 \cos \phi_1 = p_1 \cos 30° = p_1 \frac{\sqrt{3}}{2}$$

$$p_2' = p_1 \sin \phi_1 = p_1 \sin 30° = p_1 \frac{1}{2}$$

Therefore, the kinetic energies after impact are

$$T_1' = \frac{p_1'^2}{2m_1} = \frac{3}{4}\frac{p_1'^2}{2m_1} = \frac{3}{4}T = 3 \text{ MeV}$$

$$T_2' = \frac{p_2'^2}{2m_2} = \frac{1}{4}\frac{p_1'^2}{2m_1} = \frac{1}{4}T = 1 \text{ MeV} \qquad\blacksquare$$

EXAMPLE 7.7

What is the scattering angle of the center-of-mass system for Problem 7.6?

Solution:

Here Equation 7.55 gives the answer directly, namely,

$$\theta = 2\phi_1 = 60°$$ ∎

EXAMPLE 7.8

(a) Show that, for the general case of elastic scattering of a beam of particles of mass m_1 off a stationary target of particles whose mass is m_2, the opening angle ψ in the lab is given by the expression

$$\Psi = \phi_1 + \phi_2 = \frac{\pi}{2} + \frac{\phi_1}{2} - \frac{1}{2}\sin^{-1}\left(\frac{m_1}{m_2}\sin\phi_1\right)$$

(b) Suppose the beam of particles consists of protons and the target consists of helium nuclei. Calculate the opening angle for a proton scattered elastically at a lab angle $\phi_1 = 30°$.

Solution:

(a) Since particle 2 is at rest in the lab, its center-of-mass velocity $\bar{v}_2$ is equal in magnitude (and opposite in direction) to v_{cm}. For elastic collisions in the center of mass, momentum and energy conservation can be written as

$$\bar{\mathbf{p}}_1 + \bar{\mathbf{p}}_2 = \bar{\mathbf{p}}_1' + \bar{\mathbf{p}}_2' = 0$$

$$\frac{\bar{p}_1^2}{2m_1} + \frac{\bar{p}_2^2}{2m_2} = \frac{\bar{p}_1'^2}{2m_1} + \frac{\bar{p}_2'^2}{2m_2}$$

Solving for the magnitudes of the center-of-mass momenta of particle 1 in terms of particle 2, we obtain

$$\bar{p}_1 = \bar{p}_2 \qquad \bar{p}_1' = \bar{p}_2'$$

These expressions can be inserted into the energy conservation equation to obtain

$$\frac{\bar{p}_2'^2}{2\mu} = \frac{\bar{p}_2^2}{2\mu} \qquad \mu = \frac{m_1 m_2}{m_1 + m_2}$$

$$\therefore \bar{v}_2' = \bar{v}_2 = v_{cm}$$

Thus, in an elastic collision, the center-of-mass velocities of particle 2 are the same before and after the collision, and both are equal to the center-of-mass velocity. Moreover, the values of the center-of-mass velocities of particle 1 are also the same before and after the collision, and, from conservation of momentum in the center of mass, they are

$$\bar{v}_1' = \bar{v}_1 = \frac{m_2}{m_1}\bar{v}_2' = \frac{m_1}{m_2}v_{cm}$$

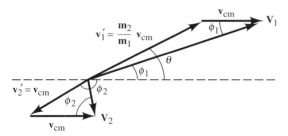

Figure 7.9   Velocity vectors in laboratory and center-of-mass frame for elastic scattering.

We can now draw a vector diagram analogous to that of Figure 7.8 but here specifically for the case of elastic scattering. From the geometry of Figure 7.9, we see that

$$\psi = \phi_1 + \phi_2$$

$$2\phi_2 = \pi - \theta$$

$$\phi_2 = \frac{\pi}{2} - \frac{\theta}{2}$$

Now, applying the law of sines to the upper triangle of the figure, we obtain

$$\frac{\frac{m_2}{m_1} v_{cm}}{\sin \phi_1} = \frac{v_{cm}}{\sin (\theta - \phi_1)}$$

$$\sin (\theta - \phi_1) = \frac{m_1}{m_2} \sin \phi_1$$

$$\therefore \theta = \phi_1 + \sin^{-1}\left(\frac{m_1}{m_2} \sin \phi_1\right)$$

Finally, substituting this last expression for θ into the one above for ϕ_2 and solving for the opening angle ψ, we obtain

$$\psi = \phi_1 + \phi_2 = \phi_1 + \left(\frac{\pi}{2} - \frac{\theta}{2}\right)$$

$$\psi = \frac{\pi}{2} + \frac{\phi_1}{2} - \frac{1}{2} \sin^{-1}\left(\frac{m_1}{m_2} \sin \phi_1\right)$$

(b) For elastic scattering of protons off helium nuclei at $\phi_1 = 30°$, $m_1/m_2 = 1/4$, and $\psi \approx 101°$.

Note: In the case where $m_1 = m_2$, $\psi = 90°$ as derived in the text.   ■

7.6   MOTION OF A BODY WITH VARIABLE MASS. ROCKET MOTION

In the case of a body whose mass changes with time, it is necessary to use care in setting up the differential equations of motion. The concept of impulse can be helpful in this type of problem.

Consider the general case of the motion of a body with changing mass. Let $\mathbf{F}_{ext}$ denote the external force acting on the body at a given time and let Δm denote the increment of the mass of the body that occurs in a short time interval Δt. Then $\mathbf{F}_{ext}\Delta t$ is the impulse delivered by the external force, and we have

$$\mathbf{F}_{ext}\Delta t = (\mathbf{p}_{total})_{t+\Delta t} - (\mathbf{p}_{total})_t \tag{7.57}$$

for the change in the total linear momentum of the system. Hence, if $\mathbf{v}$ denotes the velocity of the body and $\mathbf{V}$ the velocity of the mass increment Δm relative to the body, then we can write

$$\mathbf{F}_{ext}\Delta t = (m + \Delta m)(\mathbf{v} + \Delta \mathbf{v}) - [m\mathbf{v} + \Delta m(\mathbf{v} + \mathbf{V})]$$

This reduces to

$$\mathbf{F}_{ext}\Delta t = m\Delta \mathbf{v} + \Delta m\Delta \mathbf{v} - \mathbf{V}\Delta m \tag{7.58}$$

or, by dividing by Δt, we can write

$$\mathbf{F}_{ext} = (m + \Delta m)\frac{\Delta \mathbf{v}}{\Delta t} - \mathbf{V}\frac{\Delta m}{\Delta t} \tag{7.58a}$$

Thus, in the limit as Δt approaches zero, we have the general equation

$$\mathbf{F}_{ext} = m\dot{\mathbf{v}} - \mathbf{V}\dot{m} \tag{7.59}$$

Here the force $\mathbf{F}_{ext}$ may represent gravity, air resistance, and so on. In the case of rockets, the term $\mathbf{V}\dot{m}$ represents the thrust.

Let us apply the equation to two special cases. First, suppose that a body is moving through a fog or mist so that it collects mass as it goes. In this case the initial velocity of the accumulated matter is zero. Hence, $\mathbf{V} = -\mathbf{v}$, and we get

$$\mathbf{F}_{ext} = m\dot{\mathbf{v}} + \mathbf{v}\dot{m} = \frac{d(m\mathbf{v})}{dt} \tag{7.60}$$

for the equation of motion. It applies only if the initial velocity of the matter that is being swept up is zero. Otherwise, the general equation (Equation 7.59) must be used.

For the second case, consider the motion of a rocket. In this instance the sign of $\dot{m}$ is negative, because the rocket is losing mass in the form of ejected fuel. Hence, $\mathbf{V}\dot{m}$ is opposite to the direction of $\mathbf{V}$, the relative velocity of the ejected fuel. For simplicity we shall solve the equation of motion for the case in which the external force $\mathbf{F}_{ext}$ is zero; that is, we neglect gravity, air resistance, and so on. Then we have

$$m\dot{\mathbf{v}} = \mathbf{V}\dot{m} \tag{7.61}$$

We can now separate the variables and integrate to find **v** as follows:

$$\int d\mathbf{v} = \int \frac{\mathbf{V}\, dm}{m} \tag{7.61a}$$

If it is assumed that **V** is constant, then we can integrate between limits to find the *speed* as a function of *m:*

$$\int_{v_0}^{v} dv = -V \int_{m_0}^{m} \frac{dm}{m}$$

$$v = v_0 + V \ln \frac{m_0}{m}$$

Here m_0 is the initial mass of the rocket plus unburned fuel, m is the mass at any time, and V is the speed of the ejected fuel relative to the rocket. Owing to the nature of the logarithmic function, it is necessary to have a large fuel to payload ratio in order to attain the large speeds needed for satellite launching.

EXAMPLE 7.9   *Launching an Earth-Satellite from Cape Canaveral*

We know from Example 6.3(a) that the speed of a satellite in a circular orbit near the earth is about 8 km/s. Satellites are launched toward the east in order to take advantage of the earth's rotation. For a point on the earth near the equator the rotational speed is approximately $R_{earth}\omega_{earth}$, which is about 0.5 km/s. For most rocket fuels the effective ejection speed is of the order of 2 to 4 km/s. For example, if we take $V = 2.5$ km/s, then we find that the mass ratio calculated from Equation 7.62 is

$$\frac{m_0}{m} = \exp\left(\frac{v - v_0}{V}\right) = \exp\left(\frac{8.0 - 0.5}{2.5}\right) = e^3 = 20.1$$

to achieve orbital speed from the ground. Thus, only about 5% of the total initial mass m_0 is payload.   ∎

PROBLEMS

7.1   A system consists of three particles, each of unit mass, with positions and velocities as follows:

$$\begin{aligned} \mathbf{r}_1 &= \mathbf{i} + \mathbf{j} & \mathbf{v}_1 &= 2\mathbf{i} \\ \mathbf{r}_2 &= \mathbf{j} + \mathbf{k} & \mathbf{v}_2 &= \mathbf{j} \\ \mathbf{r}_3 &= \mathbf{k} & \mathbf{v}_3 &= \mathbf{i} + \mathbf{j} + \mathbf{k} \end{aligned}$$

Find the position and velocity of the center of mass. Find also the linear momentum of the system.

7.2 **(a)** Find the kinetic energy of the above system.

(b) Find the value of $mv_{cm}^2/2$.

(c) Find the angular momentum about the origin.

7.3 A bullet of mass m is fired from a gun of mass M. If the gun can recoil freely and the muzzle velocity of the bullet (velocity relative to the gun as it leaves the barrel) is v_0, show that the actual velocity of the bullet relative to the ground is $v_0/(1 + \gamma)$ and the recoil velocity for the gun is $-\gamma v_0/(1 + \gamma)$ where $\gamma = m/M$.

7.4 A block of wood rests on a smooth horizontal table. A gun is fired horizontally at the block and the bullet passes through the block emerging with half its initial speed just before it entered the block. Show that the fraction of the initial kinetic energy of the bullet that is lost as frictional heat is $\frac{3}{4} - \frac{1}{4}\gamma$ where γ is the ratio of the mass of the bullet to the mass of the block ($\gamma < 1$).

7.5 An artillery shell is fired at an angle of elevation of $60°$ with initial speed v_0. At the uppermost part of its trajectory, the shell bursts into two equal fragments, one of which moves directly upward, relative to the ground, with initial speed $v_0/2$. What is the direction and speed of the other fragment immediately after the burst?

7.6 A ball is dropped from a height h onto a horizontal pavement. If the coefficient of restitution is ϵ, show that the total vertical distance the ball goes before the rebounds cease is $h(1 + \epsilon^2)/(1 - \epsilon^2)$. Find also the total length of time that the ball bounces.

7.7 A small car of a mass m and initial speed v_0 collides head-on on an icy road with a truck of mass $4m$ going toward the car with initial speed $\frac{1}{2}v_0$. If the coefficient of restitution in the collision is $\frac{1}{4}$, find the speed and direction of each vehicle just after colliding.

7.8 Show that the kinetic energy of a two-particle system is $\frac{1}{2}mv_{cm}^2 + \frac{1}{2}\mu v^2$ where $m = m_1 + m_2$, v is the relative speed, and μ is the reduced mass.

7.9 If two bodies undergo a direct collision, show that the loss in kinetic energy is equal to

$$\frac{1}{2}\mu v^2(1 - \epsilon^2)$$

where μ is the reduced mass, v is the relative speed before impact, and ϵ is the coefficient of restitution.

7.10 A moving particle of mass m_1 collides elastically with a target particle of mass m_2, which is initially at rest. If the collision is head-on, show that the incident particle loses a fraction $4\mu/m$ of its original kinetic energy where μ is the reduced mass and $m = m_1 + m_2$.

7.11 Show that the angular momentum of a two-particle system is

$$\mathbf{r}_{cm} \times m\mathbf{v}_{cm} + \mathbf{R} \times \mu\mathbf{v}$$

where $m = m_1 + m_2$, μ is the reduced mass, $\mathbf{R}$ is the relative position vector, and $\mathbf{v}$ is the relative velocity of the two particles.

7.12 The observed period of the binary system Cygnus X-1, presumed to be a bright star and a black hole, is 5.6 days. If the mass of the visible component is 15 solar masses and the black hole has a mass of 8 solar masses, show that the semimajor axis of the orbit of the black hole relative to the visible star is roughly one fifth the distance from the earth to the sun.

7.13 A proton of mass m_p with initial velocity $\mathbf{v}_0$ collides with a helium atom, mass $4m_p$, that is initially at rest. If the proton leaves the point of impact at an angle of $45°$ with its original line of motion, find the final velocities of each particle. Assume that the collision is perfectly elastic.

7.14 Work Problem 7.13 for the case that the collision is inelastic and that Q is equal to one fourth of the initial energy of the proton.

7.15 Referring to Problem 7.13, find the scattering angle of the proton in the center-of-mass system.

7.16 Find the scattering angle of the proton in the center-of-mass system for Problem 7.14.

7.17 A particle of mass m with initial momentum p_1 collides with a particle of equal mass at rest. If the magnitudes of the final momenta of the two particles are p_1' and p_2', respectively, show that the energy loss of the collision is given by

$$Q = \frac{p_1' p_2'}{m} \cos \psi$$

where ψ is the angle between the paths of the two particles after colliding.

7.18 A uniform chain lies in a heap on a table. If one end is raised vertically with uniform velocity v, show that the upward force that must be exerted on the end of the chain is equal to the weight of a length $z + (v^2/g)$ of the chain where z is the length that has been uncoiled at any instant.

7.19 Find the differential equation of motion of a raindrop falling through a mist collecting mass as it falls. Assume that the drop remains spherical and that the rate of accretion is proportional to the cross-sectional area of the drop multiplied by the speed of fall. Show that if the drop starts from rest when it is infinitely small, then the acceleration is constant and equal to $g/7$.

7.20 A uniform heavy chain of length a hangs initially with a part of length b hanging over the edge of a table. The remaining part, of length $a - b$, is coiled up at the edge of the table. If the chain is released, show that the speed of the chain when the last link leaves the end of the table is $[2g(a^3 - b^3)/3a^2]^{1/2}$.

7.21 A rocket traveling through the atmosphere experiences a linear air resistance $-k\mathbf{v}$. Find the differential equation of motion when all other external forces are negligible. Integrate the equation and show that if the rocket starts from rest, the final speed is given by $v = V\alpha[1 - (m/m_0)^{1/\alpha}]$ where V is the relative speed of the exhaust fuel, $\alpha = |\dot{m}/k| = $ constant, m_0 is the initial mass of the rocket plus fuel, and m is the final mass of the rocket.

7.22 Find the equation of motion for a rocket fired vertically upward, assuming g is constant. Find the ratio of fuel to payload to achieve a final speed equal to the escape speed v_e from the earth if the speed of the exhaust gas is kv_e where k is a given constant, and the fuel burning rate is $|\dot{m}|$. Compute the numerical value of the fuel–payload ratio for $k = \frac{1}{4}$, and $|\dot{m}|$ equal to 1% of the mass of the fuel per second.

COMPUTER/CALCULATOR APPLICATIONS

7.1 Let two particles ($m_1 = m_2 = 1$ kg) repel each other with equal and opposite forces given by

$$\mathbf{F}_{12} = k\frac{b^2}{r^2}\mathbf{r}_{12} = -\mathbf{F}_{21}$$

where $b = 1$ m and $k = 1$ N. Assume that the initial positions of m_1 and m_2 are given by $(x_1, y_1)_0 = (-10, 0.5)$ m and $(x_2, y_2)_0 = (0, -0.5)$ m. Let the initial velocity of m_1 be 10 m/s in the $+x$ direction and m_2 be at rest. Numerically integrate the equations of motion for these two particles undergoing this two-dimensional "collision."

(a) Plot their trajectories up to a point where their distance of separation is 10 m.

(b) Measure the scattering angle of the incident particle and the recoil angle of the scattered particle. Is the sum of these two angles equal to 90°?

(c) What is the vector sum of their final momenta? Is it equal to the initial momentum of the incident particle?

8

MECHANICS OF RIGID BODIES.
PLANAR MOTION

". . . centre of gravity implies the more restricted concept of a solid that is only heavy, while the centre of inertia is defined by means of the inertia alone, the forces to which the solid is subject being neglected. . . . Euler also defines the moments of inertia—a concept which Huygens lacked and which considerably simplifies the language—and calculates these moments for Homogeneous bodies."

(Rene Dugas, A History of Mechanics, *Editions du Griffon, Neuchatel, Switzerland, 1955; synopsis of Leonhard Euler's comments in* Theoria motus corporum solidorum seu rigidorum, *1760)*

A rigid body may be regarded as a system of particles whose *relative* positions are fixed, or, in other words, the distance between any two particles is constant. This definition of a rigid body is idealized. In the first place, as pointed out in the definition of a particle, there are no true particles in nature. Second, real extended bodies are not strictly rigid; they become more or less deformed (stretched, compressed, or bent) when external forces are applied. We shall for the present, however, neglect such deformations. In this chapter we take up the study of rigid-body motion for the case in which the direction of the axis of rotation does not change. The general case, which involves more extensive calculation, will be treated in the next chapter.

8.1   CENTER OF MASS OF A RIGID BODY

We have already defined the center of mass (Section 7.1) of a system of particles as the point (x_{cm}, y_{cm}, z_{cm}) where

$$x_{cm} = \frac{\sum\limits_i x_i m_i}{\sum\limits_i m_i} \qquad y_{cm} = \frac{\sum\limits_i y_i m_i}{\sum\limits_i m_i} \qquad z_{cm} = \frac{\sum\limits_i z_i m_i}{\sum\limits_i m_i}$$

For a rigid extended body, we can replace the summation by an integration over the volume of the body, namely,

$$x_{cm} = \frac{\int_v \rho x \, dv}{\int_v \rho \, dv} \qquad y_{cm} = \frac{\int_v \rho y \, dv}{\int_v \rho \, dv} \qquad z_{cm} = \frac{\int_v \rho z \, dv}{\int_v \rho \, dv} \qquad (8.1)$$

where ρ is the density and dv is the element of volume.

If a rigid body is in the form of a thin shell, the equations for the center of mass become

$$x_{cm} = \frac{\int_s \rho x \, ds}{\int_s \rho \, ds} \qquad y_{cm} = \frac{\int_s \rho y \, ds}{\int_s \rho \, ds} \qquad z_{cm} = \frac{\int_s \rho z \, ds}{\int_s \rho \, ds} \qquad (8.2)$$

where ds is the element of area and ρ is the mass per unit area, the integration extending over the area of the body.

Similarly, if the body is in the form of a thin wire, we have

$$x_{cm} = \frac{\int_l \rho x \, dl}{\int_l \rho \, dl} \qquad y_{cm} = \frac{\int_l \rho y \, dl}{\int_l \rho \, dl} \qquad z_{cm} = \frac{\int_l \rho z \, dl}{\int_l \rho \, dl} \qquad (8.3)$$

In this case ρ is the mass per unit length, and dl is the element of length.

For uniform homogeneous bodies, the density factors ρ are constant in each case and, therefore, may be canceled out in each equation above.

If a body is composite, that is, if it consists of two or more parts whose centers of mass are known, then it is clear, from the definition of the center of mass, that we can write

$$x_{cm} = \frac{x_1 m_1 + x_2 m_2 + \cdots}{m_1 + m_2 + \cdots} \tag{8.4}$$

with similar equations for y_{cm} and z_{cm}. Here (x_1, y_1, z_1) is the center of mass of the part m_1, and so on.

Symmetry Considerations

If a body possesses symmetry, it is possible to take advantage of that symmetry in locating the center of mass. Thus, if the body has a plane of symmetry, that is, if each particle m_i has a mirror image of itself m_i' relative to some plane, then the center of mass lies in that plane. To prove this, let us suppose that the xy plane is a plane of symmetry. We have then

$$z_{cm} = \frac{\sum_i (z_i m_i + z_i' m_i')}{\sum_i (m_i + m_i')}$$

But $m_i = m_i'$ and $z_i = -z_i'$. Hence, the terms in the numerator cancel in pairs, and so $z_{cm} = 0$; that is, the center of mass lies in the xy plane.

Similarly, if the body has a line of symmetry, it is easy to show that the center of mass lies on that line. The proof is left as an exercise.

Solid Hemisphere

To find the center of mass of a solid homogeneous hemisphere of radius a, we know from symmetry that the center of mass lies on the radius that is normal to the plane face. Choosing coordinate axes as shown in Figure 8.1, we have that the center of mass lies

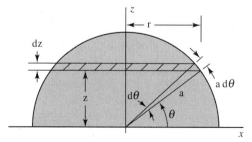

Figure 8.1   Coordinates for calculating the center of mass of a hemisphere.

on the z-axis. To calculate z_{cm} we use a circular element of volume of thickness dz and radius $= (a^2 - z^2)^{1/2}$, as shown. Thus,

$$dv = \pi(a^2 - z^2) \, dz$$

Therefore,

$$z_{cm} = \frac{\displaystyle\int_0^a \rho\pi z(a^2 - z^2) \, dz}{\displaystyle\int_0^a \rho\pi(a^2 - z^2) \, dz} = \frac{3}{8}a \tag{8.5}$$

Hemispherical Shell

For a hemispherical shell of radius a, we use the same axes as in Figure 8.1. Again, from symmetry, the center of mass is located on the z-axis. For our element of surface ds, we choose a circular strip of width $dl = a d\theta$. Hence,

$$ds = 2\pi r \, dl = 2\pi(a^2 - z^2)^{1/2} \, a d\theta$$

$$\theta = \sin^{-1}\left(\frac{z}{a}\right) \qquad d\theta = (a^2 - z^2)^{-1/2} \, dz$$

$$\therefore \ ds = 2\pi a \, dz$$

The location of the center of mass is accordingly given by

$$z_{cm} = \frac{\displaystyle\int_0^a \rho 2\pi a z \, dz}{\displaystyle\int_0^a \rho 2\pi a \, dz} = \frac{1}{2}a \tag{8.6}$$

Semicircle

To find the center of mass of a thin wire bent into the form of a semicircle of radius a, we use axes as shown in Figure 8.2. We have

$$dl = a \, d\theta$$

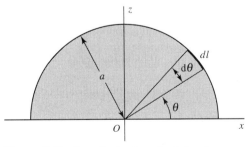

Figure 8.2 Coordinates for calculating the center of mass of a wire bent into the form of a semicircle.

and

$$z = a \sin \theta$$

Hence,

$$z_{cm} = \frac{\int_0^\pi \rho(a \sin \theta)a \, d\theta}{\int_0^\pi \rho a \, d\theta} = \frac{2}{\pi}a \tag{8.7}$$

Semicircular Lamina

In the case of a uniform semicircular lamina, the center of mass is on the z-axis (Figure 8.2). It is left for the student to verify that

$$z_{cm} = \frac{4}{3\pi}a \tag{8.8}$$

8.2   ROTATION OF A RIGID BODY ABOUT A FIXED AXIS. MOMENT OF INERTIA

The simplest type of rigid-body motion, other than pure translation, is that in which the body is constrained to rotate about a fixed axis. Let us choose the z-axis of an appropriate coordinate system as the axis of rotation. The path of a representative particle m_i located at the point (x_i, y_i, z_i) is then a circle of radius $(x_i^2 + y_i^2)^{1/2} = R_i$ centered on the z-axis. A representative cross section parallel to the xy plane is shown in Figure 8.3.

The speed v_i of particle i is given by

$$v_i = R_i\omega = (x_i^2 + y_i^2)^{1/2}\omega$$

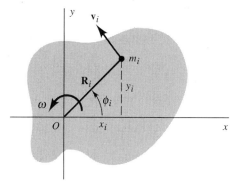

Figure 8.3   Cross section of a rigid body rotating about the z-axis. (The z-axis is out of the page.)

where ω is the angular speed of rotation. From a study of Figure 8.3, we see that the velocity has components as follows:

$$\dot{x}_i = -v_i \sin \phi_i = -\omega y_i$$
$$\dot{y}_i = v_i \cos \phi_i = \omega x_i \tag{8.9}$$
$$\dot{z}_i = 0$$

where ϕ_i is defined as shown in Figure 8.3. The above equations can also be obtained by taking the components of

$$\mathbf{v}_i = \boldsymbol{\omega} \times \mathbf{r}_i \tag{8.10}$$

where $\boldsymbol{\omega} = \mathbf{k}\omega$.

Let us calculate the kinetic energy of rotation of the body. We have

$$T_{rot} = \sum_i \frac{1}{2} m_i v_i^2 = \frac{1}{2}\left(\sum_i m_i R_i^2\right)\omega^2 = \frac{1}{2} I_z \omega^2 \tag{8.11}$$

where

$$I_z = \sum_i m_i R_i^2 = \sum_i m_i(x_i^2 + y_i^2) \tag{8.12}$$

The quantity I_z, defined by Equation 8.12, is of particular importance in the study of the motion of rigid bodies. It is called the *moment of inertia* about the z-axis.

To show how the moment of inertia further enters the picture, let us next calculate the angular momentum about the axis of rotation. Since the angular momentum of a single particle is, by definition, $\mathbf{r}_i \times m_i \mathbf{v}_i$, the z-component is

$$m_i(x_i \dot{y}_i - y_i \dot{x}_i) = m_i(x_i^2 + y_i^2)\omega = m_i R_i^2 \,\omega \tag{8.13}$$

where we have made use of Equations 8.9. The total z-component of the angular momentum, which we shall call L_z, is then given by summing over all the particles, namely,

$$L_z = \sum_i m_i R_i^2 \omega = I_z \omega \tag{8.14}$$

In Section 7.2 we found that the rate of change of angular momentum for any system is equal to the total moment of the external forces. For a body constrained to rotate about a fixed axis, taken here as the z-axis, then

$$N_z = \frac{dL_z}{dt} = \frac{d(I_z \omega)}{dt} \tag{8.15}$$

where N_z is the total moment of all the applied forces about the axis of rotation (the component of $\mathbf{N}$ along the z-axis). If the body is rigid, then I_z is constant, and we can write

$$N_z = I_z \frac{d\omega}{dt} \tag{8.16}$$

The analogy between the equations for translation and for rotation about a fixed axis is shown below:

Translation along x-axis   *Rotation about z-axis*

Linear momentum	$p_x = mv_x$	Angular momentum	$L_z = I_z\omega$
Force	$F_x = m\dot{v}_x$	Torque	$N_z = I_z\dot{\omega}$
Kinetic energy	$T = \frac{1}{2}mv^2$	Kinetic energy	$T_{rot} = \frac{1}{2}I_z\omega^2$

Thus, the moment of inertia is analogous to mass; it is a measure of the rotational inertia of a body relative to some fixed axis of rotation, just as mass is a measure of translational inertia of a body.

8.3   CALCULATION OF THE MOMENT OF INERTIA

In actual calculations of the moment of inertia $\Sigma m_i R_i^2$ for extended bodies, we can replace the summation by an integration over the body, just as we did in calculation of the center of mass. Thus, we may write for any axis

$$I = \int R^2 \ dm \tag{8.17}$$

where the element of mass dm is given by a density factor multiplied by an appropriate differential (volume, area, or length). It is important to remember that R is the perpendicular distance from the element of mass to the axis of rotation.

In the case of a composite body, it is clear, from the definition of the moment of inertia that we may write

$$I = I_1 + I_2 + \cdots \tag{8.18}$$

where I_1, I_2, and so on, are the moments of inertia of the various parts about the particular axis chosen.

Let us calculate the moments of inertia for some important special cases.

Thin Rod

For a thin uniform rod of length a and mass m, we have, for an axis perpendicular to the rod at one end [Figure 8.4(a)],

$$I_z = \int_0^a x^2\rho \ dx = \frac{1}{3}\rho a^3 = \frac{1}{3}ma^2 \tag{8.19}$$

The last step follows from the fact that $m = \rho a$.

If the axis is taken at the center of the rod [Figure 8.4(b)], we have

$$I_z = \int_{-a/2}^{a/2} x^2\rho \ dx = \frac{1}{12}\rho a^3 = \frac{1}{12}ma^2 \tag{8.20}$$

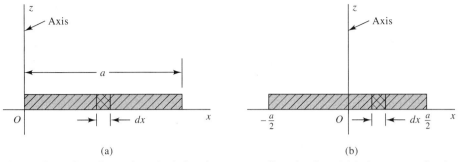

(a)   (b)

Figure 8.4   Coordinates for calculating the moment of inertia of a rod (a) about one end and (b) about the center of the rod.

Hoop or Cylindrical Shell

In the case of a thin circular hoop or cylindrical shell, for the central or *symmetry* axis all particles lie at the same distance from the axis. Thus,

$$I_{axis} = ma^2 \qquad (8.21)$$

where a is the radius and m is the mass.

Circular Disc or Cylinder

To calculate the moment of inertia of a uniform circular disc of radius a and mass m, we shall use polar coordinates. The element of mass, a thin ring of radius r and thickness dr, is given by

$$dm = \rho 2\pi r \, dr$$

where ρ is the mass per unit area. The moment of inertia about an axis through the center of the disc normal to the plane faces (Figure 8.5) is obtained as follows:

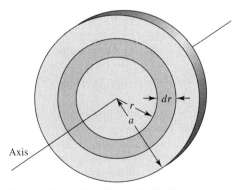

Figure 8.5   Coordinates for finding the moment of inertia of a disc.

$$I_{axis} = \int_0^a r^2 \rho \, 2\pi r \, dr = 2\pi \rho \frac{a^4}{4} = \frac{1}{2} ma^2 \qquad (8.22)$$

The last step results from the relation $m = \rho \pi a^2$.

Clearly, Equation 8.22 also applies to a uniform right-circular cylinder of radius a and mass m, the axis being the central axis of the cylinder.

Sphere

Let us find the moment of inertia of a uniform solid sphere of radius a and mass m about an axis (the z-axis) passing through the center. We shall divide the sphere into thin circular discs, as shown in Figure 8.6. The moment of inertia of a representative disc of radius y, from Equation 8.22, is $\frac{1}{2}y^2 \, dm$. But $dm = \rho \pi y^2 \, dz$; hence,

$$I_z = \int_{-a}^a \frac{1}{2}\pi \rho y^4 \, dz = \int_{-a}^a \frac{1}{2}\pi \rho (a^2 - z^2)^2 \, dz = \frac{8}{15}\pi \rho a^5 \qquad (8.23)$$

The last step above should be filled in by the student. Since the mass m is given by

$$m = \frac{4}{3}\pi a^3 \rho$$

we have

$$I_z = \frac{2}{5} ma^2 \qquad (8.24)$$

for a solid uniform sphere. Clearly also, $I_x = I_y = I_z$.

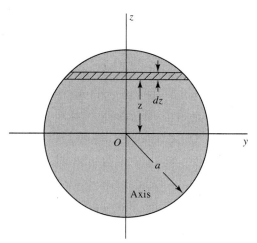

Figure 8.6 Coordinates for finding the moment of inertia of a sphere.

Spherical Shell

The moment of inertia of a thin uniform spherical shell can be found very simply by application of Equation 8.23. If we differentiate with respect to a, namely,

$$dI_z = \frac{8}{3}\pi\rho a^4 \ da$$

the result is the moment of inertia of a shell of thickness da and radius a. The mass of the shell is $4\pi a^2\rho \ da$. Hence, we can write

$$I_z = \frac{2}{3}ma^2 \tag{8.25}$$

for the moment of inertia of a thin shell of radius a and mass m. The student should verify the above result by direct integration.

EXAMPLE 8.1

Shown in Figure 8.7 is a uniform chain of length $l = 2\pi R$ and mass $m = M/2$ that was initially wrapped around a uniform, thin disc of radius R and mass M. One tiny piece of chain initially hung free, perpendicular to the horizontal axis. When the disc was released, the chain fell and unwrapped. The disc began to rotate faster and faster about its fixed z-axis, without friction. (a) Find the angular speed of the disc at the moment the chain completely unwrapped itself. (b) Solve for the case of a chain wrapped around a wheel whose mass is the same as that of the disc, but concentrated in a thin rim.

Solution:
(a) Figure 8.7 shows the disc and chain at the moment the chain unwrapped. The final angular speed of the disc is ω. Energy was conserved as the chain unwrapped.

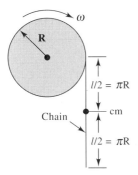

Figure 8.7 Falling chain attached to disc, free to rotate about a fixed z-axis.

Since the center of mass of the chain originally coincided with that of the disc, it fell a distance $l/2 = \pi R$, and we have

$$mg\frac{l}{2} = \frac{1}{2}I\omega^2 + \frac{1}{2}mv^2$$

$$\frac{l}{2} = \pi R \qquad v = \omega R \qquad I = \frac{1}{2}MR^2$$

Solving for ω^2 gives

$$\omega^2 = \frac{mg\dfrac{l}{2}}{\left[\dfrac{1}{2}\left(\dfrac{M}{2}\right) + \dfrac{1}{2}m\right]R^2} = \frac{mg\pi R}{\left(\dfrac{1}{2}m + \dfrac{1}{2}m\right)R^2}$$

$$= \pi\frac{g}{R}$$

(b) The moment of inertia of a wheel is $I = MR^2$. Substituting this into the above equation yields

$$\omega^2 = \pi\frac{2g}{3R}$$

Even though the mass of the wheel is the same as that of the disc, its moment of inertia is larger, since all its mass is concentrated along the rim. Thus, its angular acceleration and final angular velocity are less than that of the disc. ∎

Perpendicular-Axis Theorem for a Plane Lamina

Consider a rigid body that is in the form of a plane lamina of any shape. Let us place the lamina in the xy plane (Figure 8.8). The moment of inertia about the z-axis is given by

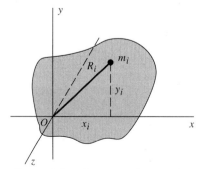

Figure 8.8   The perpendicular-axis theorem for a lamina.

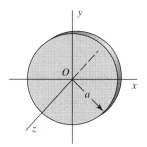

Figure 8.9 Circular disc.

$$I_z = \sum_i m_i(x_i^2 + y_i^2) = \sum_i m_i x_i^2 + \sum_i m_i y_i^2$$

But the sum $\sum m_i x_i^2$ is just the moment of inertia I_y about the y-axis, because z_i is zero for all particles. Similarly, $\sum_i m_i y_i^2$ is the moment of inertia I_x about the x-axis. The above equation can therefore be written

$$I_z = I_x + I_y \tag{8.26}$$

This is the perpendicular-axis theorem. In words: *The moment of inertia of any plane lamina about an axis normal to the plane of the lamina is equal to the sum of the moments of inertia about any two mutually perpendicular axes passing through the given axis and lying in the plane of the lamina.*

As an example of the use of this theorem, let us consider a thin circular disc in the xy plane (Figure 8.9). From Equation 8.22 we have

$$I_z = \frac{1}{2}ma^2 = I_x + I_y$$

In this case, however, we know from symmetry that $I_x = I_y$. Therefore, we must have

$$I_x = I_y = \frac{1}{4}ma^2 \tag{8.27}$$

for the moment of inertia about any axis in the plane of the disc passing through the center. The above result can also be obtained by direct integration.

Parallel-Axis Theorem for any Rigid Body

Consider the equation for the moment of inertia about some axis, say the z-axis,

$$I_z = \sum_i m_i(x_i^2 + y_i^2)$$

Now we can express x_i and y_i in terms of the coordinates of the center of mass (x_{cm}, y_{cm}, z_{cm}) and the coordinates *relative* to the center of mass ($\overline{x}_i$, $\overline{y}_i$, $\overline{z}_i$) (Figure 8.10) as follows:

$$x_i = x_{cm} + \overline{x}_i \qquad y_i = y_{cm} + \overline{y}_i \tag{8.28}$$

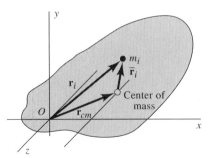

Figure 8.10   The parallel-axis theorem
for any rigid body.

We have, therefore, after substituting and collecting terms,

$$I_z = \sum_i m_i(\overline{x}_i^2 + \overline{y}_i^2) + \sum_i m_i(x_{cm}^2 + y_{cm}^2) + 2x_{cm}\sum_i m_i\overline{x}_i + 2y_{cm}\sum_i m_i\overline{y}_i \quad (8.29)$$

The first sum on the right is just the moment of inertia about an axis parallel to the z-axis and passing through the center of mass. We shall call it I_{cm}. The second sum is clearly equal to the mass of the body multiplied by the square of the distance between the center of mass and the z-axis. Let us call this distance l. That is, $l^2 = x_{cm}^2 + y_{cm}^2$.

Now, from the definition of the center of mass,

$$\sum_i m_i\overline{x}_i = \sum_i m_i\overline{y}_i = 0$$

Hence, the last two sums on the right of Equation 8.29 vanish. The final result may be written in the general form for any axis

$$I = I_{cm} + ml^2 \tag{8.30}$$

This is the *parallel-axis* theorem. It is applicable to any rigid body, solid as well as laminar.

> The theorem states, in effect, that *the moment of inertia of a rigid body about any axis is equal to the moment of inertia about a parallel axis passing through the center of mass plus the product of the mass of the body and the square of the distance between the two axes.*

Applying the perpendicular-axis theorem to a circular disc, we have, from Equations 8.22 and 8.30,

$$I = \frac{1}{2}ma^2 + ma^2 = \frac{3}{2}ma^2 \tag{8.31}$$

for the moment of inertia of a uniform circular disc about an axis *perpendicular* to the plane of the disc and passing through the edge. Furthermore, from Equations 8.27 and 8.30, we find

$$I = \frac{1}{4}ma^2 + ma^2 = \frac{5}{4}ma^2 \tag{8.32}$$

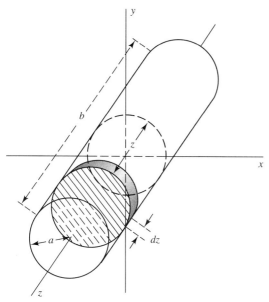

Figure 8.11 Coordinates for finding the moment of inertia of a circular cylinder.

for the moment of inertia about an axis in the plane of the disc and *tangent* to the edge.

As a second example, let us find the moment of inertia of a uniform circular cylinder of length b and radius a about an axis through the center and *perpendicular* to the central axis, namely I_x or I_y in Figure 8.11. For our element of integration, we choose a disc of thickness dz located a distance z from the xy plane. Then, from the previous result for a thin disc (Equation 8.27), together with the parallel-axis theorem, we have

$$dI_x = \frac{1}{4}a^2\,dm + z^2\,dm$$

in which $dm = \rho\pi a^2 dz$. Thus,

$$I_x = \rho\pi a^2 \int_{-b/2}^{b/2} \left(\frac{1}{4}a^2 + z^2\right) dz = \rho\pi a^2 \left(\frac{1}{4}a^2 b + \frac{1}{12}b^3\right)$$

But the mass of the cylinder is $m = \rho\pi a^2 b$, therefore,

$$I_x = I_y = m\left(\frac{1}{4}a^2 + \frac{1}{12}b^2\right) \tag{8.33}$$

Radius of Gyration

Note the similarity of Equation 8.12, the expression for the moment of inertia I_z of a rigid body about the z-axis, to the expressions for center of mass developed in Section 8.1. If we were to divide Equation 8.12 by the total mass of the rigid body, we would obtain the mass-weighted average of the square of the positions of all the mass elements away from the z-axis. Thus, moment of inertia is, in essence, the average of the squares of the radial distances away from the z-axis of all the mass elements making up the rigid body. You can understand physically why the moment of inertia must depend on the square (or, at least, some even power) of the distances away from the rotational axis; it could not be represented by a linear average over all the mass elements (or any average of the odd power of distance). If such were the case, then a body whose mass was symmetrically distributed about its rotational axis, such as a bicycle wheel, would have zero moment of inertia due to a term by term cancellation of the + and − weighted mass elements in the symmetrical distribution. An application of the slightest torque would spin up a bicycle wheel into an instantaneous frenzy, a condition that any bike racer knows is impossible.

We can formalize this discussion by defining a distance k, called the *radius of gyration*, to be this average, that is,

$$k^2 = \frac{I}{m} \qquad k = \sqrt{\frac{I}{m}} \tag{8.34}$$

Clearly, knowing the radius of gyration of any rigid body is equivalent to knowing its moment of inertia, but it better characterizes the nature of the averaging process upon which the concept of moment of inertia is based.

For example, we find for the radius of gyration of a thin rod about an axis passing through one end (see Equation 8.19)

$$k = \sqrt{\frac{\frac{1}{3}ma^2}{m}} = \frac{a}{\sqrt{3}}$$

Moments of inertia for various objects can be tabulated simply by listing the squares of their radii of gyration (Table 8.1).

8.4   THE PHYSICAL PENDULUM

A rigid body that is free to swing under its own weight about a fixed horizontal axis of rotation is known as a *physical pendulum* or *compound pendulum*. A physical pendulum is shown in Figure 8.12. Here CM is the center of mass, and O is the point on the axis of rotation that is in the vertical plane of the circular path of the center of mass.

Denoting the angle between the line OCM and the vertical line OA by θ, the moment of the gravitational force (acting at CM) about the axis of rotation is of magnitude

$$mgl \sin \theta$$

The fundamental equation of motion $N = I\dot{\omega}$ then takes the form $- mgl \sin \theta = I\ddot{\theta}$

$$\ddot{\theta} + \frac{mgl}{I} \sin \theta = 0 \tag{8.35}$$

TABLE 8.1   Values of k^2 of Various Bodies (Moment of Inertia = Mass $\times$ k^2)

Body	Axis	k^2
Thin rod, length a	Normal to rod at its center	$\dfrac{a^2}{12}$
	Normal to rod at one end	$\dfrac{a^2}{3}$
Thin rectangular lamina, sides a and b	Through the center, parallel to side b	$\dfrac{a^2}{12}$
	Through the center, normal to the lamina	$\dfrac{a^2 + b^2}{12}$
Thin circular disc, radius a	Through the center, in the plane of the disc	$\dfrac{a^2}{4}$
	Through the center, normal to the disc	$\dfrac{a^2}{2}$
Thin hoop (or ring), radius a	Through the center, in the plane of the hoop	$\dfrac{a^2}{2}$
	Through the center, normal to the plane of the hoop	a^2
Thin cylindrical shell, radius a, length b	Central longitudinal axis	a^2
Uniform solid right circular cylinder, radius a, length b	Central longitudinal axis	$\dfrac{a^2}{2}$
	Through the center, perpendicular to longitudinal axis	$\dfrac{a^2}{4} + \dfrac{b^2}{12}$
Thin spherical shell, radius a	Any diameter	$\dfrac{2}{3}a^2$
Uniform solid sphere, radius a	Any diameter	$\dfrac{2}{5}a^2$
Uniform solid rectangular parallelepiped, sides a, b, and c	Through the center, normal to face ab, parallel to edge c	$\dfrac{a^2 + b^2}{12}$

Equation 8.35 is identical in form to the equation of motion of a simple pendulum. For small oscillations, as in the case of the simple pendulum, we can replace $\sin \theta$ by θ:

$$\ddot{\theta} + \frac{mgl}{I}\theta = 0 \qquad (8.36)$$

The solution, as we know from Chapter 3, can be written

$$\theta = \theta_0 \cos(2\pi f_0 t - \epsilon) \qquad (8.37)$$

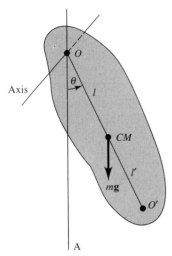

Figure 8.12   The physical pendulum.

where θ_0 is the amplitude and ϵ is a phase angle. The frequency of oscillation is given by

$$f_0 = \frac{1}{2\pi} \sqrt{\frac{mgl}{I}} \tag{8.38}$$

The period is, therefore, given by

$$T_0 = \frac{1}{f_0} = 2\pi \sqrt{\frac{I}{mgl}} \tag{8.39}$$

(To avoid confusion, we shall not use a specific symbol to designate the angular frequency $2\pi f_0$.) We can also express the period in terms of the radius of gyration k, namely,

$$T_0 = 2\pi \sqrt{\frac{k^2}{gl}} \tag{8.40}$$

Thus, the period is the same as that of a simple pendulum of length k^2/l.

Consider as an example a thin uniform rod of length a swinging as a physical pendulum about one end: $k^2 = a^2/3$, $l = a/2$. The period is then

$$T_0 = 2\pi \sqrt{\frac{a^2/3}{ga/2}} = 2\pi \sqrt{\frac{2a}{3g}}$$

which is the same as that of a simple pendulum of length $\frac{2}{3}a$.

Center of Oscillation

By use of the parallel-axis theorem, we can express the radius of gyration k in terms of the radius of gyration about the center of mass k_{cm}, as follows:

$$I = I_{cm} + ml^2$$

or

$$mk^2 = mk_{cm}^2 + ml^2$$

Canceling the m's, we get

$$k^2 = k_{cm}^2 + l^2 \tag{8.41}$$

Equation 8.40 can, therefore, be written as

$$T_0 = 2\pi \sqrt{\frac{k_{cm}^2 + l^2}{gl}} \tag{8.42}$$

Suppose that the axis of rotation of a physical pendulum is shifted to a different position O' at a distance l' from the center of mass, as shown in Figure 8.12. The period of oscillation T_0' about this new axis is given by

$$T_0' = 2\pi \sqrt{\frac{k_{cm}^2 + l'^2}{gl'}}$$

It follows that the periods of oscillation about O and about O' will be equal, provided

$$\frac{k_{cm}^2 + l^2}{l} = \frac{k_{cm}^2 + l'^2}{l'}$$

The above equation readily reduces to

$$ll' = k_{cm}^2 \tag{8.43}$$

The point O', related to O by Equation 8.43, is called the *center of oscillation* for the point O. It is clear that O is also the center of oscillation for O'. Thus, for a rod of length a swinging about one end, we have k_{cm}^2 and $a^2/12$ and $l = a/2$. Hence, from Equation 8.43, $l' = a/6$, and so the rod will have the same period when swinging about an axis located a distance $a/6$ from the center as it does for an axis passing through one end.

The "Upside-Down Pendulum." Elliptic Integrals

When the amplitude of oscillation of a pendulum is so large that the approximation $\sin \theta = \theta$ is not valid, the formula for the period (Equation 8.39) is not accurate. In Example 3.10 we obtained an improved formula for the period of a simple pendulum by using a method of successive approximations. That result also applies to the physical pendulum with l replaced by I/ml, but it is still an approximation and is completely erroneous when the amplitude approaches 180° (vertical position), Figure 8.13.

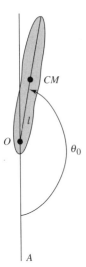

Figure 8.13 The upside-down pendulum.

To find the period for large amplitude, we start with the energy equation for the physical pendulum

$$\frac{1}{2}I\dot{\theta}^2 + mgh = E \tag{8.44}$$

where h is the vertical distance of the center of mass from the equilibrium position, that is, $h = l(1 - \cos\theta)$. Let θ_0 denote the amplitude of the pendulum's oscillation. Then $\dot{\theta} = 0$ when $\theta = \theta_0$, so that $E = mgl(1 - \cos\theta_0)$. The energy equation can then be written

$$\frac{1}{2}I\dot{\theta} + mgl(1 - \cos\theta) = mgl(1 - \cos\theta_0) \tag{8.44a}$$

Solving for $\dot{\theta}$ gives

$$\frac{d\theta}{dt} = \pm\left[\frac{2mgl}{I}(\cos\theta - \cos\theta_0)\right]^{1/2} \tag{8.44b}$$

Thus, by taking the positive root, we can write

$$t = \sqrt{\frac{I}{2mgl}}\int_0^\theta \frac{d\theta}{(\cos\theta - \cos\theta_0)^{1/2}} \tag{8.45}$$

from which we can, in principle, find t as a function of θ. Also, we note that θ increases from 0 to θ_0 in just one quarter of a complete cycle. The period T can, therefore, be expressed as

$$T = 4\sqrt{\frac{I}{2mgl}} \int_0^{\theta_0} \frac{d\theta}{(\cos\theta - \cos\theta_0)^{1/2}} \tag{8.46}$$

Unfortunately, the integrals in Equations 8.45 and 8.46 cannot be evaluated in terms of elementary functions. However, they can be expressed in terms of special functions known as *elliptic integrals*. For this purpose it is convenient to introduce a new variable of integration ϕ, which is defined as follows:

$$\sin\phi = \frac{\sin(\theta/2)}{\sin(\theta_0/2)} = \frac{1}{k}\sin(\theta/2)$$

where

$$k = \sin(\theta_0/2)$$

Thus, when $\theta = \theta_0$, we have $\sin\phi = 1$ and so $\phi = \pi/2$. The result of making the above substitutions in Equations 8.45 and 8.46 yields

$$t = \sqrt{\frac{I}{mgl}} \int_0^{\phi} \frac{d\phi}{(1 - k^2\sin^2\phi)^{1/2}} \tag{8.47}$$

$$T = 4\sqrt{\frac{I}{mgl}} \int_0^{\pi/2} \frac{d\phi}{(1 - k^2\sin^2\phi)^{1/2}} \tag{8.48}$$

The steps are left as an exercise and involve use of the identity $\cos\theta = 1 - 2\sin^2(\theta/2)$.

Tabulated values of the integrals in the above expressions can be found in various handbooks and mathematical tables. The first integral

$$\int_0^{\phi} \frac{d\phi}{(1 - k^2\sin^2\phi)^{1/2}} = F(k, \phi) \tag{8.49}$$

is called the *incomplete elliptic integral of the first kind*. In our problem, given a value of the amplitude θ_0, we can find the relationship between θ and t through a series of steps involving the definitions of k and ϕ. We are more interested in finding the period of the pendulum, which involves the second integral

$$\int_0^{\pi/2} \frac{d\phi}{(1 - k^2\sin^2\phi)^{1/2}} = F(k, \pi/2) \tag{8.50}$$

known as the *complete elliptic integral of the first kind*. [It is also variously listed as $K(k)$ or $F(k)$ in many tables.] In terms of it, the period is

$$T = 4\sqrt{\frac{I}{mgl}}\, F(k, \pi/2) \tag{8.51}$$

TABLE 8.2   Selected Values of the Complete Elliptic Integral and Corresponding Period of Oscillation of a Physical Pendulum.[1]

Amplitude, θ_0	$k = \sin(\theta_0/2)$	$F(k, \pi/2)$	Period, T
0°	0	$1.5708 = \pi/2$	T_0
10°	0.0872	1.5738	$1.0019\ T_0$
45°	0.3827	1.6336	$1.0400\ T_0$
90°	0.7071	1.8541	$1.1804\ T_0$
135°	0.9234	2.4003	$1.5281\ T_0$
178°	0.99985	5.4349	$3.5236\ T_0$
179°	0.99996	7.2660	$4.6002\ T_0$
180°	1	∞	∞

[1]For more extensive tables and other information on elliptic integrals, consult any treatise on elliptic functions, such as (1) H.B. Dwight, *Tables of Integrals and Other Mathematical Data,* The Macmillan Co., New York, 1961; and (2) M. Abramowitz and A. Stegun, *Handbook of Mathematical Functions,* Dover Publishing, New York, 1972.

Table 8.2 lists selected values of $F(k, \pi/2)$. Also listed is the period T as a factor multiplied by the period for zero amplitude: $T_0 = 2\pi(I/mgl)^{1/2}$.

Table 8.2 shows the trend as the amplitude approaches 180° at which value the elliptic integral diverges and the period becomes infinitely large. This means that, *theoretically,* a physical pendulum, such as a rigid rod, if placed exactly in the vertical position with absolutely zero initial angular velocity, would remain in that same unstable position indefinitely.

8.5   A GENERAL THEOREM CONCERNING ANGULAR MOMENTUM

In order to study the more general case of rigid-body motion, that in which the axis of rotation is not fixed, we need to develop a fundamental theorem about angular momentum. In Section 7.2 we showed that the time rate of change of angular momentum of any system is equal to the applied torque:

$$\frac{d\mathbf{L}}{dt} = \mathbf{N}$$

or, explicitly

$$\frac{d}{dt} \sum_i (\mathbf{r}_i \times m_i\mathbf{v}_i) = \sum_i (\mathbf{r}_i \times \mathbf{F}_i) \tag{8.52}$$

In the above equation all quantities are referred to some inertial coordinate system.

Let us now introduce the center of mass by expressing the position vector of each particle $\mathbf{r}_i$ in terms of the position of the center of mass $\mathbf{r}_{cm}$ and the position vector of particle i relative to the center of mass $\bar{\mathbf{r}}_i$ (as in Section 7.2), namely,

$$\mathbf{r}_i = \mathbf{r}_{cm} + \bar{\mathbf{r}}_i$$

and

$$\mathbf{v}_i = \mathbf{v}_{cm} + \bar{\mathbf{v}}_i$$

Equation 8.52 then becomes

$$\frac{d}{dt} \sum_i [(\mathbf{r}_{cm} + \bar{\mathbf{r}}_i) \times m_i(\mathbf{v}_{cm} + \bar{\mathbf{v}}_i)] = \sum_i (\mathbf{r}_{cm} + \bar{\mathbf{r}}_i) \times \mathbf{F}_i \qquad (8.53)$$

Upon expanding and using the fact that $\Sigma m_i \bar{\mathbf{r}}_i$ and $\Sigma m_i \bar{\mathbf{v}}_i$ both vanish, we find that the above equation reduces to

$$\mathbf{r}_{cm} \times \sum_i m_i \mathbf{a}_{cm} + \frac{d}{dt} \sum_i \bar{\mathbf{r}}_i \times m_i \bar{\mathbf{v}}_i = \mathbf{r}_{cm} \times \sum_i \mathbf{F}_i + \sum_i \bar{\mathbf{r}}_i \times \mathbf{F}_i \qquad (8.54)$$

where $\mathbf{a}_{cm} = \dot{\mathbf{v}}_{cm}$.

In Section 7.1 we showed that the translation of the center of mass of any system of particles obeys the equation

$$\sum_i \mathbf{F}_i = \sum_i m_i \mathbf{a}_i = m\mathbf{a}_{cm}$$

Consequently, the first term on the left of Equation 8.54 cancels the first term on the right. The final result is

$$\frac{d}{dt} \sum_i \bar{\mathbf{r}}_i \times m_i \bar{\mathbf{v}}_i = \sum_i \bar{\mathbf{r}}_i \times \mathbf{F}_i \qquad (8.55)$$

The sum on the left in the above equation is just the angular momentum of the system about the center of mass, and the sum on the right is the total moment of the external forces about the center of mass. Calling these quantities $\bar{\mathbf{L}}$ and $\bar{\mathbf{N}}$, respectively, we have

$$\frac{d\bar{\mathbf{L}}}{dt} = \bar{\mathbf{N}} \qquad (8.56)$$

This important result states that the time rate of change of angular momentum about the center of mass of any system is equal to the total moment of the external forces about the center of mass. This is true even if the center of mass is accelerating. If we choose any point other than the center of mass as a reference point, then that point must be at rest in an inertial coordinate system (except for certain special cases that we shall not attempt to discuss). An example of the use of Equation 8.56 is given in the next section.

8.6   LAMINAR MOTION OF A RIGID BODY

If the motion of a body is such that all particles move parallel to some fixed plane, then that motion is called *laminar*. In laminar motion the axis of rotation may change position, but it does not change in direction. Rotation about a fixed axis is a special case of laminar motion. The rolling of a cylinder on a plane surface is another example of laminar motion.

If a body undergoes a laminar displacement, that displacement can be specified as follows: Choose some reference point of the body, for example, the center of mass. The reference point undergoes some displacement $\Delta\mathbf{r}$. In addition, the body rotates about the reference point through some angle $\Delta\phi$. Clearly, any laminar displacement can be so specified. Consequently, laminar motion can be specified by giving the translational velocity of a convenient reference point together with the angular velocity.

The fundamental equation governing translation of a rigid body is

$$\mathbf{F} = m\ddot{\mathbf{r}}_{cm} = m\dot{\mathbf{v}}_{cm} = m\mathbf{a}_{cm} \tag{8.57}$$

where $\mathbf{F}$ represents the sum of all the external forces acting on the body, m is the mass, and $\mathbf{a}_{cm}$ is the acceleration of the center of mass.

Application of Equation 8.14 to the case of laminar motion of a rigid body yields

$$\bar{L}_C = I_{cm}\omega \tag{8.58}$$

for the component of the angular momentum about an axis C passing through the center of mass where ω is the angular speed of rotation about that axis. The fundamental equation governing the rotation of the body (Equation 8.56) then becomes

$$\frac{d\bar{L}_C}{dt} = I_{cm}\dot{\omega} = \bar{N}_C \tag{8.59}$$

where $\bar{N}_C$ is the total moment of the applied forces about the axis C.

Body Rolling Down an Inclined Plane

As an illustration of laminar motion, we shall study the motion of a round object (cylinder, ball, and so on) rolling down an inclined plane. As shown in Figure 8.14, there are three forces acting on the body. These are (1) the downward force of gravity, (2) the

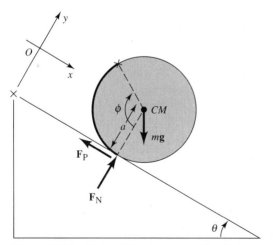

Figure 8.14   Body rolling down an inclined plane.

normal reaction of the plane: $\mathbf{F}_N$, and (3) the frictional force parallel to the plane: $\mathbf{F}_P$. Choosing axes as shown, the component equations of the translation of the center of mass are

$$m\ddot{x}_{cm} = mg \sin \theta - F_P \qquad (8.60)$$

$$m\ddot{y}_{cm} = -mg \cos \theta + F_N \qquad (8.61)$$

where θ is the inclination of the plane to the horizontal. Since the body remains in contact with the plane, we have

$$y_{cm} = \text{constant}$$

Hence,

$$\ddot{y}_{cm} = 0$$

Therefore, from Equation 8.61,

$$F_N = mg \cos \theta \qquad (8.62)$$

The only force that exerts a moment about the center of mass is the frictional force $\mathbf{F}_P$. The magnitude of this moment is $F_P a$ where a is the radius of the body. Hence, the rotational equation (Equation 8.59) becomes

$$I_{cm}\dot{\omega} = F_P a \qquad (8.63)$$

To discuss the problem further, we need to make some assumptions regarding the contact between the plane and the body. We shall solve the equations of motion for two cases.

Motion with No Slipping

If the contact is very rough so that no slipping can occur, that is, if $F_P \leq \mu_s F_N$ where μ_s is the coefficient of *static* friction, we have the following relations:

$$\dot{x}_{cm} = a\dot{\phi} = a\omega \qquad (8.64)$$

$$\ddot{x}_{cm} = a\ddot{\phi} = a\dot{\omega}$$

where ϕ is the angle of rotation. Equation 8.63 can then be written

$$\frac{I_{cm}}{a^2}\ddot{x}_{cm} = F_P \qquad (8.65)$$

Substituting the above value for F_P into Equation 8.60 yields

$$m\ddot{x}_{cm} = mg \sin \theta - \frac{I_{cm}}{a^2}\ddot{x}_{cm}$$

Solving for $\ddot{x}_{cm}$, we find

$$\ddot{x}_{cm} = \frac{mg \sin \theta}{m + (I_{cm}/a^2)} = \frac{g \sin \theta}{1 + (k_{cm}^2/a^2)} \qquad (8.66)$$

where k_{cm} is the radius of gyration about the center of mass. The body, therefore, rolls down the plane with constant linear acceleration and also with constant angular acceleration by virtue of Equations 8.64.

For example, the acceleration of a uniform cylinder ($k_{cm}^2 = a^2/2$) is

$$\frac{g \sin \theta}{1 + \dfrac{1}{2}} = \frac{2}{3} g \sin \theta$$

whereas that of a uniform sphere ($k_{cm}^2 = 2a^2/5$) is

$$\frac{g \sin \theta}{1 + \dfrac{2}{5}} = \frac{5}{7} g \sin \theta$$

Energy Considerations

The above results can also be obtained from energy considerations. In a uniform gravitational field the potential energy V of a rigid body is given by the sum of the potential energies of the individual particles, namely,

$$V = \sum_i (m_i g z_i) = m g z_{cm}$$

where z_{cm} is the vertical distance of the center of mass from some (arbitrary) reference plane. Now if the forces, other than gravity, acting on the body do no work, then the motion is conservative, and we can write

$$T + V = T + m g z_{cm} = E = \text{constant}$$

where T is the kinetic energy.

In the case of the body rolling down the inclined plane, Figure 8.14, the kinetic energy of translation is $\frac{1}{2}m \dot{x}_{cm}^2$ and that of rotation is $\frac{1}{2}I_{cm}\omega^2$, so the energy equation reads

$$\frac{1}{2}m\dot{x}_{cm}^2 + \frac{1}{2}I_{cm}\omega^2 + m g z_{cm} = E$$

But $\omega = \dot{x}_{cm}/a$ and $z_{cm} = -x_{cm} \sin \theta$. Hence,

$$\frac{1}{2}m\dot{x}_{cm}^2 + \frac{1}{2}m k_{cm}^2 \frac{\dot{x}_{cm}^2}{a^2} - m g x_{cm} \sin \theta = E$$

In the case of pure rolling motion, the frictional force does not appear in the energy equation since no mechanical energy is converted into heat unless slipping occurs. Thus, the total energy E is constant. Differentiating with respect to t and collecting terms yields

$$m\dot{x}_{cm}\ddot{x}_{cm}\left(1 + \frac{k_{cm}^2}{a^2}\right) - m g \dot{x}_{cm} \sin \theta = 0$$

Canceling the common factor $\dot{x}_{cm}$ (assuming, of course, that $\dot{x}_{cm} \neq 0$) and solving for

$\dot{x}_{cm}$, we find the same result as that obtained previously using forces and moments (Equation 8.66).

Occurrence of Slipping

Let us now consider the case in which the contact with the plane is not perfectly rough but has a certain coefficient of *sliding* friction μ_k. If slipping occurs, then the magnitude of the frictional force $\mathbf{F}_P$ is given by

$$F_P = \mu_k F_N = \mu_k mg \cos \theta \tag{8.67}$$

The equation of translation (Equation 8.60) then becomes

$$m\ddot{x}_{cm} = mg \sin \theta - \mu_k mg \cos \theta \tag{8.68}$$

and the rotational equation (Equation 8.63) is

$$I_{cm}\dot{\omega} = \mu_k mga \cos \theta \tag{8.69}$$

From Equation 8.68 we see that again the center of mass undergoes constant acceleration:

$$\ddot{x}_{cm} = g(\sin \theta - \mu_k \cos \theta) \tag{8.70}$$

and, at the same time, the angular acceleration is constant:

$$\dot{\omega} = \frac{\mu_k mga \cos \theta}{I_{cm}} = \frac{\mu_k ga \cos \theta}{k_{cm}^2} \tag{8.71}$$

Let us integrate these two equations with respect to t, assuming that the body starts from rest, that is, at $t = 0$, $\dot{x}_{cm} = 0$, $\dot{\phi} = 0$. We obtain

$$\dot{x}_{cm} = g(\sin \theta - \mu_k \cos \theta)t \tag{8.72}$$

$$\omega = \dot{\phi} = g(\mu_k a \cos \theta/k_{cm}^2)t$$

Consequently, the linear speed and the angular speed have a constant ratio, and we can write

$$\dot{x}_{cm} = \gamma a\omega$$

where

$$\gamma = \frac{\sin \theta - \mu \cos \theta}{\mu_k a^2 \cos \theta/k_{cm}^2} = \frac{k_{cm}^2}{a^2}\left(\frac{\tan \theta}{\mu_k} - 1\right) \tag{8.73}$$

Now $a\omega$ cannot be greater than $\dot{x}_{cm}$, so γ cannot be less than unity. The limiting case, that for which we have pure rolling, is given by $\dot{x}_{cm} = a\omega$, that is,

$$\gamma = 1$$

Solving for μ_k in Equation 8.73 with $\gamma = 1$, we find that the critical value of the coefficient of friction is given by

$$\mu_{crit} = \frac{\tan \theta}{1 + (a/k_{cm})^2} \tag{8.74}$$

(Actually this is the critical value for the coefficient of *static* friction μ_s.) If μ_s is greater than that given above, then the body rolls without slipping.

For example, if a ball is placed on a 45° plane, it will roll without slipping provided μ_s is greater than $\tan 45°/(1 + \frac{5}{2})$ or $\frac{2}{7}$.

EXAMPLE 8.2

A small, uniform cylinder of radius R rolls without slipping along the inside of a cylindrical surface of radius $r \gg R$ as shown in Figure 8.15. Assuming only small excursions away from the equilibrium position, show that the motion of the center of mass of the small cylinder along the x direction is simple harmonic and that the frequency of oscillation is equivalent to that of a simple pendulum of a length $l = 3(r - R)/2$.

Solution:
We can make the following approximations for the x and y positions of the center of mass of the small, inner cylinder relative to its equilibrium position

$$x = (r - R) \sin \theta \approx (r - R) \theta$$

$$\dot{x} = (r - R) \dot{\theta}$$

$$\ddot{x} = (r - R) \ddot{\theta}$$

$$y = (r - R)(1 - \cos \theta) \approx \frac{(r - R) \theta^2}{2} \approx 0$$

$$\dot{y} \approx (r - R) \theta\dot{\theta} \approx 0$$

$$\ddot{y} \approx (r - R)(\dot{\theta}^2 + \theta\ddot{\theta}) \approx 0$$

The y coordinate of the center of mass of the cylinder (and its derivatives) depend on products of small quantities. Hence, they are all approximately zero.

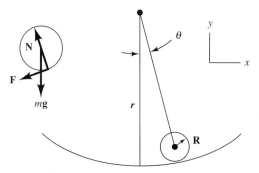

Figure 8.15   Small cylinder free to roll without slipping on the inside of a large cylindrical surface.

The forces acting on the cylinder are shown in the insert of Figure 8.15. **N** is the normal force exerted on the inner cylinder by the surface upon which it rolls. **F** is the force of static friction between the two cylindrical surfaces. Using the above approximations, we can write Newton's second law of motion for the y coordinate as

$$N \cos \theta \; - \; F \sin \theta \; - \; mg \; = \; m\ddot{y} \; \approx \; 0$$

The second term above is small and can be neglected with respect to the first term. This can be seen as follows: The largest that F can be is $\mu_s N$, a condition that holds true only when the inner cylinder is about to slip. Since μ_s is always less than 1, F is less than N. Substituting this upper bound for F into the above equation, along with the first-order, small angle approximations for both $\sin \theta$ and $\cos \theta$ ($\sin \theta \approx \theta$ and $\cos \theta \approx 1$), we get

$$N \; - \; \mu_s N \theta \; - \; mg \; \approx \; 0$$

$$\therefore N \; \approx \; \frac{mg}{1 \; - \; \mu_s \theta} \; \approx \; mg$$

The last step follows by neglecting the small product $\mu_s \theta$ with respect to 1. Therefore, the normal force **N** is essentially equal to the weight of the cylinder primarily because motion along the y coordinate is negligible. The second law of motion for the x coordinate is

$$- F \cos \theta \; - \; N \sin \theta \; = \; m\ddot{x}$$

$$- F \; - \; mg \; \theta \; \approx \; m\ddot{x}$$

We can estimate the frictional force **F** by calculating the total torque about the center of mass of the cylinder and setting the result equal to the rate of change of the angular momentum about the center of mass, according to Equation 8.56, even though the cylinder is accelerating

$$FR \; = \; I_{cm}\dot{\omega} \; = \; I_{cm}\frac{\dot{v}_{cm}}{R} \; = \; I_{cm}\frac{(r \; - \; R) \, \ddot{\theta}}{R} \; \approx \; I_{cm}\frac{\ddot{x}}{R}$$

$$F \; \approx \; \frac{I_{cm}}{R^2} \, \ddot{x}$$

Substituting this for F, along with the moment of inertia of a cylinder, $mR^2/2$, into the x coordinate equation of motion, yields

$$-\frac{I_{cm}}{R^2}\ddot{x} \; - \; mg\theta \; = \; m\ddot{x}$$

$$\frac{3}{2}\ddot{x} \; + \; g\frac{x}{(r \; - \; R)} \; = \; 0$$

$$\ddot{x} \; + \; \frac{g}{\dfrac{3}{2}(r \; - \; R)}x \; = \; 0$$

Thus, the cylinder executes simple harmonic motion along the x direction with the same frequency as that of a pendulum whose length is $3(r - R)/2$.   ■

EXAMPLE 8.3

Solve Example 8.2 using energy considerations, only this time obtain an equation of motion for the θ coordinate of the center of mass of the cylinder, thus explicitly demonstrating the equivalence of this system to that of a simple pendulum of length $3(r - R)/2$.

Solution:
The key to a solution using energy considerations hinges on the fact that energy is conserved by this system. Since there is no relative motion between the two surfaces (no slipping), friction does not lead to any loss of energy. Moreover, since the energy of a system is nothing other than the first integral of the equation of motion, we should be able to generate the equation of motion for the θ coordinate by first calculating the energy as a function of θ, then taking its time derivative and setting the result equal to zero.

First, we write down an expression for the kinetic energy

$$T = \frac{1}{2}mv_{cm}^2 + \frac{1}{2}I_{cm}\omega^2$$

Using the geometrical approximations from Example 8.2, we can express the kinetic energy as a function of θ and $\dot{\theta}$.

$$T = \frac{1}{2}m(r - R)^2\dot{\theta}^2 + \frac{1}{2}\left(\frac{1}{2}mR^2\right)\frac{(r - R)^2\dot{\theta}^2}{R^2}$$
$$= \frac{3}{4}m(r - R)^2\dot{\theta}^2$$

The potential energy of the center of mass is

$$V = mg(r - R)(1 - \cos\theta)$$
$$\approx \frac{1}{2}mg(r - R)\theta^2$$

The total energy of the system is, thus, quadratic in $\dot{\theta}$ and θ

$$E = T + V = \frac{3}{4}m(r - R)^2\dot{\theta}^2 + \frac{1}{2}mg(r - R)\,\theta^2$$

and setting the time derivative of the total energy to zero leads us directly to the desired equation of motion

$$\dot{E} = 0 = \frac{3}{2}m(r - R)^2\dot{\theta}\ddot{\theta} + mg(r - R)\theta\dot{\theta}$$

$$\therefore \ddot{\theta} + \frac{g}{\frac{3}{2}(r - R)}\theta = 0$$

The problem was much more tractable using energy methods.   ■

8.7 IMPULSE AND COLLISIONS INVOLVING RIGID BODIES

In the previous chapter we considered the case of an impulsive force acting on a particle. In this section we shall extend the notion of impulsive force to the case of laminar motion of a rigid body. First, we know that the translation of the body, assuming constant mass, is governed by the general equation $\mathbf{F} = m d\mathbf{v}_{cm}/dt$, so that if $\mathbf{F}$ is an impulsive type of force, the change of linear momentum of the body is given by

$$\int \mathbf{F}\ dt = \hat{\mathbf{P}} = m\Delta\mathbf{v}_{cm} \tag{8.75}$$

Thus, the result of an impulse $\hat{\mathbf{P}}$ is to produce a sudden change in the velocity of the center of mass by an amount

$$\Delta\mathbf{v}_{cm} = \frac{\hat{\mathbf{P}}}{m} \tag{8.76}$$

Second, the rotational part of the motion of the body obeys the equation $N = \dot{L} = I d\omega/dt$, so the change in angular momentum is

$$\int N\ dt = I\Delta\omega \tag{8.77}$$

The integral $\int N\ dt$ is called the *rotational impulse*. Let us use the symbol $\hat{L}$ to designate it. Now if the primary impulse $\hat{\mathbf{P}}$ is applied to the body in such a way that its line of action is a distance l from the reference axis about which the angular momentum is calculated, then $N = Fl$, and we have

$$\hat{L} = \hat{P}l \tag{8.78}$$

Consequently, the change in angular velocity produced by an impulse $\hat{\mathbf{P}}$ acting on a rigid body in laminar motion is given by

$$\Delta\omega = \frac{\hat{P}l}{I} \tag{8.79}$$

For the general case of free laminar motion, the reference axis must be taken through the center of mass, and the moment of inertia $I = I_{cm}$. On the other hand, if the body is constrained to rotate about a fixed axis, then the rotational equation alone suffices to determine the motion, and I is the moment of inertia about the fixed axis.

In collisions involving rigid bodies, the forces, and therefore the impulses, that the bodies exert on one another during the collision are always equal and opposite. Thus, the principles of conservation of linear and angular momentum apply.

Center of Percussion: The "Baseball Bat Theorem"

As an example of the above theory, let us discuss the collision of a ball of mass m, treated as a particle, with a rigid body (bat) of mass M. For simplicity we shall assume that the body is initially at rest on a smooth horizontal surface and is free to move in laminar-type motion. Let $\hat{\mathbf{P}}$ denote the impulse delivered to the body by the ball. Then the equations for translation are

$$\hat{\mathbf{P}} = M\mathbf{v}_{cm} \tag{8.80}$$

$$-\hat{\mathbf{P}} = m\mathbf{v}_1 - m\mathbf{v}_0 \tag{8.81}$$

where $\mathbf{v}_0$ and $\mathbf{v}_1$ are, respectively, the initial and final velocities of the ball and $\mathbf{v}_{cm}$ is the velocity of the mass center of the body after the impact. Clearly, the above two equations imply conservation of linear momentum.

Since the body is initially at rest, the rotation about the center of mass, as a result of the impact, is given by

$$\omega = \frac{\hat{P}l}{I_{cm}} \tag{8.82}$$

in which l is the distance OC from the center of mass C to the line of action of $\hat{\mathbf{P}}$, as shown in Figure 8.16. Let us now consider a point O' located a distance l' from the

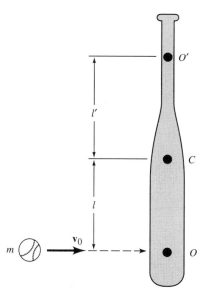

Figure 8.16   Baseball colliding with a bat.

center of mass such that the line CO' is the extension of OC, as shown. The (scalar) velocity of O' is obtained by combining the translational and rotational parts, namely,

$$v_{O'} = v_{cm} - \omega l' = \frac{\hat{P}}{M} - \frac{\hat{P}l}{I_{cm}} l' = \hat{P}\left(\frac{1}{M} - \frac{ll'}{I_{cm}}\right) \tag{8.83}$$

In particular, the velocity of O' will be zero if the quantity in parentheses vanishes, that is, if

$$ll' = \frac{I_{cm}}{M} = k_{cm}^2 \tag{8.84}$$

where k_{cm} is the radius of gyration of the body about its center of mass. In this case the point O' is the instantaneous center of rotation of the body just after impact. O is called the *center of percussion* about O'. The two points are related in the same way as the centers of oscillation, defined previously in our analysis of the physical pendulum (Equation 8.43).

Anyone who has played baseball knows that if the ball hits the bat in just the right spot there will be no "sting" upon impact. This "right spot" is just the center of percussion about the point at which the bat is held.

PROBLEMS

8.1 Find the center of mass of each of the following:
  (a) A thin wire bent into the form of a block "⬚" with each segment of equal length b
  (b) A quadrant of a uniform circular lamina of radius b
  (c) The area bounded by parabola $y = x^2/b$ and the line $y = b$
  (d) The volume bounded by paraboloid of revolution $z = (x^2 + y^2)/b$ and the plane $z = b$
  (e) A solid uniform right circular cone of height b

8.2 The linear density of a thin rod is given by $\rho = cx$ where c is a constant and x is the distance measured from one end. If the rod is of length b, find the center of mass.

8.3 A solid uniform sphere of radius a has a spherical cavity of radius $a/2$ centered at a point $a/2$ from the center of the sphere. Find the center of mass.

8.4 Find the moments of inertia of each of the objects in Problem 8.1 about their symmetry axes.

8.5 Find the moment of inertia of the sphere in Problem 8.3 about an axis passing through the center of the sphere and the center of the cavity.

8.6 Show that the moment of inertia of a solid uniform octant of a sphere of radius a is $(2/5)ma^2$ about an axis along one of the straight edges. (*Note:* This is the same formula as that for a solid sphere of the same radius.)

8.7 Show that the moments of inertia of a solid uniform rectangular parallelepiped, elliptic cylinder, and ellipsoid are, respectively, $(m/3)(a^2 + b^2)$, $(m/4)(a^2 + b^2)$, and $(m/5)(a^2 + b^2)$, where m is the mass, and $2a$ and $2b$ are the principal diameters of the solid at right angles to the axis of rotation, the axis being through the center in each case.

8.8 Show that the period of a physical pendulum is equal to $2\pi(d/g)^{1/2}$, where d is the distance between the point of suspension O and the center of oscillation O'.

8.9 A circular hoop of radius a swings as a physical pendulum about a point on the circumference. Find the period of oscillation for small amplitude if the axis of rotation is

(a) Normal to the plane of the hoop

(b) In the plane of the hoop

8.10 A uniform solid ball has a few turns of light string wound around it. If the end of the string is held steady and the ball is allowed to fall under gravity, what is the acceleration of the center of the ball? (Assume the string remains vertical.)

8.11 Two people are holding the ends of a uniform plank of length l and mass m. Show that if one person suddenly lets go, the load supported by the other person suddenly drops from $mg/2$ to $mg/4$. Show also that the initial downward acceleration of the free end is $\frac{3}{2}g$.

8.12 A uniform solid ball contains a hollow spherical cavity at its center, the radius of the cavity being $\frac{1}{2}$ the radius of the ball. Show that the acceleration of the ball rolling down a rough inclined plane is just $\frac{98}{101}$ of that of a uniform solid ball with no cavity. (*Note:* This suggests a method for nondestructive testing.)

8.13 Two weights of mass m_1 and m_2 are tied to the ends of a light inextensible cord. The cord passes over a rough pulley of radius a and moment of inertia I. Find the accelerations of the weights, assuming $m_1 > m_2$ and neglecting friction in the axle of the pulley.

8.14 A uniform right-circular cylinder of radius a is balanced on the top of a perfectly rough fixed cylinder of radius $b(b > a)$, the axes of the two cylinders being parallel. If the balance is slightly disturbed, show that the rolling cylinder leaves the fixed one when the line of centers makes an angle with the vertical of $\cos^{-1}(4/7)$.

8.15 A uniform ladder leans against a smooth vertical wall. If the floor is also smooth, and the initial angle between the floor and the ladder is θ_0, show that the ladder, in sliding down, will lose contact with the wall when the angle between the floor and the ladder is $\sin^{-1}(\frac{2}{3}\sin\theta_0)$.

8.16 At Cape Canaveral a Saturn V rocket stands in a vertical position ready for launch. Unfortunately, before firing, a slight disturbance causes the rocket to fall over. Find the horizontal and vertical components of the reaction on the launch pad as functions of the angle θ between the rocket and the vertical at any instant. Show from this that the rocket will tend to slide backward for $\theta < \cos^{-1}(2/3)$ and forward for $\theta > \cos^{-1}(2/3)$. (Assume the rocket to be a thin uniform rod.)

8.17 A ball is initially projected, without rotation, at a speed v_0 up a rough inclined plane of inclination θ and coefficient of sliding friction μ_k. Find the position of the ball as a function of time and determine the position of the ball when pure rolling begins. Assume that μ_k is greater than $\frac{2}{7}\tan\theta$.

8.18 A billiard ball of radius a is initially spinning about a horizontal axis with angular speed ω_0 and with zero forward speed. If the coefficient of sliding friction between the ball and the billiard table is μ_k, find the distance the ball travels before slipping ceases to occur.

8.19 A thin uniform plank of length l lies at rest on a horizontal sheet of ice. If the plank is given a kick at one end in a direction normal to the plank, show that the plank will begin to rotate about a point located a distance $l/6$ from the center.

8.20 Show that the edge (cushion) of a billiard table should be at a height of $7/10$ of the diameter of the billiard ball in order that no reaction occurs between the table surface and the ball when the ball strikes the cushion.

8.21 A ballistic pendulum is made of a long plank of length l and mass m. It is free to swing about one end O and is initially at rest in a vertical position. A bullet of mass m' is fired horizontally into the pendulum at a distance l' below O, the bullet coming to rest in the plank. If the resulting amplitude of oscillation of the pendulum is θ_0, find the speed of the bullet.

8.22 Two uniform rods AB and BC of equal mass m and equal length l are smoothly joined at B. The system is initially at rest on a smooth horizontal surface, the points A, B, and C lying in a straight line. If an impulse $\hat{\mathbf{P}}$ is applied at A at right angles to the rod, find the initial motion of the system. (*Hint:* Isolate the rods.)

COMPUTER/CALCULATOR APPLICATIONS

8.1 Shown below is a table of the density of the Sun as a function of distance from the solar core. The densities are given as fractions relative to the core density ρ_c, and the distances are given as fractions of the solar radius $R_\odot$.

Distance ($r/R_\odot$)	Density ($\rho/\rho c$)
0.00	1.000
0.01	0.991
0.02	0.967
0.03	0.933
0.04	0.891
0.05	0.844
0.06	0.794
0.07	0.743
0.08	0.692
0.09	0.641
0.10	0.592
0.20	0.231
0.30	0.0780
0.40	0.0244
0.50	7.30×10^{-3}
0.60	2.11×10^{-3}
0.70	5.99×10^{-4}
0.80	1.67×10^{-4}
0.90	4.59×10^{-5}
1.00	1.25×10^{-5}

The mass of the sun is $M_\odot = 1.989 \times 10^{30}$ kg, and its radius is $R_\odot = 6.96 \times 10^5$ km.

(a) Using the above data, estimate the core density of the Sun by numerical integration. (*Hint: It may be necessary to interpolate to do the integration.*)

(b) Estimate the moment of inertia of the Sun.

(c) Assuming that the period of rotation for the Sun is 25 days, estimate its rotational angular momentum.

(d) Suppose the Sun "imploded" such that its entire mass were suddenly concentrated within a spherical ball of radius 10 km and uniform density. What would be its new period of rotation?

9

MOTION OF RIGID BODIES
IN THREE DIMENSIONS

'The body can no longer participate in the diurnal motion which actuates our sphere. Indeed, although because of its short length, its axis appears to preserve its original direction relatively to terrestrial objects, the use of a microscope is suffi-cient to establish an apparent and continuous motion which follows the motion of the celestial sphere exactly. . . . As the original direction of this axis is disposed ar-bitrarily in all azimuths about the vertical, the observed deviations can be, at will, given all the values contained between that of the total deviation and that of this total deviation as reduced by the sine of the latitude. . . . In one fell swoop, with a deviation in the desired direction, a new proof of the rotation of the Earth is ob-tained; this with an instrument reduced to small dimensions, easily transportable, and which mirrors the continuous motion of the Earth itself. . . ."

(J. B. L. Foucalt, Comptes rendus de l' Academie Sciences, *Vol 35, 27-Sep-1852)*

In the motion of a rigid body constrained either to rotate about a fixed axis or to move parallel to a fixed plane, the direction of the axis does not change. In the more general cases of rigid-body motion, which we take up in this chapter, the direction of the axis may vary. Compared to the previous chapter, the analysis here is considerably more involved. In fact, even in the case of a freely rotating body on which no external forces whatever are acting, the motion, as we shall see, is not simple.

9.1 ROTATION OF A RIGID BODY ABOUT AN ARBITRARY AXIS. MOMENTS AND PRODUCTS OF INERTIA. ANGULAR MOMENTUM AND KINETIC ENERGY

We begin the study of the general motion of a rigid body with some mathematical pre-liminaries. First, we shall give a calculation of the moment of inertia about an axis whose direction is arbitrary. The axis passes through a fixed point O, Figure 9.1(a), taken as the origin of our coordinate system. We shall apply the fundamental definition

$$I = \sum_i m_i R_i^2$$

where R_i is the perpendicular distance from the particle of mass m_i to the axis of rotation. The direction of the axis of rotation is defined by the unit vector **n**. Then

$$R_i = |r_i \sin \theta_i| = |\mathbf{r}_i \times \mathbf{n}|$$

in which θ_i is the angle between $\mathbf{r}_i$ and **n,** and

$$\mathbf{r}_i = \mathbf{i}x_i + \mathbf{j}y_i + \mathbf{k}z_i$$

is the position vector of the ith particle. Let the direction cosines of the axis be $\cos \alpha$, $\cos \beta$, and $\cos \gamma$ (See Figures 9.1). Then

$$\mathbf{n} = \mathbf{i} \cos \alpha + \mathbf{j} \cos \beta + \mathbf{k} \cos \gamma \qquad (9.1)$$

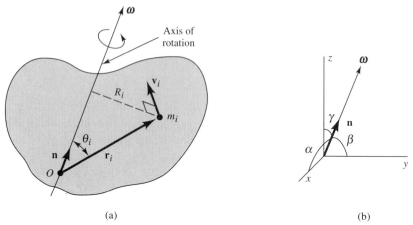

(a)   (b)

Figure 9.1a   The velocity vector of a representative particle of a rotating rigid body.
Figure 9.1b α, β and γ are the angles that the angular velocity vector $\boldsymbol{\omega}$ (or the vector $\mathbf{n}$) makes with the x, y, and z axes respectively.

and so

$$R_i^2 = |\mathbf{r}_i \times \mathbf{n}|^2$$

$$= (y_i \cos \gamma - z_i \cos \beta)^2 + (z_i \cos \alpha - x_i \cos \gamma)^2 + (x_i \cos \beta - y_i \cos \alpha)^2$$

Upon rearranging terms, we can write

$$R_i^2 = (y_i^2 + z_i^2) \cos^2 \alpha + (z_i^2 + x_i^2) \cos^2 \beta + (x_i^2 + y_i^2) \cos^2 \gamma$$
$$- 2y_i z_i \cos \gamma \cos \beta - 2z_i x_i \cos \alpha \cos \gamma - 2x_i y_i \cos \alpha \cos \beta$$

The moment of inertia about our general axis of rotation is then given by the rather lengthy expression

$$I = \sum_i m_i(y_i^2 + z_i^2) \cos^2 \alpha + \sum_i m_i(z_i^2 + x_i^2) \cos^2 \beta$$
$$+ \sum_i m_i(x_i^2 + y_i^2) \cos^2 \gamma - 2\sum_i m_i y_i z_i \cos \gamma \cos \beta \qquad (9.2)$$
$$- 2\sum_i m_i z_i x_i \cos \alpha \cos \gamma - 2\sum_i m_i x_i y_i \cos \alpha \cos \beta$$

As we shall see later, the formula can be simplified. First, we immediately recognize the sums involving the squares of the coordinates as the moments of inertia of the body about the three coordinate axes. We shall use a slightly modified notation for them as follows:

$$\sum_i m_i(y_i^2 + z_i^2) = I_{xx} \qquad \textit{moment of inertia about the x-axis}$$

$$\sum_i m_i(z_i^2 + x_i^2) = I_{yy} \qquad \textit{moment of inertia about the y-axis}$$

$$\sum_i m_i(x_i^2 + y_i^2) = I_{zz} \qquad \textit{moment of inertia about the z-axis}$$

The sums involving the products of the coordinates are new to us. They are called *products of inertia*. These quantities will be designated as follows:

$$-\sum_i m_i x_i y_i = I_{xy} = I_{yx} \quad xy \text{ product of inertia}$$

$$-\sum_i m_i y_i z_i = I_{yz} = I_{zy} \quad yz \text{ product of inertia}$$

$$-\sum_i m_i z_i x_i = I_{zx} = I_{xz} \quad zx \text{ product of inertia}$$

Notice that our definition includes the minus sign. (In some textbooks the minus sign is not included.) Products of inertia have the same physical dimensions as moments of inertia, namely, mass $\times$ (length)^2, and their values are determined by the mass distribution and orientation of the body relative to the coordinate axes. They make their appearance here because we wish to allow the axis of rotation to have an arbitrary direction, whereas in the previous chapter one of the coordinate axes was taken as the axis of rotation. In order for the moments and products of inertia to be constant quantities, it will generally be necessary to employ a coordinate system that is fixed to the body and rotates with it.

In actually computing the moments and products of inertia of an extended rigid body, we replace the summations by integrations over the volume, as we have done previously:

$$I_{zz} = \int (x^2 + y^2) \, dm$$

$$I_{xy} = -\int xy \, dm$$

with similar expressions for the other I's. We have already found the moments of inertia for a number of cases in the previous chapter. It is important to remember that the values of the moments and products of inertia depend on the choice of the coordinate system.

Using the above notation, the general expression (Equation 9.2) for the moment of inertia about an arbitrary axis becomes

$$I = I_{xx} \cos^2 \alpha + I_{yy} \cos^2 \beta + I_{zz} \cos^2 \gamma + 2I_{yz} \cos \gamma \cos \beta \qquad (9.3)$$
$$+ 2I_{zx} \cos \alpha \cos \gamma + 2I_{xy} \cos \alpha \cos \beta$$

Although Equation 9.3 seems rather cumbersome for obtaining the moment of inertia, it is nevertheless useful for certain applications. Furthermore, the calculation is included here in order to show how the products of inertia enter into the general problem of rigid-body dynamics.

At this point, we would like to express the moment of inertia of a rigid body in the more compact notational form of tensors and/or equivalent matrices. Such a representation conveys more than just economy and elegance. Expressions for certain kinematic variables are more easily remembered, and it provides us with more powerful techniques for solving complicated problems in rotational motion of rigid bodies.

If we examine the expressions developed above for the moment of inertia of a rigid body about an arbitrary axis, they look like they could be written as the components I_{ij}

of a symmetric, 3×3 matrix. Let us define the quantity $\mathbf{I}$ to be the *moment of inertia tensor* whose components in matrix form are the values given above

$$\mathbf{I} = \begin{bmatrix} I_{xx} & I_{xy} & I_{xz} \\ I_{yx} & I_{yy} & I_{yz} \\ I_{zx} & I_{zy} & I_{zz} \end{bmatrix}$$

Note that $I_{ij} = I_{ji}$ and the matrix is symmetric. Furthermore, since the vector $\mathbf{n}$ can be represented in matrix notation by the column vector

$$\mathbf{n} = \begin{bmatrix} \cos \alpha \\ \cos \beta \\ \cos \gamma \end{bmatrix}$$

we can express the moment of inertia about an axis aligned with the vector $\mathbf{n}$ in matrix notation as

$$I = \tilde{\mathbf{n}}\mathbf{I}\mathbf{n} = [\cos \alpha \; \cos \beta \; \cos \gamma] \begin{bmatrix} I_{xx} & I_{xy} & I_{xz} \\ I_{yx} & I_{yy} & I_{yz} \\ I_{zx} & I_{zy} & I_{zz} \end{bmatrix} \begin{bmatrix} \cos \alpha \\ \cos \beta \\ \cos \gamma \end{bmatrix} \qquad (9.3a)$$

where $\tilde{\mathbf{n}}$ means "$\mathbf{n}$ transpose." The transpose of a matrix is obtained by simply "flipping the matrix over about its diagonal," that is, exchanging the row elements for the column elements. In the case above $\tilde{\mathbf{n}}$ is a row vector whose elements are equal to those of the column vector $\mathbf{n}$. Equation 9.3a is identical to Equation 9.3. However, take note of its notational compactness and elegance.

Angular Momentum Vector

In Chapter 7 we showed that the time rate of change of the angular momentum of a system of particles is equal to the total moment of all the external forces acting on the system. This rotational equation of motion is expressed by Equation 7.14

$$\frac{d\mathbf{L}}{dt} = \mathbf{N}$$

The angular momentum of a system of particles about some coordinate origin is given by Equation 7.15

$$\mathbf{L} = \sum_i \mathbf{r}_i \times m_i \mathbf{v}_i$$

These equations also apply to a rigid body, which is nothing other than a system of particles whose relative positions are fixed or whose distances of separation remain constant. However, before we can apply the rotational equation of motion to a rigid body, we must be able to calculate its angular momentum about an arbitrary axis.

First, we note that the rotational velocity of any constituent particle of the rigid body is given by the cross product

$$\mathbf{v}_i = \boldsymbol{\omega} \times \mathbf{r}_i$$

The total angular momentum of the rigid body is the sum of the angular momenta of each particle about the coordinate origin

$$\mathbf{L} = \sum_i [m_i \mathbf{r}_i \times \mathbf{v}_i] = \sum_i [m_i \mathbf{r}_i \times (\boldsymbol{\omega} \times \mathbf{r}_i)]$$

This expression contains a vector triple product which, using Equation 1.23, can be reduced to

$$[\mathbf{r}_i \times (\boldsymbol{\omega} \times \mathbf{r}_i)] = r_i^2 \boldsymbol{\omega} - \mathbf{r}_i(\mathbf{r}_i \cdot \boldsymbol{\omega})$$

Hence, the angular momentum of the rigid body can be written as

$$\mathbf{L} = \sum_i m_i r_i^2 \boldsymbol{\omega} - \sum_i m_i \mathbf{r}_i(\mathbf{r}_i \cdot \boldsymbol{\omega}) \tag{9.4}$$

We could easily evaluate the x, y, z components of the angular momentum vector using this equation. However, in keeping with the philosophy initiated above, we will cast this equation into tensor form, defining a tensor in the process

$$\mathbf{L} = \left(\sum_i m_i r_i^2 \right) \boldsymbol{\omega} - \left(\sum_i m_i \mathbf{r}_i \mathbf{r}_i \right) \cdot \boldsymbol{\omega}$$
$$= \left[\left(\sum_i m_i r_i^2 \mathbf{1} \right) - \left(\sum_i m_i \mathbf{r}_i \mathbf{r}_i \right) \right] \cdot \boldsymbol{\omega} \tag{9.4a}$$

where the vector $\boldsymbol{\omega}$ in the first term has been written as

$$\mathbf{1} \cdot \boldsymbol{\omega} = \boldsymbol{\omega}$$

which may be viewed as the definition of the *unit tensor*

$$\mathbf{1} = \mathbf{ii} + \mathbf{jj} + \mathbf{kk}$$

You can confirm the above identity by carrying out the dot product operation on the vector $\boldsymbol{\omega}$, that is,

$$\begin{aligned} \mathbf{1} \cdot \boldsymbol{\omega} &= (\mathbf{ii} + \mathbf{jj} + \mathbf{kk}) \cdot \boldsymbol{\omega} \\ &= \mathbf{i}(\mathbf{i} \cdot \boldsymbol{\omega}) + \mathbf{j}(\mathbf{j} \cdot \boldsymbol{\omega}) + \mathbf{k}(\mathbf{k} \cdot \boldsymbol{\omega}) \\ &= \mathbf{i}\omega_x + \mathbf{j}\omega_y + \mathbf{k}\omega_z \\ &= \boldsymbol{\omega} \end{aligned} \tag{9.4b}$$

Both the unit tensor and the second term in brackets in Equation 9.4a contain a product of vectors that we have never seen before. This type of vector product (for example, **ab**) is called a *dyad product*. It is a tensor defined by its dot product operation upon another vector **c** in the same way that we defined the unit tensor above.

$$(\mathbf{ab}) \cdot \mathbf{c} = \mathbf{a}(\mathbf{b} \cdot \mathbf{c})$$

This dot product yields a vector. The operation may be expressed in matrix form as

$$(\mathbf{ab}) \cdot \mathbf{c} = \begin{bmatrix} a_x b_x & a_x b_y & a_x b_z \\ a_y b_x & a_y b_y & a_y b_z \\ a_z b_x & a_z b_y & a_z b_z \end{bmatrix} \begin{bmatrix} c_x \\ c_y \\ c_z \end{bmatrix}$$

$$= \begin{bmatrix} a_x(b_x c_x + b_y c_y + b_z c_z) \\ a_y(b_x c_x + b_y c_y + b_z c_z) \\ a_z(b_x c_x + b_y c_y + b_z c_z) \end{bmatrix} = \mathbf{a}(\mathbf{b} \cdot \mathbf{c})$$

(9.5)

If we were to "dot" Equation 9.5 from the left with another vector **d**, the result would be a simple scalar, that is,

$$\mathbf{d} \cdot (\mathbf{ab}) \cdot \mathbf{c} = (\mathbf{d} \cdot \mathbf{a})(\mathbf{b} \cdot \mathbf{c})$$

We leave it as an exercise for the reader to obtain this result using matrices.

In three dimensions a tensor has nine components. The components may be generated in the following way:

$$T_{ij} = \mathbf{i} \cdot \mathbf{T} \cdot \mathbf{j}$$

(9.5a)

You should be able to convince yourself that the components of the dyad product **ab** contained in the matrix of Equation 9.5 can be generated by the operation given in Equation 9.5a.

Using these definitions, we can see that the term in brackets in Equation 9.4a

$$\mathbf{I} = \sum_i m_i r_i^2 \mathbf{1} - \sum_i m_i \mathbf{r}_i \mathbf{r}_i$$

(9.6)

is the previously defined moment of inertia tensor whose components were given in Equation 9.2. We can demonstrate this equivalence by calculating the components as follows:

$$\mathbf{i} \cdot \mathbf{I} \cdot \mathbf{i} = \mathbf{i} \cdot \left[\sum_i (m_i r_i^2 (\mathbf{ii} + \mathbf{jj} + \mathbf{kk}) - m_i \mathbf{r}_i \mathbf{r}_i) \right] \cdot \mathbf{i}$$

$$I_{xx} = \sum_i m_i r_i^2 - m_i x_i^2 = \sum_i m_i(y_i^2 + z_i^2)$$

$$\mathbf{i} \cdot \mathbf{I} \cdot \mathbf{j} = \mathbf{i} \cdot \left[\sum_i (m_i r_i^2 (\mathbf{ii} + \mathbf{jj} + \mathbf{kk}) - m_i \mathbf{r}_i \mathbf{r}_i) \right] \cdot \mathbf{j}$$

$$I_{xy} = -\sum_i m_i x_i y_i \text{ and so on}$$

Thus, using tensor notation the angular momentum vector can be written as

$$\mathbf{L} = \mathbf{I} \cdot \boldsymbol{\omega}$$

(9.7)

An important fact should now be apparent: *The direction of the angular momentum vector is not necessarily aligned along the axis of rotation; **L** and **ω** are not necessarily* *parallel.* For example let $\boldsymbol{\omega}$ be directed along the x-axis ($\omega_x = \omega$, $\omega_y = 0$, $\omega_z = 0$). In this case the above expression reduces to

$$\mathbf{L} = \mathbf{I} \cdot \boldsymbol{\omega}$$
$$= \mathbf{i}(\omega_x I_{xx} + \omega_y I_{xy} + \omega_z I_{xz}) + \mathbf{j}(\omega_x I_{yx} + \omega_y I_{yy} + \omega_z I_{yz})$$
$$+ \mathbf{k}(\omega_x I_{zx} + \omega_y I_{zy} + \omega_z I_{zz})$$
$$\mathbf{L} = \mathbf{i}\omega I_{xx} + \mathbf{j}\omega I_{xy} + \mathbf{k}\omega I_{xz}$$

Thus, $\mathbf{L}$ may have components perpendicular to the x-axis (axis of rotation). Note that the component of angular momentum along the axis of rotation is $L_x = \omega I_{xx}$ in agreement with the results of the last chapter.

EXAMPLE 9.1

Find the moment of inertia of a uniform square lamina of side a and mass m about a diagonal.

Solution:
Let us choose coordinate axes as shown in Figure 9.2 with the lamina lying in the xy plane with a corner at the origin. Then, from the last chapter, we have $I_{xx} = I_{yy} = ma^2/3$ and $I_{zz} = I_{xx} + I_{yy} = 2ma^2/3$. Now $z = 0$ for all points in the lamina;

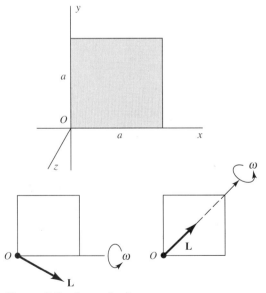

Figure 9.2 Square lamina.

therefore, the xz and yz products of inertia vanish: $I_{xz} = I_{yz} = 0$. The xy product of inertia is found by integrating as follows:

$$I_{xy} = I_{yx} = -\int_0^a \int_0^a xy\rho \ dx dy = -\rho \int_0^a \frac{a^2}{2} y dy = -\rho \frac{a^4}{4}$$

where ρ is the mass per unit area, that is $\rho = m/a^2$. Therefore, we get

$$I_{xy} = -\frac{1}{4} ma^2$$

We now have the essential ingredients necessary to construct the moment of inertia tensor of the square plate related to the xyz coordinate system indicated in Figure 9.2. It is

$$\mathbf{I} = \begin{bmatrix} \dfrac{ma^2}{3} & \dfrac{-ma^2}{4} & 0 \\ \dfrac{-ma^2}{4} & \dfrac{ma^2}{3} & 0 \\ 0 & 0 & \dfrac{2ma^2}{3} \end{bmatrix} = ma^2 \begin{bmatrix} \dfrac{1}{3} & -\dfrac{1}{4} & 0 \\ -\dfrac{1}{4} & \dfrac{1}{3} & 0 \\ 0 & 0 & \dfrac{2}{3} \end{bmatrix}$$

In order to calculate the moment of inertia of the plate about its diagonal according to Equation 9.3a, we need the components of a unit vector **n** directed along the diagonal. The angles that this vector makes with the coordinate axes are $\alpha = \beta = 45°$ and $\gamma = 90°$. The direction cosines are then $\cos \alpha = \cos \beta = 1/\sqrt{2}$ and $\cos \gamma = 0$. Equation 9.3a then gives

$$I = \tilde{\mathbf{n}}\mathbf{I}\mathbf{n} = \begin{bmatrix} \dfrac{1}{\sqrt{2}} & \dfrac{1}{\sqrt{2}} & 0 \end{bmatrix} \begin{bmatrix} \dfrac{1}{3} & -\dfrac{1}{4} & 0 \\ -\dfrac{1}{4} & \dfrac{1}{3} & 0 \\ 0 & 0 & \dfrac{2}{3} \end{bmatrix} \begin{bmatrix} \dfrac{1}{\sqrt{2}} \\ \dfrac{1}{\sqrt{2}} \\ 0 \end{bmatrix} ma^2$$

$$= \begin{bmatrix} \dfrac{1}{\sqrt{2}} & \dfrac{1}{\sqrt{2}} & 0 \end{bmatrix} \begin{bmatrix} \dfrac{1}{12\sqrt{2}} \\ \dfrac{1}{12\sqrt{2}} \\ 0 \end{bmatrix} ma^2 = \dfrac{1}{12} ma^2$$

for the moment of inertia about a diagonal. This result could also be obtained by direct integration. The student may wish to verify this as an exercise. See also Problem 9.2(c).   ■

EXAMPLE 9.2

Find the angular momentum about the origin of the above square plate when it is rotating with angular speed ω about (a) the x-axis and (b) the diagonal through the origin.

Solution:

(a) For rotation about the x-axis we have

$$\omega_x = \omega \qquad \omega_y = 0 \qquad \omega_z = 0$$

The total angular momentum is $\mathbf{L} = \mathbf{I} \cdot \boldsymbol{\omega}$ or, in matrix notation,

$$\mathbf{L} = \mathbf{I}\boldsymbol{\omega} = ma^2 \begin{bmatrix} \dfrac{1}{3} & -\dfrac{1}{4} & 0 \\ -\dfrac{1}{4} & \dfrac{1}{3} & 0 \\ 0 & 0 & \dfrac{2}{3} \end{bmatrix} \begin{bmatrix} \omega \\ 0 \\ 0 \end{bmatrix} = ma^2\omega \begin{bmatrix} \dfrac{1}{3} \\ -\dfrac{1}{4} \\ 0 \end{bmatrix}$$

with the final result expressed in matrix form as a column vector.

(b) The components of $\boldsymbol{\omega}$ about the diagonal are

$$\omega_x = \omega_y = \omega \cos 45° = \frac{\omega}{\sqrt{2}} \qquad \omega_z = 0$$

Therefore,

$$\mathbf{L} = ma^2 \begin{bmatrix} \dfrac{1}{3} & -\dfrac{1}{4} & 0 \\ -\dfrac{1}{4} & \dfrac{1}{3} & 0 \\ 0 & 0 & \dfrac{2}{3} \end{bmatrix} \begin{bmatrix} \dfrac{\omega}{\sqrt{2}} \\ \dfrac{\omega}{\sqrt{2}} \\ 0 \end{bmatrix} = \frac{ma^2\omega}{\sqrt{2}} \begin{bmatrix} \dfrac{1}{12} \\ \dfrac{1}{12} \\ 0 \end{bmatrix}$$

Notice that in case (a) the angular momentum vector $\mathbf{L}$ does not point in the same direction as the angular velocity vector $\boldsymbol{\omega}$ but points downward as shown in Figure 9.2. However, in case (b) the two vectors point in the same direction as shown in Figure 9.2.

The magnitude of the angular momentum vector for case (a) is given by $(\mathbf{L} \cdot \mathbf{L})^{1/2}$. Using matrix notation, we obtain

$$L^2 = \tilde{\mathbf{L}}\mathbf{L} = (ma^2\omega)^2 \begin{bmatrix} \dfrac{1}{3} & -\dfrac{1}{4} & 0 \end{bmatrix} \begin{bmatrix} \dfrac{1}{3} \\ -\dfrac{1}{4} \\ 0 \end{bmatrix} = (ma^2\omega)^2\left(\dfrac{1}{9} + \dfrac{1}{16}\right)$$

$$= (ma^2\omega)^2 \frac{25}{144} \qquad \therefore L = ma^2\omega \frac{5}{12}$$

and for case (b)

$$L = ma^2\omega \frac{1}{12}$$

∎

Rotational Kinetic Energy of a Rigid Body

We next calculate the kinetic energy of rotation of our general rigid body of Figure 9.1. As in our calculation of the angular momentum, we use the fact that the velocity of a representative particle is given by $\mathbf{v}_i = \boldsymbol{\omega} \times \mathbf{r}_i$. The rotational kinetic energy is, therefore, given by the summation

$$T_{rot} = \sum_i \frac{1}{2} m_i \mathbf{v}_i \cdot \mathbf{v}_i = \frac{1}{2} \sum_i (\boldsymbol{\omega} \times \mathbf{r}_i) \cdot m_i \mathbf{v}_i$$

Now in any triple scalar product we can exchange the dot and the cross: $(\mathbf{A} \times \mathbf{B}) \cdot \mathbf{C} = \mathbf{A} \cdot (\mathbf{B} \times \mathbf{C})$. (See Section 1.7.) Hence,

$$T_{rot} = \frac{1}{2} \sum_i \boldsymbol{\omega} \cdot (\mathbf{r}_i \times m_i \mathbf{v}_i) = \frac{1}{2} \boldsymbol{\omega} \cdot \sum_i (\mathbf{r}_i \times m_i \mathbf{v}_i)$$

But, by definition, the sum $\sum_i (\mathbf{r}_i \times m_i \mathbf{v}_i)$ is the angular momentum $\mathbf{L}$. Thus, we can write

$$T_{rot} = \frac{1}{2} \boldsymbol{\omega} \cdot \mathbf{L} \tag{9.8}$$

for the kinetic energy of rotation of a rigid body. We recall from Chapter 7 that the translational kinetic energy of any system is equal to the expression $\frac{1}{2}\mathbf{v}_{cm} \cdot \mathbf{p}$ where $\mathbf{p} = m\mathbf{v}_{cm}$ is the linear momentum of the system and $\mathbf{v}_{cm}$ is the velocity of the mass center. For a rigid body the total kinetic energy is accordingly

$$T = T_{rot} + T_{trans} = \frac{1}{2} \boldsymbol{\omega} \cdot \mathbf{L} + \frac{1}{2} \mathbf{v}_{cm} \cdot \mathbf{p} \tag{9.9}$$

where $\mathbf{L}$ is the angular momentum about the center of mass.

Using the results of the previous section, we can express the rotational kinetic energy of a rigid body in terms of the moment of the inertia tensor as

$$T_{rot} = \frac{1}{2} \boldsymbol{\omega} \cdot \mathbf{I} \cdot \boldsymbol{\omega}$$

or we can carry out the calculation explicitly in matrix notation

$$T_{rot} = \frac{1}{2} \tilde{\boldsymbol{\omega}} \, \mathbf{I} \, \boldsymbol{\omega} = \frac{1}{2} [\omega_x \; \omega_y \; \omega_z] \begin{bmatrix} I_{xx} & I_{xy} & I_{xz} \\ I_{yx} & I_{yy} & I_{yz} \\ I_{zx} & I_{zy} & I_{zz} \end{bmatrix} \begin{bmatrix} \omega_x \\ \omega_y \\ \omega_z \end{bmatrix}$$

$$= \frac{1}{2} \, (I_{xx}\omega_x^2 + I_{yy}\omega_y^2 + I_{zz}\omega_z^2 \tag{9.10}$$

$$+ \; 2I_{xz}\omega_x\omega_y + 2I_{xz}\omega_x\omega_z + 2I_{yz}\omega_y\omega_z)$$

EXAMPLE 9.3

Find the rotational kinetic energy of the square plate in Example 9.2.

Solution:
For the case (a) rotation about the x-axis, we have

$$T = \frac{1}{2}\tilde{\boldsymbol{\omega}}\,\mathbf{I}\,\boldsymbol{\omega} = \frac{1}{2}ma^2\omega^2[1\ 0\ 0]\begin{bmatrix} \frac{1}{3} & -\frac{1}{4} & 0 \\ -\frac{1}{4} & \frac{1}{3} & 0 \\ 0 & 0 & \frac{2}{3} \end{bmatrix}\begin{bmatrix} 1 \\ 0 \\ 0 \end{bmatrix} = \frac{1}{6}ma^2\omega^2$$

For case (b), rotation about a diagonal, we have

$$T = \frac{1}{2}ma^2\omega^2\left[\frac{1}{\sqrt{2}}\ \frac{1}{\sqrt{2}}\ 0\right]\begin{bmatrix} \frac{1}{3} & -\frac{1}{4} & 0 \\ -\frac{1}{4} & \frac{1}{3} & 0 \\ 0 & 0 & \frac{2}{3} \end{bmatrix}\begin{bmatrix} \frac{1}{\sqrt{2}} \\ \frac{1}{\sqrt{2}} \\ 0 \end{bmatrix}$$

$$= \frac{1}{2}ma^2\omega^2\left[\frac{1}{\sqrt{2}}\ \frac{1}{\sqrt{2}}\ 0\right]\begin{bmatrix} \frac{1}{12\sqrt{2}} \\ \frac{1}{12\sqrt{2}} \\ 0 \end{bmatrix} = \frac{1}{24}ma^2\omega^2$$

9.2 PRINCIPAL AXES OF A RIGID BODY. DYNAMIC BALANCING

A considerable simplification in the above-derived mathematical formulas for rigid-body motion results if we employ a coordinate system such that the products of inertia all vanish. It turns out that such a coordinate system does in fact exist for any rigid body and for any point taken as the origin. The axes of this coordinate system are said to be *principal axes* for the body at the point O, the origin of the coordinate system in question. (Often we shall choose O to be the center of mass.)

Explicitly, if the coordinate axes are principal axes of the body, then $I_{xy} = I_{xz} = I_{yz} = 0$. In this case we shall employ the following notation:

$$\begin{aligned} I_{xx} &= I_1 & \omega_x &= \omega_1 & \mathbf{i} &= \mathbf{e}_1 \\ I_{yy} &= I_2 & \omega_y &= \omega_2 & \mathbf{j} &= \mathbf{e}_2 \\ I_{zz} &= I_3 & \omega_z &= \omega_3 & \mathbf{k} &= \mathbf{e}_3 \end{aligned}$$

The three moments of inertia I_1, I_2, and I_3 are known as the *principal moments* of the rigid body at the point O. In a coordinate system whose axes are aligned with the principal axes, the moment of inertia tensor takes on a particularly simple, diagonal form

$$\mathbf{I} = \begin{bmatrix} I_1 & 0 & 0 \\ 0 & I_2 & 0 \\ 0 & 0 & I_3 \end{bmatrix} \tag{9.10a}$$

Thus, the problem of finding the principal axes of a rigid body is equivalent to the mathematical problem of diagonalizing a 3×3 matrix. The moment of inertia tensor is always expressible as a square, symmetric matrix and any such matrix can always be diagonalized. *Thus, a set of principal axes exists for any rigid body at any point in space.* We will argue this point on the basis of more physical considerations in the following section.

The moment of inertia, angular momentum, and rotational kinetic energy of a rigid body about any arbitrary rotational axis all take on fairly simple forms in a coordinate system whose axes are aligned with the principal axes of the rigid body. Let $\mathbf{n}$ be a unit vector designating the direction of the axis of rotation of a rigid body, and let its components relative to the principal axes be given by the direction cosines ($\cos \alpha$, $\cos \beta$, $\cos \gamma$). The moment of inertia about that axis is

$$I = \tilde{\mathbf{n}} \mathbf{I} \mathbf{n} = [\cos \alpha \ \cos \beta \ \cos \gamma] \begin{bmatrix} I_1 & 0 & 0 \\ 0 & I_2 & 0 \\ 0 & 0 & I_3 \end{bmatrix} \begin{bmatrix} \cos \alpha \\ \cos \beta \\ \cos \gamma \end{bmatrix}$$

$$= I_1 \cos^2 \alpha + I_2 \cos^2 \beta + I_3 \cos^2 \gamma \tag{9.11}$$

The angular velocity $\boldsymbol{\omega}$ points in the same direction as $\mathbf{n}$, and its components relative to the principal axes are (ω_1, ω_2, ω_3). The total angular momentum $\mathbf{L} = \mathbf{I} \cdot \boldsymbol{\omega}$, in this frame of reference, can be written in matrix notation as

$$\mathbf{L} = \mathbf{I}\boldsymbol{\omega} = \begin{bmatrix} I_1 & 0 & 0 \\ 0 & I_2 & 0 \\ 0 & 0 & I_3 \end{bmatrix} \begin{bmatrix} \omega_1 \\ \omega_2 \\ \omega_3 \end{bmatrix} = \begin{bmatrix} I_1\omega_1 \\ I_2\omega_2 \\ I_3\omega_3 \end{bmatrix}$$

$$\mathbf{L} = \mathbf{e}_1 I_1 \omega_1 + \mathbf{e}_2 I_2 \omega_2 + \mathbf{e}_3 I_3 \omega_3 \tag{9.12}$$

And, finally, the kinetic energy of rotation T_{rot} is

$$T_{rot} = \frac{1}{2} \tilde{\boldsymbol{\omega}} \mathbf{I} \boldsymbol{\omega} = \frac{1}{2} [\omega_1 \ \omega_2 \ \omega_3] \begin{bmatrix} I_1 & 0 & 0 \\ 0 & I_2 & 0 \\ 0 & 0 & I_3 \end{bmatrix} \begin{bmatrix} \omega_1 \\ \omega_2 \\ \omega_3 \end{bmatrix}$$

$$= \frac{1}{2} (I_1 \omega_1^2 + I_2 \omega_2^2 + I_3 \omega_3^2) \tag{9.13}$$

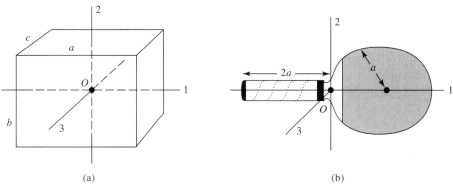

(a)   (b)

Figure 9.3 Principal axes for a (a) uniform rectangular block and (b) ping-pong paddle.

Existence of Principal Axes

Let us now investigate the question of finding the principal axes. First, if the body possesses some symmetry, then it is usually possible to choose a set of coordinate axes by inspection such that one or more of the three products of inertia consists of two parts of equal magnitude and opposite algebraic sign and therefore vanishes. For example, the rectangular block and the symmetric laminar body (ping-pong paddle) have the principal axes at O as indicated in Figure 9.3.

A body does not have to be symmetric, however, in order that the products of inertia vanish for an appropriately chosen coordinate system. For example, consider a plane lamina of any shape (Figure 9.4). If the xy plane is the plane of the lamina, then $z = 0$ for all parts of the body and so $I_{zy} = -\int zy\,dm$ and $I_{zx} = -\int zx\,dm$ both vanish. Now, relative to any given origin in the plane of the lamina, we can easily prove that there always exists a set of axes such that $I_{xy} = -\int xy\,dm$ also vanishes. To show this, we observe that the integral changes sign as the Oxy system is rotated through an angle of $90°$, because the lamina passes from one quadrant to the next, as shown. Consequently,

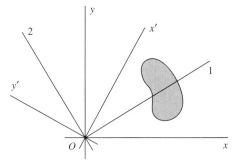

Figure 9.4 Showing the existence of principal axes for a laminar body of any shape.

the integral must vanish for some angle between $0°$ and $90°$. This angle defines a set of coordinate axes for which all products of inertia vanish. By definition this is a set of principal axes.

It can be argued in a similar way that for any rigid body there always exists, at any given point, at least one set of three mutually orthogonal axes such that all products of inertia vanish; that is, there always exists at least one set of principal axes at the point in question. A method of determining principal axes is discussed below.

EXAMPLE 9.4

(a) For the rectangular block shown in Figure 9.3(a), the principal moments at the center of mass O are clearly just those indicated in Table 8.1, namely,

$$I_1 = \frac{m}{12}(b^2 + c^2)$$

$$I_2 = \frac{m}{12}(a^2 + c^2)$$

$$I_3 = \frac{m}{12}(a^2 + b^2)$$

in which m is the mass of the block and a, b, and c are the edge lengths. The moment of inertia about any axis passing through O is then given by Equation 9.11. Note in particular that for a *cubical* block ($a = b = c$) the three principal moments at O are all equal. In this case since $\cos^2 \alpha + \cos^2 \beta + \cos^2 \gamma = 1$, then the moment of inertia about any axis through the center of the block is independent of the direction of the axis and is given by $I = \frac{1}{6}ma^2$.

(b) To find the principal moments of the ping-pong paddle at the point O indicated in Figure 9.3(b), we shall assume for simplicity that the paddle is a circular lamina of radius a and mass $m/2$ attached to a thin rod for a handle of mass $m/2$ and length $2a$. We borrow from the results of Section 8.3. The principal moments are each calculated by adding the moments about the respective axes for the two parts, namely,

$$I = I_{rod} + I_{disc}$$

$$I_1 = 0 + \frac{1}{4}\frac{m}{2}a^2 = \frac{1}{8}ma^2$$

$$I_2 = \frac{1}{3}\frac{m}{2}(2a)^2 + \frac{5}{4}\frac{m}{2}a^2 = \frac{31}{24}ma^2$$

$$I_3 = \frac{1}{3}\frac{m}{2}(2a)^2 + \frac{3}{2}\frac{m}{2}a^2 = \frac{17}{12}ma^2$$

We note that $I_3 = I_1 + I_2$, since the object is assumed to be laminar. ■

Suppose that a body is rotating about one of its principal axes, say the 1-axis. Then $\omega = \omega_1$, $\omega_2 = \omega_3 = 0$. The expression for the angular momentum (Equation 9.12) then reduces to just one term, namely,

$$\mathbf{L} = \mathbf{e}_1 I_1 \omega_1$$

or, equivalently

$$\mathbf{L} = I_1 \boldsymbol{\omega}$$

Thus, in this circumstance the angular momentum vector is in the same direction as the angular velocity vector or axis of rotation. We have, therefore, the following important fact: *The angular momentum vector is either in the same direction as the axis of rotation, or is not, depending on whether the axis of rotation is, or is not, a principal axis.*

Dynamic Balancing

The above rule finds application in the case of a rotating device such as an automobile wheel or fan blade. If the device is *statically balanced,* the center of mass lies on the axis of rotation. To be *dynamically balanced* the axis of rotation must also be a principal axis so that, as the body rotates, the angular momentum vector **L** will lie along the axis. Otherwise, if the rotational axis is not a principal one, the angular momentum vector varies in direction: It describes a cone as the body rotates (Figure 9.5). Then, since $d\mathbf{L}/dt$ is equal to the applied torque, there must be a torque exerted on the body. The direction of this torque is at right angles to the axis. The result is a reaction on the bearings. Thus, in the case of a dynamically unbalanced wheel or rotator, there may be violent vibration and wobbling, even if the wheel is statically balanced.

Determination of the Other Two Principal Axes When One Is Known

In many instances a body possesses sufficient symmetry so that at least one principal axis can be found by inspection; that is, the axis can be chosen so as to make two of the

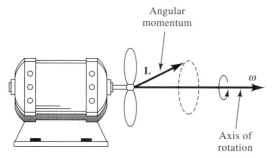

Figure 9.5 A rotating fan blade. The angular momentum vector **L** describes a cone about the axis of rotation if the blade is not dynamically balanced.

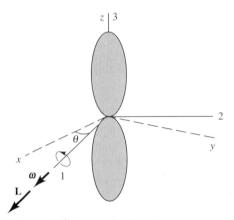

Figure 9.6 Determination of two principal axes (1 and 2) when the third one (z) is known.

three products of inertia vanish. If such is the case, then the other two principal axes can be determined as follows. Figure 9.6 is a front view of the fan blade in Figure 9.5. The z-axis is a symmetry axis and coincides with the third principal axis of the fan blade. Thus, for such cases, we have

$$I_{zx} = I_{zy} = 0 \qquad I_{zz} = I_3 \neq 0$$

The other two principal axes are each perpendicular to the z-axis. They must lie in the xy plane. Suppose the body is rotating about one of these two, as yet, unknown principal axes. If so, the rotating object is dynamically balanced as illustrated by the fan blade in Figure 9.6. The angular momentum vector **L** lies in the same direction as the angular velocity vector **ω**, thus

$$\mathbf{L} = I_1 \boldsymbol{\omega} = I_1 \begin{bmatrix} \omega_x \\ \omega_y \\ 0 \end{bmatrix}$$

where I_1 is one of the two principal moments of inertia in question. Furthermore, in matrix notation, the angular momentum **L,** in the xyz frame of reference, is given by

$$\mathbf{L} = I\boldsymbol{\omega} = \begin{bmatrix} I_{xx} & I_{xy} & 0 \\ I_{xy} & I_{yy} & 0 \\ 0 & 0 & I_3 \end{bmatrix} \begin{bmatrix} \omega_x \\ \omega_y \\ 0 \end{bmatrix}$$

(Remember, the products of inertia about the z-axis are zero.) Thus, equating components of the angular momentum given by these two expressions gives

$$I_{xx}\omega_x + I_{xy}\omega_y = I_1\omega_x \qquad (9.14a)$$

$$I_{xy}\omega_x + I_{yy}\omega_y = I_1\omega_y \qquad (9.14b)$$

Let θ denote the angle between the x-axis and the principal axis I_1 about which the body is rotating (See Figure 9.6). Then $\omega_y/\omega_x = \tan\theta$, so, upon dividing by ω_x, we have

$$I_{xx} + I_{xy} \tan\theta = I_1$$

$$I_{xy} + I_{yy} \tan\theta = I_1 \tan\theta$$

Elimination of I_1 between the two equations yields

$$(I_{yy} - I_{xx}) \tan\theta = I_{xy}(\tan^2\theta - 1)$$

from which θ can be found. In this calculation it is helpful to employ the trigonometric identity $\tan 2\theta = 2\tan\theta/(1 - \tan^2\theta)$. This gives

$$\tan 2\theta = \frac{2I_{xy}}{I_{xx} - I_{yy}} \qquad (9.15)$$

In the interval $0°$ to $180°$ there are two values of θ, differing by $90°$, that satisfy Equation 9.15, and these give the directions of the two principal axes in the xy plane.

Note that in the case $I_{xx} = I_{yy}$, $\tan 2\theta = \infty$ so that the two values of θ are $45°$ and $135°$. (This is the case for the square lamina of Example 9.1 when the origin is at a corner.) Also, if $I_{xy} = 0$ the equation is satisfied by the two values $\theta = 0°$ and $\theta = 90°$; that is, the x and y-axes are already principal axes.

EXAMPLE 9.5 *Balancing a Crooked Wheel*

Suppose an automobile wheel, through some defect or accident, has its axis of rotation (axle) slightly bent relative to the symmetry axis of the wheel. The situation can be remedied by use of counterbalance weights suitably located on the rim so as to make the axle a principal axis for the total system: wheel plus weights. For simplicity we shall treat the wheel as a thin uniform circular disc of radius a and mass m, Figure 9.7. We choose $Oxyz$ axes such that the disc lies in the yz plane, with the x-axis as the symmetry axis of the disc. The axis of rotation (axle) is taken as the 1-axis inclined by an angle θ relative to the x-axis and lying in the xy plane, as shown. Two balancing weights, each of mass m', are attached to the wheel by means of light supports of length b. The weights both lie in the xy plane, as indicated. The wheel will be dynamically balanced if the 123 coordinate axes are principal axes for the total system.

Now, from symmetry relative to the xy plane, we see that the z-axis is a principal axis for the wheel plus weights: z is zero for the weights, and the xy plane divides the wheel into two equal parts having opposite signs for the products zx and zy. Consequently, we can use Equation 9.15 to find the relationship between θ and the other parameters.

From the previous chapter we know that for the wheel alone the moments of inertia about the x- and the y-axes are $\frac{1}{2}ma^2$ and $\frac{1}{4}ma^2$, respectively. Thus, for the wheel plus weights

$$I_{xx} = \frac{1}{2}ma^2 + 2m'a^2$$

$$I_{yy} = \frac{1}{4}ma^2 + 2m'b^2$$

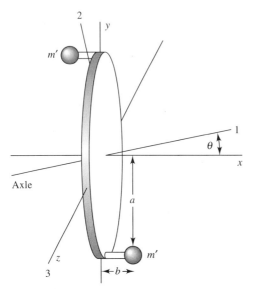

Figure 9.7   Principal axes for a bent wheel
with balancing weights.

Now the xy product of inertia for the wheel alone is zero, and so we need consider
only the weights for finding I_{xy} for the system, namely,

$$I_{xy} = -\Sigma x_i y_i m_i' = -[(-b)am' + b(-a)m'] = 2abm'$$

Notice that this is a positive quantity for our choice of coordinate axes. Equation
9.15 then gives the inclination of the 1-axis:

$$\tan 2\theta = \frac{2I_{xy}}{I_{xx} - I_{yy}} = \frac{4abm'}{\dfrac{1}{4}ma^2 + 2m'(a^2 - b^2)}$$

If, as is typical, θ is very small, and also m' is small compared to m, then we can
express the above relation in approximate form by neglecting the second term in
the denominator and using the fact that $\tan u \approx u$ for small u. The result is

$$\theta \approx 8\frac{bm'}{am}$$

As a numerical example, let $\theta = 1° = 0.017$ rad, $a = 7$ in., $b = 2$ in.,
$m = 20$ lb. Solving for m', we find

$$m' = \theta\frac{am}{8b} = 0.017\frac{7 \text{ in.} \times 20 \text{ lb}}{8 \times 2 \text{ in.}} = 0.15 \text{ lb} = 2.4 \text{ oz}$$

for the required balance weights.   ■

Determining Principal Axes by Diagonalizing the Moment of Inertia Matrix

There are a number of ways to diagonalize a real, symmetric matrix.[1] Before doing so, we reiterate the first statement made at the beginning of Section 9.2 regarding the choice of a coordinate system in which all the products of inertia vanish. The axes of such a coordinate system are the principal axes, and in such a coordinate system the moment of inertia tensor is represented by a diagonal matrix. If this tensor is "dotted" with one of the unit vectors designating one of these axes, the result is equivalent to a simple multiplication of the unit vector by a scalar quantity, that is,

$$\mathbf{I} \cdot \mathbf{e}_i = \lambda_i \mathbf{e}_i \qquad (9.15a)$$

The scalar quantities λ_i are just the principal moments of inertia about their respective principal axes. The problem of finding the principal axes is one of finding those vectors $\mathbf{e}_i$ that satisfy the condition

$$(\mathbf{I} - \lambda \mathbf{1}) \cdot \mathbf{e}_i = 0 \qquad (9.15b)$$

In general this condition is not satisfied for any arbitrary set of orthonormal unit vectors $\mathbf{e}_i$. It is satisfied only by a set of unit vectors aligned with the principal axes of the rigid body. Any arbitrary xyz coordinate system can always be rotated such that the coordinate axes line up with the principal axes. The unit vectors specifying these coordinate axes will then satisfy the above condition. This condition is equivalent to the vanishing of the following determinant[2]

$$|\mathbf{I} - \lambda \mathbf{1}| = 0 \qquad (9.15c)$$

Explicitly, this equation reads

$$\begin{vmatrix} I_{xx} - \lambda & I_{xy} & I_{xz} \\ I_{yx} & I_{yy} - \lambda & I_{yz} \\ I_{zx} & I_{zy} & I_{zz} - \lambda \end{vmatrix} = 0 \qquad (9.15d)$$

It is a cubic in λ, namely,

$$-\lambda^3 + A\lambda^2 + B\lambda + C = 0 \qquad (9.15e)$$

in which A, B, and C are functions of the I's. The three roots, λ_1, λ_2 and λ_3 are the three principal moments of inertia.

We now have the principal moments of inertia, but the task of specifying the components of the unit vectors representing the principal axes in terms of our initial coordinate system remains to be solved. Here we can make use of the fact that when the rigid body rotates about one of its principal axes, the angular momentum vector is in the same direction as the angular velocity vector. Let the angles of one of the principal axes rela-

---

[1] For example, see J. Mathews and R. L. Walker, *Mathematical Methods of Physics*, W. A. Benjamin, New York, 1970.

[2] *Ibid.*

tive to the initial xyz coordinate system be α_1, β_1, and γ_1, and let the body rotate about this axis. Therefore, a unit vector pointing in the direction of this principal axis has components ($\cos \alpha_1$, $\cos \beta_1$, $\cos \gamma_1$). The angular momentum is given by

$$\mathbf{L} = \mathbf{I} \cdot \boldsymbol{\omega} = \lambda_1 \boldsymbol{\omega}$$

in which λ_1, the first principal moment of the three (λ_1, λ_2, and λ_3), is obtained by solving Equation 9.15e. In matrix form the above equation reads

$$\lambda_1 \omega \begin{bmatrix} \cos \alpha_1 \\ \cos \beta_1 \\ \cos \gamma_1 \end{bmatrix} = \begin{bmatrix} I_{xx} & I_{xy} & I_{xz} \\ I_{yx} & I_{yy} & I_{yz} \\ I_{zx} & I_{zy} & I_{zz} \end{bmatrix} \omega \begin{bmatrix} \cos \alpha_1 \\ \cos \beta_1 \\ \cos \gamma_1 \end{bmatrix}$$

Note that we have extracted the common factor ω from each one of its components, thus directly exposing the desired principal axis unit vector. The resultant equation is equivalent to the condition expressed by Equation 9.15a, namely, that the dot product of the moment of inertia tensor with a principal axis unit vector is tantamount to multiplying that vector by the scalar quantity λ_1, that is,

$$\mathbf{I} \cdot \mathbf{e_1} = \lambda_1 \mathbf{e_1}$$

This vector equation can be written in matrix form as

$$\begin{bmatrix} (I_{xx} - \lambda_1) & I_{xy} & I_{xz} \\ I_{yx} & (I_{yy} - \lambda_1) & I_{yz} \\ I_{zx} & I_{zy} & (I_{zz} - \lambda_1) \end{bmatrix} \begin{bmatrix} \cos \alpha_1 \\ \cos \beta_1 \\ \cos \gamma_1 \end{bmatrix} = 0 \qquad (9.15f)$$

The direction cosines may be found by solving the above equations. The solutions are not independent. They are subject to the constraint

$$\cos^2 \alpha_1 + \cos^2 \beta_1 + \cos^2 \gamma_1 = 1$$

In other words the resultant vector $\mathbf{e_1}$ specified by these components is a unit vector. The other two vectors may be found by repeating the above process for the other two principal moments λ_2 and λ_3.

EXAMPLE 9.6

Find the principal moments of inertia of a square plate about a corner.

Solution:
We choose the same xyz system as initially chosen in Example 9.1. We have all the moments of inertia relative to those axes. They are the same as in Example 9.1. The vanishing of the determinant expressed by Equation 9.15d reads

$$\begin{vmatrix} \left(\dfrac{1}{3} - \lambda\right) & -\dfrac{1}{4} & 0 \\ -\dfrac{1}{4} & \left(\dfrac{1}{3} - \lambda\right) & 0 \\ 0 & 0 & \left(\dfrac{2}{3} - \lambda\right) \end{vmatrix} ma^2 = 0$$

Note: We have extracted a common factor ma^2, which will leave us with only the desired numerical coefficients for each value of λ. We must then put the ma^2 factor back in to get the final values for the principal moments.

Evaluating the above determinantal equation gives

$$\left[\left(\frac{1}{3} - \lambda\right)^2 - \left(\frac{1}{4}\right)^2\right]\left(\frac{2}{3} - \lambda\right) = 0$$

The second factor gives

$$\lambda_3 = \frac{2}{3}(ma^2)$$

The first factor gives

$$\frac{1}{3} - \lambda = \pm\frac{1}{4}$$

$$\lambda_1 = \frac{1}{12}(ma^2)$$

$$\lambda_2 = \frac{7}{12}(ma^2)$$ ∎

EXAMPLE 9.7

Find the directions of the principal axes of a square plate about a corner.

Solution:
Equations 9.15f give

$$\left(\frac{1}{3} - \lambda\right)\cos\alpha - \frac{1}{4}\cos\beta = 0$$

$$-\frac{1}{4}\cos\alpha + \left(\frac{1}{3} - \lambda\right)\cos\beta = 0$$

$$\left(\frac{2}{3} - \lambda\right)\cos\gamma = 0$$

We would guess that at least one of the principal axes (say, the third axis) is perpendicular to the plane of the square plate, that is, $\gamma_3 = 0°$ and $\alpha_3 = \beta_3 = 90°$. We would also guess from looking at the last equation that the principal moment about this axis would be $\lambda_3 = 2/3\ (ma^2)$. Such choices would ensure that the last equation above would automatically vanish as would the first two, since both $\cos\alpha_3$ and $\cos\beta_3$ would be identically zero. The remaining axes can be determined by inserting the other two principal moments into the above equations. Thus, if we set $\lambda_1 = 1/12\ (ma^2)$, we obtain the conditions that

$$\cos\alpha_1 - \cos\beta_1 = 0 \qquad \cos\gamma_1 = 0$$

which can be satisfied only by $\alpha_1 = \beta_1 = 45°$ and $\gamma_1 = 90°$. Now, if we insert the final principal moment $\lambda_2 = 7/12\ (ma^2)$ into the above equation, we obtain the conditions that can be satisfied if $\alpha_2 = 135°$, $\beta_2 = 45°$, and $\gamma_2 = 90°$.

$$\cos \alpha_2 + \cos \beta_2 = 0 \qquad \cos \gamma_2 = 0$$

Thus, two of the principal axes lie in the plane of the plate, one being the diagonal, the other perpendicular to the diagonal. The third is normal to the plate. The moment of inertia matrix in this coordinate representation is, thus,

$$\mathbf{I} = \begin{bmatrix} \dfrac{1}{12} & 0 & 0 \\ 0 & \dfrac{7}{12} & 0 \\ 0 & 0 & \dfrac{2}{3} \end{bmatrix} ma^2$$

and the corresponding principal axes in the original coordinate system are given by the vectors

$$\mathbf{e}_1 = \frac{1}{\sqrt{2}} \begin{bmatrix} 1 \\ 1 \\ 0 \end{bmatrix} \qquad \mathbf{e}_2 = \frac{1}{\sqrt{2}} \begin{bmatrix} -1 \\ 1 \\ 0 \end{bmatrix} \qquad \mathbf{e}_3 = \begin{bmatrix} 0 \\ 0 \\ 1 \end{bmatrix}$$

It can be seen that the principal axes can be obtained by a simple $45°$ rotation of the original coordinate system in the counter clockwise direction about the z-axis.　■

9.3　EULER'S EQUATIONS OF MOTION OF A RIGID BODY

We come now to what we may call the essential physics of the present chapter, namely, the actual three-dimensional rotation of a rigid body under the action of external forces. As we have learned from Chapter 7, the fundamental equation governing the rotational part of the motion of any system, referred to an inertial coordinate system, is

$$\mathbf{N} = \frac{d\mathbf{L}}{dt}$$

in which $\mathbf{N}$ is the net applied torque and $\mathbf{L}$ is the angular momentum. For a rigid body, we have seen that $\mathbf{L}$ is most simply expressed if the coordinate axes are principal axes for the body. Thus, in general, we must employ a coordinate system that is fixed in the body and rotates with it. That is, the angular velocity of the body and the angular velocity of the coordinate system are one and the same. (There is an exception: If two of the three principal moments I_1, I_2, and I_3 are equal to each other, then the coordinate axes need not be fixed in the body in order to be principal axes. This case will be considered later.) In any case, our coordinate system is not an inertial one.

Referring to the theory of rotating coordinate systems developed in Chapter 5, we know that the time rate of change of the angular momentum vector in a fixed (inertial) system versus a rotating system is given by the formula

$$\left(\frac{d\mathbf{L}}{dt}\right)_{fixed} = \left(\frac{d\mathbf{L}}{dt}\right)_{rot} + \boldsymbol{\omega} \times \mathbf{L}$$

Thus, the equation of motion in the rotating system is

$$\mathbf{N} = \left(\frac{d\mathbf{L}}{dt}\right)_{rot} + \boldsymbol{\omega} \times \mathbf{L} \tag{9.16}$$

where

$$\dot{\mathbf{L}} = \mathbf{I} \cdot \dot{\boldsymbol{\omega}} \tag{9.16a}$$

$$\boldsymbol{\omega} \times \mathbf{L} = \boldsymbol{\omega} \times (\mathbf{I} \cdot \boldsymbol{\omega})$$

The latter cross product above can be written as the determinant

$$\boldsymbol{\omega} \times (\mathbf{I} \cdot \boldsymbol{\omega}) = \begin{vmatrix} \mathbf{e}_1 & \mathbf{e}_2 & \mathbf{e}_3 \\ \omega_1 & \omega_2 & \omega_3 \\ I_1\omega_1 & I_2\omega_2 & I_3\omega_3 \end{vmatrix}$$

where the components of $\boldsymbol{\omega}$ are taken along the directions of the principal axes. Thus, Equation 9.16 can be written in matrix form as

$$\begin{bmatrix} N_1 \\ N_2 \\ N_3 \end{bmatrix} = \begin{bmatrix} I_1\dot{\omega}_1 \\ I_2\dot{\omega}_2 \\ I_3\dot{\omega}_3 \end{bmatrix} + \begin{bmatrix} \omega_2\omega_3(I_3 - I_2) \\ \omega_3\omega_1(I_1 - I_3) \\ \omega_2\omega_2(I_2 - I_1) \end{bmatrix} \tag{9.16b}$$

These are known as Euler's equations for the motion of a rigid body in components along the principal axes of the body.

Body Constrained to Rotate About a Fixed Axis

As a first application of Euler's equations, we take up the special case of a rigid body that is constrained to rotate about a fixed axis with constant angular velocity. Then

$$\dot{\omega}_1 = \dot{\omega}_2 = \dot{\omega}_3 = 0$$

and Euler's equations reduce to

$$N_1 = \omega_2\omega_3(I_3 - I_2)$$
$$N_2 = \omega_3\omega_1(I_1 - I_3) \tag{9.17}$$
$$N_3 = \omega_1\omega_2(I_2 - I_1)$$

These give the components of the torque that must be exerted on the body by the constraining support.

In particular, suppose that the axis of rotation is a principal axis, say the 1-axis. Then $\omega_2 = \omega_3 = 0$, $\omega = \omega_1$. In this case all three components of the torque vanish:

$$N_1 = N_2 = N_3 = 0$$

That is, there is no torque at all. This agrees with our discussion concerning dynamic balancing in the previous section.

9.4   FREE ROTATION OF A RIGID BODY. GEOMETRIC DESCRIPTION OF THE MOTION

Let us consider the case of a rigid body that is free to rotate in any direction about a certain point O. There are no torques acting on the body. This is the case of free rotation and is exemplified, for example, by a body supported on a smooth pivot at its center of mass. Another example is that of a rigid body moving freely under no forces or falling freely in a uniform gravitational field so that there are no torques. The point O in this case is the center of mass.

With zero torque the angular momentum of the body, as seen from the outside, must remain constant in direction and magnitude according to the general principle of conservation of angular momentum. However, with respect to rotating axes fixed in the body, the direction of the angular momentum vector may change, although its magnitude must remain constant. This fact can be expressed by the equation

$$\mathbf{L} \cdot \mathbf{L} = \text{constant} \tag{9.18}$$

In terms of components referred to the principal axes of the body, Equation 9.18 reads

$$I_1^2\omega_1^2 + I_2^2\omega_2^2 + I_3^2\omega_3^2 = L^2 = \text{constant} \tag{9.18a}$$

As the body rotates, the components of ω may vary, but they must always satisfy the above equation.

A second relation is obtained by considering the kinetic energy of rotation. Again, since there is zero torque, the total rotational kinetic energy must remain constant. This may be expressed as

$$\boldsymbol{\omega} \cdot \mathbf{L} = 2T_{rot} = \text{constant} \tag{9.19}$$

or, equivalently in terms of components,

$$I_1\omega_1^2 + I_2\omega_2^2 + I_3\omega_3^2 = 2T_{rot} = \text{constant} \tag{9.19a}$$

We now see that the components of $\boldsymbol{\omega}$ must simultaneously satisfy two different equations expressing the constancy of kinetic energy and of magnitude of angular momentum. (These two equations can also be obtained by use of Euler's equations. See Problem 9.7.) These are the equations of two ellipsoids whose principal axes coincide with the principal axes of the body. The first ellipsoid (Equation 9.18a) has principal diameters in the ratios $I_1^{-1}:I_2^{-1}:I_3^{-1}$. The second ellipsoid (Equation 9.19a) has principal diameters in the ratios $I_1^{-1/2}:I_2^{-1/2}:I_3^{-1/2}$. It is known as the *Poinsot ellipsoid*. As the body rotates, the extremity of the angular velocity vector thus describes a curve that is the intersection of the two ellipsoids. This is illustrated in Figure 9.8.

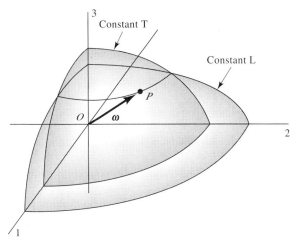

Figure 9.8 Intersecting ellipsoids of constant L and T for a rigid body undergoing torque-free rotation. (Only one octant is shown for clarity.)

From the equations of the intersecting ellipsoids, it can be shown that in the case where the initial axis of rotation coincides with one of the principal axes of the body, then the curve of intersection diminishes to a point. In other words the two ellipsoids just touch at a principal diameter, and the body rotates steadily about this axis. This is true, however, only if the initial rotation is about the axis of either the largest or the smallest moment of inertia. If it is about the intermediate axis, say the 2-axis where $I_3 > I_2 > I_1$, then the intersection of the two ellipsoids is not a point, but a curve that goes entirely around both, as illustrated in Figure 9.9. In this case the rotation is unstable, since the axis of rotation precesses all around the body. See Problem 9.19. (If the initial axis of rotation is almost, but not exactly, along one of the two stable axes, then the angular velocity vector describes a tight cone about the corresponding axis.) These facts can easily be illustrated by tossing an oblong block, a book, or a ping-pong paddle into the air.

9.5 FREE ROTATION OF A RIGID BODY WITH AN AXIS OF SYMMETRY. ANALYTICAL TREATMENT. THE EULERIAN ANGLES

Although the geometric description of the motion of a rigid body given in the preceding section is helpful in visualizing free rotation under no torques, the method does not immediately give numerical values. We now proceed to augment that description with an analytical approach based on the direct integration of Euler's equations.

We shall solve Euler's equations for the special case in which the body possesses an axis of symmetry, so that two of the three principal moments of inertia are equal.

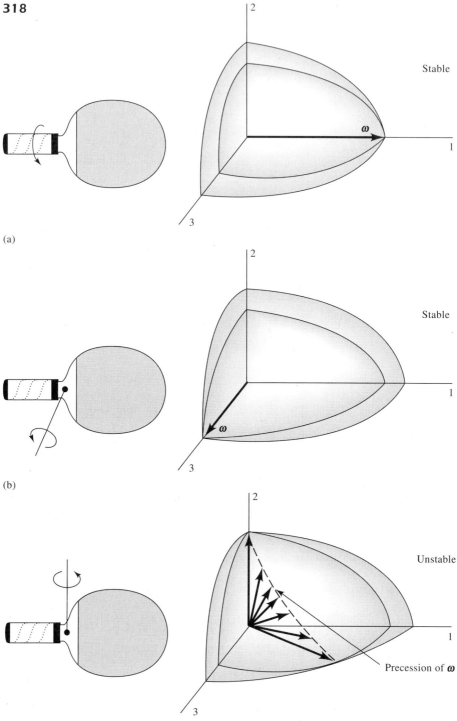

(a)

(b)

(c)

Figure 9.9 Ellipsoids of constant L and constant T for a rigid body rotating freely about the axis of (a) least, (b) greatest, and (c) intermediate moment of inertia.

Let us choose the 3-axis as the axis of symmetry. We introduce the following notation:

$I_s = I_3$ *moment of inertia about the symmetry axis*

$I = I_1 = I_2$ *moment about the axes normal to the symmetry axis*

For the case of zero torque, Euler's equations then read

$$I\dot{\omega}_1 + \omega_2\omega_2(I_s - I) = 0$$
$$I\dot{\omega}_2 + \omega_3\omega_1(I - I_s) = 0 \tag{9.20}$$
$$I_s\dot{\omega}_3 = 0$$

From the last equation it follows that

$$\omega_3 = \text{constant}$$

Let us now define a constant Ω as

$$\Omega = \omega_3\frac{I_s - I}{I} \tag{9.21}$$

Then the first two of Equations 9.20 may be written

$$\dot{\omega}_1 + \Omega\omega_2 = 0 \tag{9.22}$$
$$\dot{\omega}_2 - \Omega\omega_1 = 0 \tag{9.23}$$

To separate the variables in the above pair of equations, we differentiate the first with respect to t and obtain

$$\ddot{\omega}_1 + \Omega\dot{\omega}_2 = 0$$

Upon solving for $\dot{\omega}_2$ and inserting the result into Equation 9.23, we find

$$\ddot{\omega}_1 + \Omega^2\omega_1 = 0$$

This is the equation for simple harmonic motion. A solution is

$$\omega_1 = \omega_0 \cos \Omega t \tag{9.24}$$

in which ω_0 is a constant of integration. To find ω_2, we differentiate the above equation with respect to t and insert the result into Equation 9.22. We can then solve for ω_2 to obtain

$$\omega_2 = \omega_0 \sin \Omega t \tag{9.25}$$

Thus, ω_1 and ω_2 vary harmonically in time with angular frequency Ω, and their phases differ by $\pi/2$. It follows that the projection of $\boldsymbol{\omega}$ on the 1, 2 plane describes a circle of radius ω_0 at the angular frequency Ω.

We can summarize the above results as follows: In the free rotation of a rigid body with an axis of symmetry, the angular velocity vector describes a conical motion (precesses) about the symmetry axis. It describes a surface called the *body cone*. (See Figure 9.12.) The angular frequency of this precession is the constant Ω defined by Equation

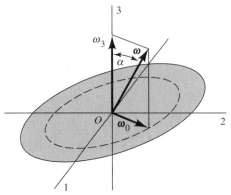

Figure 9.10   Angular velocity vector and its components for the free precession of a disc.

9.21. Let α denote the angle between the symmetry axis (3-axis) and the axis of rotation (direction of $\boldsymbol{\omega}$) as shown in Figure 9.10. Then $\omega_3 = \omega \cos \alpha$, and so

$$\Omega = \left(\frac{I_s}{I} - 1 \right) \omega \cos \alpha \qquad (9.26)$$

giving the rate of precession of the angular velocity vector about the axis of symmetry. (Some specific examples are discussed at the end of the present section.)

We can now see the connection between the above analysis of the torque-free rotation of a rigid body and the geometric description of the previous section. The circular path of radius ω_0 traced out by the extremity of the angular velocity vector is just the intersection of the two ellipsoids of Figure 9.8.

DESCRIPTION OF THE ROTATION OF A RIGID BODY RELATIVE TO A FIXED COORDINATE SYSTEM. THE EULERIAN ANGLES

In the foregoing analysis of the free rotation of a rigid body, the precessional motion was relative to a set of principal axes fixed in the body and rotating with it. In order to describe the motion relative to an observer outside the body, we must use a fixed coordinate system. In Figure 9.11(a) the coordinate system $Oxyz$ has a fixed orientation in space. The numbered system $O123$ is our previously defined set of principal axes, which is fixed in the body and rotates with it. A third system $Ox'y'z'$ is defined as follows: The z'-axis coincides with the 3-axis or symmetry axis of the body, and the x'-axis is the line of intersection of the 1, 2 plane with the fixed xy plane. The angle between the x- and the x'-axes is denoted by ϕ, and the angle between the z- and z'- or 3-axes is θ. The turning of the body about the 3-axis is determined by the angle between the 1-axis and the x'-axis, denoted by ψ as shown. The three angles ϕ, θ, and ψ completely define the orientation of the body in space and are called the *Eulerian angles*.

Now, from a study of the figure we see that $\boldsymbol{\omega}$, the angular velocity of the body (numbered system) consists of a turning about the z'-axis or 3-axis with angular rate $\dot{\psi}$

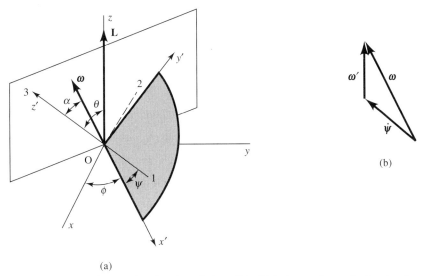

(a)

Figure 9.11 Diagram showing the relation of the Eulerian angles to the fixed and the rotating coordinate axes.

superimposed on the rotation of the primed system $Ox'y'z'$. Let us call $\boldsymbol{\omega}'$ the angular velocity of the primed coordinate system. $\boldsymbol{\omega}'$ is not shown in the figure because it would clutter the diagram. [It is shown in Figure 9.11(b).] However, it should be clear that the actual rotation of the primed system consists of two parts: an angular rate $\dot{\theta}$ about the x'-axis, and an angular rate $\dot{\phi}$ about the fixed z-axis. The components of $\boldsymbol{\omega}'$ in the $Ox'y'z'$ system are, therefore,

$$\omega'_{x'} = \dot{\theta}$$
$$\omega'_{y'} = \dot{\phi}_{y'} = \dot{\phi} \sin \theta \qquad (9.27)$$
$$\omega'_{z'} = \dot{\phi}_{z'} = \dot{\phi} \cos \theta$$

Now $\boldsymbol{\omega}$ differs from $\boldsymbol{\omega}'$ only in the turning with angular rate $\dot{\psi}$ about the z'-axis. Therefore, for the components of $\boldsymbol{\omega}$ in the primed system, we can write

$$\omega_{x'} = \dot{\theta}$$
$$\omega_{y'} = \dot{\phi} \sin \theta \qquad (9.28)$$
$$\omega_{z'} = \dot{\phi} \cos \theta + \dot{\psi}$$

We can now express the components of $\boldsymbol{\omega}$ in the $O123$ system. They are as follows:

$$\omega_1 = \omega_{x'} \cos \psi + \omega_{y'} \sin \psi = \dot{\theta} \cos \psi + \dot{\phi} \sin \theta \sin \psi$$
$$\omega_2 = \omega_{x'}(-\sin \psi) + \omega_{y'} \cos \psi = -\dot{\theta} \sin \psi + \dot{\phi} \sin \theta \cos \psi \qquad (9.29)$$
$$\omega_3 = \omega_{z'} = \dot{\phi} \cos \theta + \dot{\psi}$$

(We shall not need to use Equations 9.29 at present but will refer to them later.)

Now in the present case in which there is zero torque acting on the body, the angular momentum vector $\mathbf{L}$ is constant in magnitude and direction in the fixed system $Oxyz$. Let us choose the z-axis to be the direction of $\mathbf{L}$. This is known as the *invariable line*. From Figure 9.11 we see that the components of $\mathbf{L}$ in the primed system are

$$L_{x'} = 0$$
$$L_{y'} = L \sin \theta \qquad (9.30)$$
$$L_{z'} = L \cos \theta$$

We again restrict ourselves to the case of a body with an axis of symmetry, the 3-axis. Since the x' and y'-axes lie in the 1, 2 plane, and the z'-axis coincides with the 3-axis, then the primed axes are also principal axes. In fact, the principal moments are the same: $I_1 = I_2 = I_{x'x'} = I_{y'y'} = I$ and $I_3 = I_{z'z'} = I_s$.

Now consider the first of Equations 9.28 and 9.30 giving the x' component of the angular velocity and angular momentum of the body, namely zero. From these we see that $\dot\theta = 0$. Hence, θ is constant, and $\boldsymbol{\omega}$, having no x'-component, must lie in the $y'z'$ plane as shown. Let α denote the angle between the angular velocity vector $\boldsymbol{\omega}$ and the z'-axis. Then, in addition to Equations 9.28 and 9.30, we also have the following:

$$\omega_{y'} = \omega \sin \alpha \qquad \omega_{z'} = \omega \cos \alpha \qquad (9.31)$$
$$L_{y'} = I\omega \sin \alpha \qquad L_{z'} = I_s\omega \cos \alpha$$

It readily follows that

$$\frac{L_{y'}}{L_{z'}} = \tan \theta = \frac{I}{I_s} \tan \alpha \qquad (9.32)$$

giving the relation between the angles θ and α.

According to the above result, θ is less than or greater than α, depending on whether I is less than I_s or greater than I_s, respectively. In other words, the angular momentum vector lies between the symmetry axis and the axis of rotation in the case of a flattened body ($I < I_s$), whereas in the case of an elongated body ($I > I_s$), the axis of rotation lies between the axis of symmetry and the angular momentum vector. The two cases are illustrated in Figure 9.12. In either case, as the body rotates, the axis of symmetry (z'-axis or 3-axis) describes a conical motion or precesses about the constant angular momentum vector $\mathbf{L}$. At the same time the axis of rotation ($\boldsymbol{\omega}$ vector) precesses about $\mathbf{L}$ with the same frequency. The surface traced out by $\boldsymbol{\omega}$ about $\mathbf{L}$ is called the *space cone*, as indicated.

Referring to Figure 9.11, we see that the angular speed of rotation of the $y'z'$ plane about the z-axis is equal to the time rate of change of the angle ϕ. Thus, $\dot\phi$ is the angular rate of precession of the symmetry axis and also of the axis of rotation about the invariable line ($\mathbf{L}$ vector) as viewed from outside the body. This precession appears as a "wobble" such as that seen in an imperfectly thrown football or discus. From the second of Equations 9.28 and the first of Equations 9.31, we have $\dot\phi \sin \theta = \omega \sin \alpha$ or

$$\dot\phi = \omega \frac{\sin \alpha}{\sin \theta} \qquad (9.33)$$

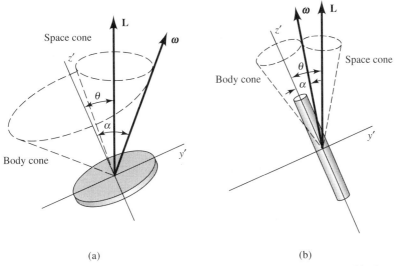

(a)   (b)

Figure 9.12   Free rotation of a (a) disc and (b) rod. The space cones and body cones are shown dotted.

for the rate of precession. Equation 9.33 can be put into a somewhat more useful form by using the relation between α and θ given by Equation 9.32. After a little algebra, we obtain

$$\dot{\phi} = \omega \left[1 + \left(\frac{I_s^2}{I^2} - 1 \right) \cos^2 \alpha \right]^{1/2} \tag{9.34}$$

for the wobble rate in terms of the angular speed ω of the body about its axis of rotation and the inclination α of the axis of rotation to the symmetry axis of the body.

  To summarize our analysis of the free rotation of a rigid body with an axis of symmetry, there are three basic angular rates: the magnitude ω of the angular velocity, the precession of angular rate Ω of the axis of rotation (direction of $\boldsymbol{\omega}$) about the symmetry axis of the body, and the precession (wobble) of angular rate $\dot{\phi}$ of the symmetry axis about the invariable line (constant angular momentum vector).

EXAMPLE 9.8   *Precession of a Frisbee*

  As an example of the above theory we consider the case of a thin disc, or any symmetric and fairly "flat" object such as a china plate or a Frisbee. The perpendicular-axis theorem for principal axes is $I_1 + I_2 = I_3$, and, for a symmetric body $I_1 = I_2$, so that $2I_1 = I_3$. In our present notation this is $2I = I_s$, so the ratio

$$\frac{I_s}{I} = 2$$

to a good approximation. If our object is thrown into the air in such a way that the angular velocity $\boldsymbol{\omega}$ is inclined to the symmetry axis by an angle α, then Equation 9.26 gives

$$\Omega = \omega \cos \alpha$$

for the rate of precession of the rotational axis about the symmetry axis.

For the precession of the symmetry axis about the invariable line, the wobble as seen from the outside, Equation 9.34 yields

$$\dot{\phi} = \omega(1 + 3 \cos^2 \alpha)^{1/2}$$

In particular, if α is quite small so that $\cos \alpha$ is very nearly unity, then we have approximately

$$\Omega \simeq \omega$$

$$\dot{\phi} \simeq 2\omega \qquad \blacksquare$$

Thus, the wobble rate is very nearly twice the angular speed of rotation.

EXAMPLE 9.9   *Free Precession of the Earth*

In the motion of the earth, it is known that the axis of rotation is very slightly inclined with respect to the geographic pole defining the axis of symmetry. The angle α is about 0.2 sec of arc (shown exaggerated in Figure 9.13). It is also known that the ratio of the moments of inertia I_s/I is about 1.00327 as determined from the earth's oblateness. From Equation 9.26 we have, therefore,

$$\Omega = 0.00327\omega$$

Then, since $\omega = 2\pi/\text{day}$, the period of the above precession is calculated to be

$$\frac{2\pi}{\Omega} = \frac{1}{0.00327} \text{ days} = 305 \text{ days}$$

The observed period of precession of the earth's axis of rotation about the pole is about 440 days. The disagreement between the observed and calculated values is attributed to the fact that the earth is not perfectly rigid.

With regard to the precession of the earth's symmetry axis as viewed from space, Equation 9.34 gives

$$\dot{\phi} = 1.00327\omega$$

The associated period of the earth's wobble is, thus,

$$\frac{2\pi}{\dot{\phi}} = \frac{2\pi}{\omega} \frac{1}{1.00327} \simeq 0.997 \text{ day}$$

This free precession of the earth's axis in space is superimposed upon a very much longer gyroscopic precession of 26,000 years, the latter resulting from the fact that there is actually a torque exerted on the earth (because of its oblateness) by the sun

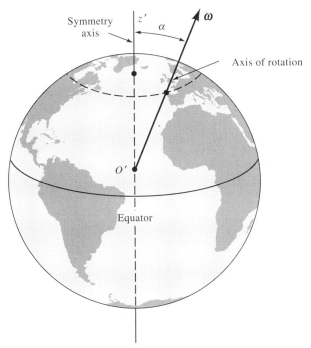

Figure 9.13 Showing the symmetry axis and the rotational axis of the earth. (The angle α is greatly exaggerated.)

and the moon. The fact that the period of gyroscopic precession is so much longer than that of the free precession justifies the neglect of the external torques in calculating the period of the free precession. ■

9.6 GYROSCOPIC PRECESSION. MOTION OF A TOP

In this section we shall study the motion of a symmetrical rigid body that is free to turn about a fixed point and on which there is exerted a torque, instead of no torque, as in the case of free precession. The case is exemplified by a simple gyroscope (or top).

The notation for our coordinate axes is shown in Figure 9.14(a). For clarity, only the z'-, y'-, and z-axes are shown in Figure 9.14(b), the x'-axis being normal to the paper. The origin O is the fixed point about which the body turns.

The torque about O resulting from the weight is of magnitude $mgl \sin \theta$, l being the distance from O to the center of mass C. This torque is about the x'-axis, so that

$$N_{x'} = mgl \sin \theta$$
$$N_{y'} = 0 \qquad\qquad (9.35)$$
$$N_{z'} = 0$$

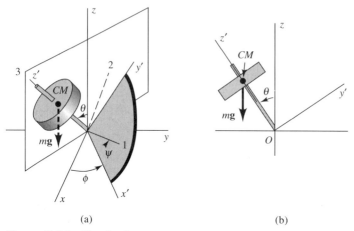

(a)   (b)

Figure 9.14   The simple gyroscope.

The components of the angular velocity of the body $\boldsymbol{\omega}$ are given by Equations 9.28. Hence, the angular momentum of the gyroscope has the following components in the primed system:

$$L_{x'} = I_{x'x'}\omega_{x'} = I\dot{\theta}$$

$$L_{y'} = I_{y'y'}\omega_{y'} = I\dot{\phi}\sin\theta \tag{9.36}$$

$$L_{z'} = I_{z'z'}\omega_{z'} = I_s(\dot{\phi}\cos\theta + \dot{\psi}) = I_s S$$

Here we use the same notation for the moments of inertia as in the previous section, and in the last equation above we have abbreviated the quantity $\omega_{z'} = \dot{\phi}\cos\theta + \dot{\psi}$ by the letter S, called the *spin*.

The fundamental equation of motion in the primed system is

$$\mathbf{N} = \left(\frac{d\mathbf{L}}{dt}\right)_{rot} + \boldsymbol{\omega}' \times \mathbf{L} \tag{9.37}$$

in which the components of $\mathbf{N}$, $\mathbf{L}$, and $\boldsymbol{\omega}'$, are given by Equations 9.35, 9.36, and 9.27, respectively. Consequently, the component equations of motion are found to be the following:

$$mgl\sin\theta = I\ddot{\theta} + I_s S\dot{\phi}\sin\theta - I\dot{\phi}^2\cos\theta\sin\theta \tag{9.37a}$$

$$0 = I\frac{d}{dt}(\dot{\phi}\sin\theta) - I_s S\dot{\theta} + I\dot{\theta}\dot{\phi}\cos\theta \tag{9.37b}$$

$$0 = I_s \dot{S} \tag{9.37c}$$

The last equation shows that S, the spin of the body about the symmetry axis, remains constant. Also, of course, the component of the angular momentum along that axis is constant

$$L_{z'} = I_s S = \text{constant} \tag{9.38}$$

The second equation is then equivalent to

$$0 = \frac{d}{dt}(I\dot{\phi} \sin^2 \theta + I_s S \cos \theta) \tag{9.39}$$

so that

$$I\dot{\phi} \sin^2 \theta + I_s S \cos \theta = B = \text{constant} \tag{9.40}$$

Steady Precession of a Gyroscope

At this point we shall discuss a simple special case of gyroscopic motion, namely that of steady precession in a horizontal plane. This is the common "demonstration" case in which the axis remains horizontal and precesses at a constant rate around a vertical line, the z-axis in our notation. Then we have $\theta = 90° = $ constant, $\dot{\theta} = \ddot{\theta} = 0$. Equation 9.37a then reduces to the simple relation

$$mgl = I_s S\dot{\phi} \tag{9.41}$$

Now it is easy to see that the quantity mgl is just the (scalar) torque about the x'-axis. Furthermore, the horizontal (vector) component of the angular momentum has a magnitude of $I_s S$, and it describes a circle in the horizontal plane. Consequently, the extremity of the **L** vector has a velocity (time rate of change) of magnitude $I_s S \dot{\phi}$ and a direction that is parallel to the x'-axis. Thus, Equation 9.41 is simply a statement of the general relation $\mathbf{N} = d\mathbf{L}/dt$ for the special case in point.

The more general case of steady precession in which the angle θ is constant but has a value other than 90° is still handled by use of Equation 9.37a, which gives, upon setting $\ddot{\theta} = 0$ and canceling the common factor $\sin \theta$,

$$mgl = I_s S\dot{\phi} - I\dot{\phi}^2 \cos \theta \tag{9.42}$$

This is a quadratic equation in the unknown $\dot{\phi}$. Solving it yields two roots

$$\dot{\phi} = \frac{I_s S \pm (I_s^2 S^2 - 4mglI \cos \theta)^{1/2}}{2I \cos \theta} \tag{9.43}$$

Thus, for a given value of θ, there are two possible rates of steady precession of the gyroscope: a fast precession (upper sign) and a slow precession (lower sign). Which of the two occurs depends on the initial conditions. Usually, it is the slower one that takes place in the motion of a simple top or gyroscope. In either case, the quantity in parentheses must be zero or positive for a physically possible solution, that is,

$$I_s^2 S^2 \geq 4mglI \cos \theta \tag{9.44}$$

Sleeping Top

Anyone who has played with a top knows that if the top is set spinning sufficiently fast and is started in a vertical position, the axis of the top will remain steady in the upright position, a condition called *sleeping*. This corresponds to a constant value of zero for θ

in the above equations. Since $\dot{\phi}$ must be real, we conclude that the criterion for stability of the sleeping top is given by

$$I_s^2 S^2 \geq 4mgIl \tag{9.45}$$

If the top slows down through friction so that the above condition no longer holds, then it will begin to fall and will eventually topple over.

EXAMPLE 9.10

A toy gyroscope has a mass of 100 g and is made in the form of a uniform disc of radius $a = 2$ cm fastened to a light spindle, the center of the disc being 2 cm from the pivot. If the gyroscope is set spinning at a rate of 20 revolutions per second, find the period for steady horizontal precession.

Solution:
Using cgs units we have $I_s = \frac{1}{2}ma^2 = \frac{1}{2} \times 100$ g $\times (2$ cm$)^2 = 200$ g cm^2. For the spin we must convert revolutions per second to radians per second, that is, $S = 20 \times 2\pi$ rad/s. Equation 9.41 then gives the precession rate

$$\dot{\phi} = \frac{mgl}{I_s S} = \frac{100 \text{ g} \times 980 \text{ cms}^{-2} \times 2 \text{ cm}}{200 \text{ g cm}^2 \times 40 \times 3.142 \text{ s}^{-1}} = 7.8 \text{ s}^{-1}$$

in radians per second. The associated period is then

$$\frac{2\pi}{\dot{\phi}} = \frac{2 \times 3.142}{7.8 \text{ s}^{-1}} = 0.81 \text{ s}$$

EXAMPLE 9.11

Find the minimum spin of the above gyroscope in order that it can sleep in the vertical position.

Solution:
We need, in addition to the above values, the moment of inertia I about the x'- or y'-axes. By the parallel-axis theorem, we have $I = I_{x'x'} = I_{y'y'} = \frac{1}{4}ma^2 + ml^2 = \frac{1}{4} 100$ g $\times (2$ cm$)^2 + 100$ g $\times (2$ cm$)^2 = 500$ g cm^2. From Equation 9.45, we can then write

$$S \geq \frac{2}{I_s}(mgIl)^{1/2} = \frac{2}{200}(200 \times 980 \times 2 \times 500)^{1/2} \text{ s}^{-1} = 140 \text{ s}^{-1}$$

or, in revolutions per second, the minimum spin is

$$S = \frac{140}{2\pi} = 22.3 \text{ rps}$$

∎

The Energy Equation and Nutation

If there are no frictional forces acting on the gyroscope to dissipate its energy, then the total energy $T_{rot} + V$ remains constant:

$$\frac{1}{2}(I\omega_{x'}^2 + I\omega_{y'}^2 + I_s S^2) + mg\,h = E$$

or equivalently, in terms of the Eulerian angles,

$$\frac{1}{2}(I\dot{\theta}^2 + I\dot{\phi}^2 \sin^2 \theta + I_s S^2) + mgl \cos \theta = E \qquad (9.46)$$

From Equation 9.40, we can solve for $\dot{\phi}$ and substitute into Equation 9.46. The result is

$$\frac{1}{2}I\dot{\theta}^2 + \frac{(B - I_s S \cos \theta)^2}{2I \sin^2 \theta} + \frac{1}{2}I_s S^2 + mgl \cos \theta = E \qquad (9.47)$$

which is entirely in terms of θ. This equation allows us, in principle, to find θ as a function of t by integration. Let us make the substitution

$$u = \cos \theta$$

Thus $\dot{u} = -(\sin \theta)\dot{\theta} = -(1 - u^2)^{1/2}\dot{\theta}$. We find that Equation 9.47 then becomes

$$\dot{u}^2 = (1 - u^2)(2E - I_s S^2 - 2mglu)I^{-1} - (B - I_s Su)^2 I^{-2} \qquad (9.47a)$$

or

$$\dot{u}^2 = f(u) \qquad (9.47b)$$

from which u (hence, θ) can be found as a function of t by integration:

$$t = \int \frac{du}{\sqrt{f(u)}} \qquad (9.48)$$

Here $f(u)$ is a cubic polynomial, and the integration can be carried out in terms of elliptic functions.[3]

We need not actually perform the integration, however, to discuss the general properties of the motion. We see that $f(u)$ must be positive in order that t be real. Hence, the limits of the motion in θ are determined by the roots of the equation $f(u) = 0$. Since θ must lie between 0° and 90°, then u must take values between 0 and $+1$. A plot of $f(u)$ is shown in Figure 9.15(a) for the case in which there are two distinct roots u_1 and u_2 between 0 and $+1$. The corresponding values of θ, namely θ_1 and θ_2, are then the limits of the vertical motion. The axis of the top oscillates back and forth between these two values of θ as the top precesses about the vertical, Figure 9.16. This oscillation is called *nutation*. If we have a double root, that is, if $u_1 = u_2$, then there is no nutation, and the top precesses steadily, Figure 9.15(b). The sleeping case is shown in Figure 9.15c.

---

[3] See reference cited in footnote 1 (Table 8.2).

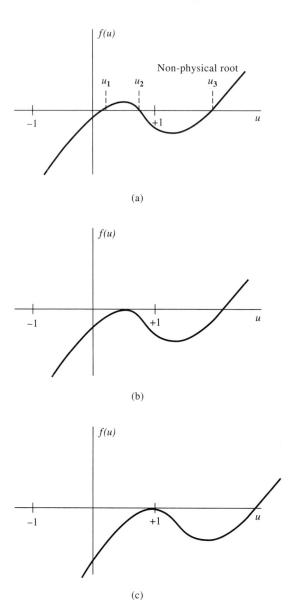

Figure 9.15   Graphs of the function $f(u)$ for a spinning top or simple gyroscope. (a) Two distinct roots: regular nutation. (b) Two equal roots: steady precession. (c) Condition for the sleeping top.

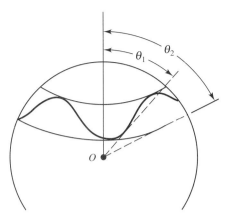

Figure 9.16   Illustrating nutation.

9.7   THE GYROCOMPASS

Let us consider the motion of a gyroscope that is mounted on a gimball support that constrains the spin axis to remain horizontal, but the axis is otherwise free to turn in any direction. The situation is diagrammed in Figure 9.17, which is taken from Figure 9.14 except that now $\theta = 90°$ and the unprimed axes are labeled to correspond to directions on the earth's surface as shown. The gyroscope is centrally mounted so that $l = 0$.

We know from Chapter 5 that the earth's angular velocity, here denoted by $\boldsymbol{\omega}_e$, has components $\omega_e \cos \lambda$ (north) and $\omega_e \sin \lambda$ (vertical) where λ is the latitude. In the primed coordinate system we can then write

$$\boldsymbol{\omega}_e = \mathbf{i}'\omega_e \cos \lambda \cos \phi + \mathbf{j}'\omega_e \sin \lambda + \mathbf{k}'\omega_e \cos \lambda \sin \phi \qquad (9.49)$$

Now the primed system is turning about the vertical with angular rate $\dot{\phi}$ so the angular velocity of the primed system is

$$\boldsymbol{\omega}' = \boldsymbol{\omega}_e + \mathbf{j}'\dot{\phi} = \mathbf{i}'\omega_e \cos \lambda \cos \phi + \mathbf{j}'(\dot{\phi} + \omega_e \sin \lambda) \qquad (9.50)$$
$$+ \mathbf{k}'\omega_e \cos \lambda \sin \phi$$

Similarly, the gyroscope itself is turning about the z'-axis at a rate $\dot{\psi}$ superimposed on the above components so that the angular velocity of the gyroscope, referred to the primed system, is

$$\boldsymbol{\omega} = \boldsymbol{\omega}' + \mathbf{k}'\dot{\psi} = \mathbf{i}'\omega_e \cos \lambda \cos \phi + \mathbf{j}'(\dot{\phi} + \omega_e \sin \lambda) \qquad (9.51)$$
$$+ \mathbf{k}'(\dot{\psi} + \omega_e \cos \lambda \sin \phi)$$

The principal moments of inertia of the gyroscope are, as before, $I_1 = I_2 = I, I_3 = I_s$. Hence, the angular momentum can be expressed as

$$\mathbf{L} = \mathbf{i}'I\omega_e \cos \lambda \cos \phi + \mathbf{j}'I(\dot{\phi} + \omega_e \sin \lambda) + \mathbf{k}'I_s S \qquad (9.52)$$

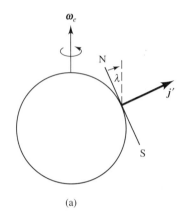

(a)

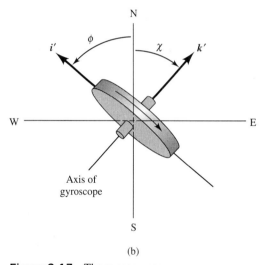

(b)

Figure 9.17   The gyrocompass.

where, in the last term, the total spin S is

$$S = \dot{\psi} + \omega_e \cos \lambda \sin \phi$$

Now, since the gyroscope is free to turn about both the vertical (y'-axis) and the spin or z'-axis, the applied torque required to keep the axis horizontal must be about the x'-axis: $\mathbf{N} = \mathbf{i}'N$. The equation of motion

$$\mathbf{N} = \left(\frac{d\mathbf{L}}{dt}\right)_{rot} + \boldsymbol{\omega}' \times \mathbf{L} \tag{9.53}$$

thus has components referred to the primed system as follows:

$$N = I\frac{d}{dt}(\omega_e \cos \lambda \cos \phi) + (\boldsymbol{\omega}' \times \mathbf{L})_{x'} \tag{9.53a}$$

$$0 = I\frac{d}{dt}(\dot{\phi} + \omega_e \sin \lambda) + (\boldsymbol{\omega}' \times \mathbf{L})_{y'} \tag{9.53b}$$

$$0 = I_s\frac{dS}{dt} + (\boldsymbol{\omega}' \times \mathbf{L})_{z'} \tag{9.53c}$$

From the expressions for $\boldsymbol{\omega}'$ and $\mathbf{L}$, we find that $(\boldsymbol{\omega}' \times \mathbf{L})_{z'} = 0$, so the last equation becomes $dS/dt = 0$. Thus, S is constant. Further, we find that the second equation becomes

$$0 = I\ddot{\phi} + I\omega_e^2 \cos^2 \lambda \cos \phi \sin \phi - I_s S\omega_e \cos \lambda \cos \phi \tag{9.54}$$

It is convenient at this point to express the angle ϕ in terms of its complement $\chi = 90°$ $- \phi$, so $\cos \phi = \sin \chi$ and $\ddot{\phi} = -\ddot{\chi}$. Furthermore, we can neglect the term involving ω_e^2 in the above equation because $S \gg \omega_e$ in the present case. Consequently, the equation reduces to

$$\ddot{\chi} + \left(\frac{I_s S\omega_e \cos \lambda}{I}\right) \sin \chi = 0 \tag{9.54a}$$

This is similar to the differential equation for a pendulum. The variable χ oscillates about the value $\chi = 0$, and the presence of any damping will cause the axis of the gyroscope to "seek" and eventually settle down to a north–south direction. For small amplitude the period of the oscillation is

$$T_0 = 2\pi \left(\frac{I}{I_s \omega_e S \cos \lambda}\right)^{1/2} \tag{9.55}$$

Since the ratio I_s/I is very nearly 2 for any flat-type symmetric object, the period of oscillation is essentially independent of the mass and dimensions of the gyroscope. Further, since ω_e is very small, the spin S must be fairly large in order to have a reasonably small period. For example, let $S = 60$ s$^{-1} = 2\pi \times 60$ rad/s. Then, for a flat-type gyroscope, we find for a latitude of 45° N

$$T_0 = 2\pi \left(\frac{24 \times 60 \times 60}{2 \times 2\pi \times 60 \times 2\pi \times 0.707}\right)^{1/2} \text{s} = \left(\frac{12 \times 60}{0.707}\right)^{1/2} \text{s} = 31.9 \text{ s}$$

or about 1/2 min. In the above calculation we have used the fact that $\omega_e = 2\pi/(24 \times 60 \times 60)$ rad/s so that the factors 2π all cancel.

9.8 GENERAL MOTION OF A RIGID BODY. ROLLING WHEEL

As we have seen, the general motion of any system can be resolved into two parts: (1) the translational motion of the center of mass, governed by the equation $\mathbf{F} = m d\mathbf{v}_{cm}/dt$, and (2) the rotation about the center of mass, determined by $\mathbf{N} = d\mathbf{L}/dt$. As an example

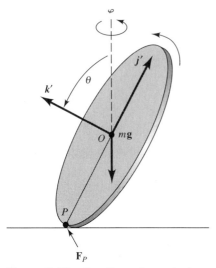

Figure 9.18 Coordinates for analyzing
the motion of a rolling wheel.

we shall discuss the motion of a wheel or disc rolling on a perfectly rough horizontal
surface. The external forces are the weight $m\mathbf{g}$ acting at the center and the reaction $\mathbf{F}_P$
at the point of contact, Figure 9.18. The general equations of motion are then $\mathbf{F}_P + m\mathbf{g}$
$= m d\mathbf{v}_{cm}/dt$ and $\mathbf{r}_{OP} \times \mathbf{F}_P = d\mathbf{L}/dt$. Elimination of $\mathbf{F}_P$ between the two equations yields
the single equation

$$\mathbf{r}_{OP} \times \left(m\frac{d\mathbf{v}_{cm}}{dt} - m\mathbf{g} \right) = \frac{d\mathbf{L}}{dt} \tag{9.56}$$

If $\boldsymbol{\omega}$ is the angular velocity of the wheel, then the velocity of the mass center is given by

$$\mathbf{v}_{cm} = \boldsymbol{\omega} \times \mathbf{r}_{PO} = \boldsymbol{\omega} \times (-\mathbf{r}_{OP})$$

Now, in the primed coordinate system $\mathbf{r}_{OP} = -\mathbf{j}'a$, and $\mathbf{g} = -g(\mathbf{j}' \sin \theta + \mathbf{k}'$
$\cos \theta)$, in which a is the radius and θ is the inclination of the wheel's axis to the vertical as shown. Further, we have a rotating coordinate system, so we must write
$d/dt = d/dt)_{rot} + \boldsymbol{\omega}' \times$ where $\boldsymbol{\omega}'$ is the angular velocity of the primed system. Equation 9.56 then becomes

$$-ma^2\mathbf{j}' \times \left[\left(\frac{d\boldsymbol{\omega}}{dt} \right)_{rot} \times \mathbf{j}' + \boldsymbol{\omega}' \times (\boldsymbol{\omega} \times \mathbf{j}') \right]$$

$$- mga[\mathbf{j}' \times (\mathbf{j}' \sin \theta + \mathbf{k}' \cos \theta)] = \left(\frac{d\mathbf{L}}{dt} \right)_{rot} + \boldsymbol{\omega}' \times \mathbf{L} \tag{9.57}$$

We shall not attempt to solve the above general equation of motion. Rather, we wish
to discuss a special case, namely, that for which the wheel stays very nearly vertical and
the direction of the rolling motion is constant or nearly constant: *steady rolling*. This

means that θ remains close to $90°$ so that the complement $\chi = 90° - \theta$ and the angle ϕ both remain small. Under these assumptions $\sin \chi = \chi$ and $\sin \phi = \phi$, approximately, and the general expressions for the components of $\boldsymbol{\omega'}$, $\boldsymbol{\omega}$, and $\mathbf{L}$, as given by Equations 9.27, 9.28, and 9.36, simplify to give

$$\boldsymbol{\omega'} = -\mathbf{i}\dot{\chi} + \mathbf{j'}\dot{\phi} \tag{9.58}$$

$$\boldsymbol{\omega} = -\mathbf{i'}\dot{\chi} + \mathbf{j'}\dot{\phi} + \mathbf{k'}S \tag{9.59}$$

$$\mathbf{L} = -\mathbf{i'}I\dot{\chi} + \mathbf{j'}I\dot{\phi} + \mathbf{k'}I_sS \tag{9.60}$$

Inserting these into Equation 9.57 and performing the indicated operations and dropping higher-order terms in the small quantities χ and ϕ, the following result is obtained:

$$ma^2(\mathbf{i'}\ddot{\chi} - \mathbf{k'}\dot{S} - \mathbf{i'}S\dot{\phi}) - mgai'\chi = \mathbf{i'}(I\ddot{\chi} + T_sS\dot{\phi}) \tag{9.61}$$
$$+ \mathbf{j'}(I\ddot{\phi} + I_sS\dot{\chi}) + \mathbf{k'}I_s\dot{S}$$

Equating the three components gives

$$ma^2(\ddot{\chi} - S\dot{\phi}) - mga\chi = -I\ddot{\chi} + I_sS\dot{\phi} \tag{9.61a}$$

$$0 = I\ddot{\phi} + I_sS\dot{\chi} \tag{9.61b}$$

$$-ma^2\dot{S} = I_s\dot{S} \tag{9.61c}$$

The last of the three equations shows that $\dot{S} = 0$ so S is constant. The second equation can then be integrated to yield $I\dot{\phi} + I_sS\chi = 0$ provided we assume that $\chi = \phi = 0$ for the initial condition. Then $\dot{\phi} = -I_sS\chi/I$, which, inserted into the first equation, gives the following separated differential equation for χ:

$$I(I + ma^2)\ddot{\chi} + [I_s(I_s + ma^2)S^2 - Imga]\chi = 0 \tag{9.62}$$

It follows that the assumed stable rolling will take place if the quantity in brackets is positive. Thus, the stability criterion is

$$S^2 > \frac{Imga}{I_s(I_s + ma^2)} \tag{9.63}$$

EXAMPLE 9.12   *Rolling Penny*

How fast must a penny ($a = 0.95$ cm) roll in order to remain upright?

Solution:
Assuming the penny to be a uniform thin lamina, we have $I_s = 2I = \frac{1}{2}ma^2$, so the criterion for stable rolling is

$$S^2 > \frac{mga}{2\left(\dfrac{1}{2}ma^2 + ma^2\right)} = \frac{g}{3a}$$

Since the rolling speed is $v = v_{cm} = aS$, the above criterion can be expressed alternatively as

$$v^2 > \frac{ga}{3}$$

Thus,

$$v > \left(\frac{980 \text{ cm s}^{-2} \times 0.95 \text{ cm}}{3}\right)^{1/2} = 17.6 \text{ cm/s}$$

PROBLEMS

9.1   A thin uniform rectangular plate (lamina) is of mass m and dimensions $2a$ by a. Choose a coordinate system $Oxyz$ such that the plate lies in the xy plane with origin at a corner, the long dimension being along the x-axis. Find the following:

(a) The moments and products of inertia

(b) The moment of inertia about the diagonal through the origin

(c) The angular momentum about the origin if the plate is spinning with angular rate ω about the diagonal through the origin

(d) The kinetic energy in part (c)

9.2   A rigid body consists of three thin uniform rods, each of mass m and length $2a$, held mutually perpendicular at their midpoints. Choose a coordinate system with axes along the rods.

(a) Find the angular momentum and kinetic energy of the body if it rotates with angular velocity $\boldsymbol{\omega}$ about an axis passing through the origin and the point $(1, 1, 1)$.

(b) Show that the moment of inertia is the same for any axis passing through the origin.

(c) Show that the moment of inertia of a uniform square lamina is that given in Example 9.1 for any axis passing through the center of the lamina and lying in the plane of the lamina.

9.3   Find a set of principal axes for the lamina of Problem 9.1 in which the origin is

(a) At a corner

(b) At the center of the lamina

9.4   A uniform block of mass m and dimensions a by $2a$ by $3a$ spins about a long diagonal with angular velocity $\boldsymbol{\omega}$. Using a coordinate system with origin at the center of the block,

(a) Find the kinetic energy.

(b) Find the angle between the angular velocity vector and the angular momentum vector about the origin.

9.5   A thin uniform rod of length l and mass m is constrained to rotate with constant angular velocity $\boldsymbol{\omega}$ about an axis passing through the center O of the rod and making an angle α with the rod.

(a) Show that the angular momentum $\mathbf{L}$ about O is perpendicular to the rod and is of magnitude $(ml^2\omega/12) \sin \alpha$.

(b) Show that the torque vector $\mathbf{N}$ is perpendicular to the rod and to $\mathbf{L}$ and is of magnitude $(ml^2\omega^2/12) \sin \alpha \cos \alpha$.

9.6   Find the magnitude of the torque that must be exerted on the block in Problem 9.4 if the angular velocity $\boldsymbol{\omega}$ is constant in magnitude and direction.

9.7 A rigid body of arbitrary shape rotates freely under zero torque. By means of Euler's equations show that both the rotational kinetic energy and the magnitude of the angular momentum are constant, as stated in Section 9.4. [*Hint:* For $\mathbf{N} = 0$, multiply Euler's equations (Equation 9.16b) by ω_1, ω_2, and ω_3, respectively, and add the three equations. The result indicates the constancy of kinetic energy. Next, multiply by $I_1\omega_1$, $I_2\omega_2$, and $I_3\omega_3$, respectively, and add. The result shows that L^2 is constant.]

9.8 A lamina of arbitrary shape rotates freely under zero torque. Use Euler's equations to show that the sum $\omega_1^2 + \omega_2^2$ is constant if the 1, 2 plane is the plane of the lamina. This means that the projection of $\boldsymbol{\omega}$ on the plane of the lamina is constant in magnitude, although the component ω_3 normal to the plane is not necessarily constant. (*Hint:* Use the perpendicular-axis theorem.) What kind of lamina gives $\omega_3 = $ constant as well?

9.9 A square plate of side a and mass m is thrown into the air so that it rotates freely under zero torque. The rotational period $2\pi/\omega$ is 1 s. If the axis of rotation makes an angle of 45° with the symmetry axis of the plate, find the period of the precession of the axis of rotation about the symmetry axis and the period of wobble of the symmetry axis about the invariable line for two cases:

(a) A thin plate

(b) a thick plate of thickness $a/4$

9.10 A rigid body having an axis of symmetry rotates freely about a fixed point under no torques. If α is the angle between the axis of symmetry and the instantaneous axis of rotation, show that the angle between the axis of rotation and the invariable line (the $\mathbf{L}$ vector) is

$$\tan^{-1}\left[\frac{(I_s - I)\tan\alpha}{I_s + I\tan^2\alpha}\right]$$

where I_s (the moment of inertia about the symmetry axis) is greater than I (the moment of inertia about an axis normal to the symmetry axis).

9.11 Since the greatest value of the ratio $I_s/I = 2$ (symmetrical lamina), show from the result of Problem 9.10 that the angle between $\boldsymbol{\omega}$ and $\mathbf{L}$ cannot exceed $\tan^{-1}(1/\sqrt{8})$ or about 19.5° and that the corresponding value of α is $\tan^{-1}\sqrt{2}$ or about 54.7°.

9.12 Find the angle between $\boldsymbol{\omega}$ and $\mathbf{L}$ for the two cases in Problem 9.9.

9.13 Find the same angle for the earth.

9.14 A space platform in the form of a thin circular disc of radius a and mass m (flying saucer) is initially rotating steadily with angular velocity $\boldsymbol{\omega}$ about its symmetry axis. A meteorite strikes the platform at the edge, imparting an impulse $\hat{\mathbf{P}}$ to the platform. The direction of $\hat{\mathbf{P}}$ is parallel to the axis of the platform, and the magnitude of $\hat{\mathbf{P}}$ is equal to $ma\omega/4$. Find the resulting values of the precessional rate Ω, the wobble rate $\dot{\phi}$, and the angle α between the symmetry axis and the new axis of rotation.

9.15 A Frisbee is thrown into the air in such a way that it has a definite wobble. If air friction exerts a frictional torque $-c\boldsymbol{\omega}$ on the rotation of the Frisbee, show that the component of $\boldsymbol{\omega}$ in the direction of the symmetry axis decreases exponentially with time. Show also that the angle α between the symmetry axis and the angular velocity vector $\boldsymbol{\omega}$ decreases with time if I_s is larger than I, which is the case for a flat-type object. Thus, the degree of wobble steadily diminishes if there is air friction.

9.16 A simple gyroscope consists of a heavy circular disc of mass m and radius a mounted at the center of a thin rod of mass $m/2$ and length a. If the gyroscope is set spinning at a given rate S, and with the axis at an angle of 45° with the vertical, there are two possible values of the precession rate $\dot{\phi}$ such that the gyroscope precesses steadily at a constant value of $\theta = 45°$.

(a) Find the two numerical values of $\dot{\phi}$ when $S = 900$ rpm and $a = 10$ cm.

(b) How fast must the gyroscope spin in order to sleep in the vertical position? Express the results in revolutions per minute.

9.17 A pencil is set spinning in an upright position. How fast must the spin be in order that the pencil will remain in the upright position? Assume that the pencil is a uniform cylinder of length a and diameter b. Find the value of the spin in revolutions per second for $a = 20$ cm and $b = 1$ cm.

9.18 A bicycle wheel of diameter 30 in rolls along the ground. How fast must it roll in order to remain upright? Assume that half the mass of the wheel is on the periphery (rim), one fourth of the mass is in the spokes, and the remainder is concentrated at the center (hub). Compare the result with that obtained if the spokes and hub are neglected.

9.19 A rigid body rotates freely under zero torque. By differentiating the first of Euler's equations with respect to t, and eliminating $\dot{\omega}_2$ and $\dot{\omega}_3$ by means of the second and third of Euler's equations, show that the following result is obtained:

$$\ddot{\omega}_1 + K_1\omega_1 = 0$$

in which the function K_1 is given by

$$K_1 = -\omega_2^2[(I_3 - I_2)(I_2 - I_1)/I_1I_3] + \omega_3^2[(I_3 - I_2)(I_3 - I_1)/I_1I_2]$$

Two similar pairs of equations are obtained by cyclic permutations: $1 \to 2$, $2 \to 3$, $3 \to 1$. In the above expression for K_1 both quantities in brackets are *positive constants* if $I_1 < I_2 < I_3$, or if $I_1 > I_2 > I_3$. Discuss the question of the growth of ω_1 (stability) if initially ω_1 is very small and (a) $\omega_2 = 0$ and ω_3 is large: initial rotation is very nearly about the 3-axis, and (b) $\omega_3 = 0$ and ω_2 is large: initial rotation is nearly about the 2-axis. (*Note:* This is an analytical method of deducing the stability criteria illustrated in Figure 9.9.)

9.20 A rigid body consists of six particles, each of mass m, fixed to the ends of three light rods of length $2a$, $2b$, and $2c$, respectively, the rods being held mutually perpendicular to one another at their midpoints.

(a) Show that a set of coordinate axes defined by the rods are principal axes, and write down the inertia tensor for the system in these axes.

(b) Use matrix algebra to find the angular momentum and the kinetic energy of the system when it is rotating with angular velocity $\boldsymbol{\omega}$ about an axis passing through the origin and the point (a, b, c).

9.21 Work Problems 9.1 and 9.4 using matrix methods.

9.22 A uniform rectangular block of dimensions $2a$ by $2b$ by $2c$ and mass m spins about a long diagonal. Find the inertia tensor for a coordinate system with origin at the center of the block and with axes normal to the faces. Find also the angular momentum and the kinetic energy. Find also the inertia tensor for axes with origin at one corner.

COMPUTER/CALCULATOR APPLICATIONS

9.1 Consider the spinning top discussed in Examples 9.10 and 9.11. Suppose that it is set spinning at 35 rev/s and initially its spin axis is held fixed at an angle $\theta_0 = 60°$. The axis is then released and the top starts to topple over. As it falls, its axis starts to precess as well as to nutate between two limiting polar angles θ_1 and θ_2.

(a) Calculate the two limits θ_1 and θ_2.

(b) Estimate the period of nutation analytically. *(Hint: Make approximations, where necessary, in the expression given in the text for $\dot{u}$ and then integrate.)*

(c) Estimate the average period of precession analytically. *(Hint: Make approximations in the expression for $\dot{\phi}$ given in the text.)*

(d) Find $\cos \theta(t)$ and $\phi(t)$ by numerically integrating the appropriate equations of motion over a time interval somewhat greater than one nutational period.

(e) Letting the function $x(t) = \cos \theta_1 - \cos \theta(t)$, where θ_1 is the smaller of the two angular limits of the nutational motion, plot $x(t)$ versus $\phi(t)$ over the same time interval as in part (d). From this plot, calculate both the nutational period and the average precessional period. Compare the results obtained from the plot with those from parts (a), (b), and (c).

Joseph Louis de Lagrange.
(Stock Montage, Inc.)

10

LAGRANGIAN MECHANICS

". . . to reduce the theory of mechanics, and the art of solving the associated problems, to general formulae, whose simple development provides all the equations necessary for the solution of each problem. . . . to unite, and present from one point of view, the different principles which have, so far, been found to assist in the solution of problems in mechanics; by showing their mutual dependence and making a judgement of their validity and scope possible. . . . No diagrams will be found in this work. The methods that I explain in it require neither constructions nor geometrical or mechanical arguments, but only the algebraic operations inherent to a regular and uniform process. Those who love Analysis will, with joy, see mechanics become a new branch of it and will be grateful to me for having extended its field."

(Joseph Louis de Lagrange, Avertissement for Mechanique Analytique, *1788)*

Another way of looking at mechanics, other than from the Newtonian perspective, was developed in continental Europe somewhat contemporaneously with the efforts of Newton. This work was championed by Wilhelm von Leibniz (with whom Newton was deeply embroiled in a bitter quarrel regarding who should get credit for the development of calculus). It was based upon mathematical operations with the scalar quantities of energy, as opposed to the vector quantities of force and acceleration. The development was to take more than a century to complete and would occupy the talents of many of the world's greatest minds. Following Leibniz, progress with the new mechanics was made chiefly by Johann Bernoulli. In 1717 he established the principle of "virtual work" to describe the equilibrium of static systems. This principle was extended by D'Alembert to include the motion of dynamical systems. The development culminated with the work of James Joseph Lagrange, who used the virtual work principle and its D'Alembertian extension as a foundation for the derivation of the dynamical equations of motion that, in his honor, now bear his name.

We shall not take Lagrange's approach here in developing his equations of motion. The interested reader can find such an exposition in many excellent texts.[1] Instead, we shall take another approach, originally pursued with the goal of solving problems that run the entire gamut of physics, not merely those limited to the domain of classical mechanics. It is an approach that stems from the deep philosophical belief that the physical universe operates according to laws of nature that are based on a principle of economy. They should be simple and elegant in form. This belief has gripped many of the most brilliant physicists and mathematicians throughout history, among them, Euler, Gauss, Einstein, Bernoulli, and Rayleigh to name a few. The basic idea is that "mother nature," given choices, always dictates that objects making up the physical universe follow paths through space and time based on extrema principles. For example, moving bodies "seek out" trajectories that are geodesics, namely, the shortest distance between two points on a given geometrical surface; a ray of light follows a path that minimizes (or, interestingly enough, maximizes) its transit time; and ensembles of particles assume equilibrium configurations that minimize their energy.

It may or may not be true that such a hypothesis conveys some deep meaning about the workings of nature. It is an issue that provides fodder for philosophers and theologians alike. From the physicist's point of view, the proof, so to speak, is in the pudding. Elegant and beautiful though our laws of nature may be, we must insist ultimately upon their experimental verification. The laws that we select to depict the reality of nature must stand up to scientific scrutiny. Failure to live up to the standards of this requirement relegated many an elegant hypothesis to the junk heap.

However, the hypothesis of global economy, which we shall introduce here, has withstood the assaults of all experimental battering rams (indeed, it leads to Newton's laws of motion). It was first announced in 1834 by the brilliant Irish mathematician Sir

---

[1] For a discussion of the principle of virtual work, see, for example, N. G. Chateau, *Theoretical Mechanics,* Springer-Verlag, Berlin, 1989. For the development of the Lagrange equations from D'Alembert's principle, see, for example, (1) H. Goldstein, *Classical Mechanics,* Addison-Wesley, Reading, MA (1965); (2) F. A. Scheck, *Mechanics—From Newton's Laws to Deterministic Chaos,* Springer-Verlag, Berlin, 1990.

William Rowan Hamilton. We will use this hypothesis, now called *Hamilton's varia-tional principle,* to show that in the specific case of a body falling in a uniform gravita-tional field, the motion it predicts is the same as that predicted by Newton's second law. We will then use it to generate the Lagrange equations of motion for any general, con-servative system. Next, we will use the resulting Lagrange equations to generate the differential equations of motion for several physical systems. Finally, we will demon-strate the equivalence of the Lagrangian and Newtonian approaches to solving problems in mechanics.

10.1   HAMILTON'S VARIATIONAL PRINCIPLE. AN EXAMPLE

Hamilton's variational principle states that the integral

$$\int_{t_1}^{t_2} L \, dt$$

taken along a path of the possible motion of a physical system is an extremum when evaluated along the path of motion that is the one actually taken. $L = T - V$ is the *Lagrangian* of the system, or the difference between its kinetic and potential energy. In other words, out of the myriad of ways in which a system could change its configuration during a time interval $t_2 - t_1$, the actual motion that does occur is the one that either maximizes or minimizes the above integral. This statement can be expressed mathe-matically as

$$\delta \int_{t_1}^{t_2} L \, dt = 0 \qquad (10.1)$$

in which δ is an operation that represents a variation of any particular system parameter by an infinitesimal amount away from that value taken by the parameter when the above integral is an extremum. For example, the δ that occurs explicitly in Equation 10.1 represents a variation in the entire integral about its extremum value. Such a variation is obtained by varying the coordinates and velocities of a dynamical system away from the values actually taken as the system evolves in time from t_1 to t_2, under the constraint that the variation in all parameters is zero at the endpoints of the motion at t_1 and t_2. That is, the variation of the system parameters between t_1 and t_2 is completely arbitrary under the provisos that the motion must be completed during that time interval and that all system parameters must assume their unvaried values at the beginning and end of the motion.

Let us apply Hamilton's variational principle to the case of a particle dropped from rest in a uniform gravitational field. We will see that the integral in Equation 10.1 is an extremum when the path taken by the object is the one for which the particle obeys Newton's second law. Let the height of the particle above ground at any time t be de-noted by y and its speed by $\dot{y}$. Then δy and $\delta \dot{y}$ represent small, virtual displacements of y and $\dot{y}$ away from the true position and speed of the particle at any time t during its actual motion. See Figure 10.1. The potential energy of the particle is mgy and its kinetic

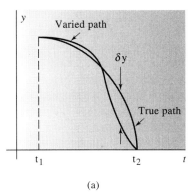

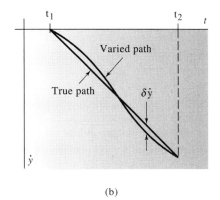

(a)   (b)

Figure 10.1   (a) Variation of the coordinate of a particle from its true path taken in free-fall. (b) Variation in the speed of a particle from the true value taken during free-fall.

energy is $m\dot{y}^2/2$. The Lagrangian is $L = m\dot{y}^2/2 - mgy$. The variation in the integral of the Lagrangian is given by

$$\delta \int_{t_1}^{t_2} L\, dt = \delta \int_{t_1}^{t_2} \left[\frac{m\dot{y}^2}{2} - mgy\right] dt = \int_{t_1}^{t_2} (m\dot{y}\delta\dot{y} - mg\delta y)\, dt \qquad (10.2)$$

The variation in the speed can be transformed into a coordinate variation by noting that

$$\delta\dot{y} = \frac{d}{dt}\,\delta y$$

Integrating the first term in Equation 10.2 by parts gives

$$\int_{t_1}^{t_2} m\dot{y}\delta\dot{y}\, dt = \int_{t_1}^{t_2} m\dot{y}\frac{d}{dt}\delta y\, dt = m\dot{y}\delta y\Big]_{t_1}^{t_2} - \int_{t_1}^{t_2} m\ddot{y}\delta y\, dt$$

The integrated term on the right-hand side is identically zero since the parameters of the admissible paths of motion do not vary at the endpoints of the motion. Hence, we obtain

$$\delta\int_{t_1}^{t_2} L\, dt = \int_{t_1}^{t_2} (-m\ddot{y} - mg)\,\delta y\, dt = 0 \qquad (10.3)$$

Since δy represents a completely arbitrary variation of the parameter y away from its true value throughout the motion of the particle (except at the endpoints where the variation is constrained to be zero), the only way in which Equation 10.3 can be zero under such conditions is for the term in parentheses to be identically zero at all times. Thus,

$$-mg - m\ddot{y} = 0 \qquad (10.4)$$

which, as advertised, is Newton's second law of motion for a particle falling in a uniform gravitational field.

Why the Lagrangian?

The above example showed that Hamilton's variational principle can be used to generate the same differential equation of motion for a particle falling freely in a gravitational field that could have been obtained from Newton's second law. The variational principle can be generalized to yield a set of dynamical equations of motion, called *Lagrange's equations,* that also lead to the correct differential equations of motion for any physical system. We will show this later. Therefore, Hamilton's variational principle, the resulting Lagrange equations, and Newton's second law are equivalent ways of presenting the laws of mechanics. This equivalence is the ultimate justification for using the Lagrangian function in Hamilton's variational principle above. To ask the question, Why the Lagrangian and not some other function?, is equivalent to asking the question, Why $\mathbf{F} = m\mathbf{a}$ and not ma^2? The answer is that $\mathbf{F} = m\mathbf{a}$ leads to a correct description of the motion of particles whereas $\mathbf{F} = ma^2$ does not. Experimental observations agree with predictions based on $\mathbf{F} = m\mathbf{a}$. Given that Lagrange's equations are equivalent to $\mathbf{F} = m\mathbf{a}$, the same must be true for them as well.

One might well ask why we should even bother with the Lagrangian method if all we get out of it is nothing other than equations that could be obtained directly from Newton's laws. The answer concerns the ease with which the correct equations of motion for rather complicated physical situations can be generated using the Lagrangian technique. The Lagrange equations of motion that arise from Hamilton's variational principle deal only with scalar variables as opposed to Newton's laws, which are intrinsically vectorial in nature (a situation we sidestepped in the above example by considering only one-dimensional motion). In many situations, generating the correct differential equations of motion using Newton's laws requires an inordinate amount of skill and physical insight. On the other hand use of Lagrange's equations can make the required effort seem almost trivially mechanistic by comparison.

10.2 GENERALIZED COORDINATES

Coordinates are used to define the position in space of an ensemble of particles. In general we can select any set of coordinates we like in order to describe the motion of a physical system. There are certain choices, however, that prove to be more economical than others because of the existence of geometrical constraints that restrict the allowable configuration of any system.

For example, consider the motion of the pendulum in Figure 10.2. It is constrained to move in the *xy* plane along an arc of radius *r*. We could choose to describe the configuration of the pendulum by means of its position vector

$$\mathbf{r} = x\mathbf{i} + y\mathbf{j} + z\mathbf{k}$$

Clearly, such a choice would be tantamount to lunacy. It ignores the fact that there are two conditions of constraint to which the pendulum must adhere, namely

$$z = 0 \qquad r^2 - (x^2 + y^2) = 0$$

Only one scalar coordinate is really needed to specify the position of the pendulum. At

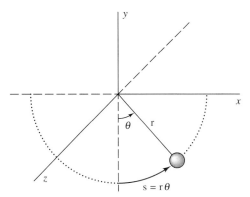

Figure 10.2 Pendulum swinging in the *xy* plane.

first sight either *x* or *y* might work, but in order to resolve a possible left–right ambiguity, we would probably choose *x* to define the position uniquely. We would then assume that the implied value of *y* is always negative. Given the *x*-value of the pendulum, *y* and *z* would then be determined completely by the conditions of constraint. This choice would still prove an awkward one for describing the configuration of the pendulum.

A better choice would be the arc length displacement s ($= r\theta$) or, equivalently, the angular displacement θ of the pendulum away from the vertical. Either of these choices would be better, since only a single number would be needed to tell us the whereabouts of the pendulum. The important point here is that the pendulum really has only *one degree of freedom;* that is, it can move only one way and that way is along an arc of radius *r*. There exists only a single, independent coordinate necessary to depict its configuration uniquely. *Generalized coordinates* are any collection of independent coordinates q_i (not connected by any equations of constraint) that just suffice to specify uniquely the configuration of a system of particles. The required number of generalized coordinates is equal to the system's number of degrees of freedom. If fewer than this number is chosen to describe the system's configuration, the result will be indeterminate; if a greater number is chosen, then some of the coordinates must be determinable from the others by conditions of constraint.

For example, a single particle able to move freely in three-dimensional space exhibits three degrees of freedom and requires three coordinates to specify its configuration. There exist no equations of constraint connecting the coordinates of a single free particle. Two free particles would require six coordinates to specify the configuration completely, but two particles connected by a rigid straight line like a dumbbell (see Figure 10.3) would require only five coordinates. Let us see why. The position of particle 1 could be specified by the coordinates (x_1, y_1, z_1), whereas the position of particle 2 could be specified by the coordinates (x_2, y_2, z_2) as in the free particle case. However, there exists an equation of constraint connecting the coordinates

$$d^2 - [(x_1 - x_2)^2 + (y_1 - y_2)^2 + (z_1 - z_2)^2] = 0$$

namely, the distance between the two particles is fixed and equal to *d*.

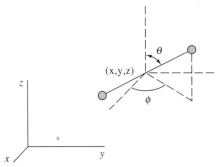

Figure 10.3   Generalized coordinates for two particles connected by an infinitesimally thin rigid rod.

Suppose, in specifying the position of the system, we picked the above six coordinates, one by one. We would not have complete freedom of choice in the selection process, since the choice of the sixth coordinate would be forced upon us after the first five had been made. It would make more sense to choose initially only five independent coordinates, say (X, Y, Z, θ, ϕ), unconnected by any equations of constraint, where (X, Y, Z) are the coordinates of the center of mass and (θ, ϕ) are the zenith and azimuthal angles, which describe the orientation of the dumbbell relative to the vertical ($\theta = 0°$ when particle 1 is directly above particle 2 and $\phi = 0°$ when the projection of the line from particle 2 to particle 1 onto the xy plane points parallel to the x-axis).

As a final example, consider the situation of a particle constrained to move along the surface of a sphere. Again, the coordinates (x, y, z) do not constitute an independent set. They are connected by the constraint

$$R^2 - (x^2 + y^2 + z^2) = 0$$

where R is the radius of the sphere. The particle has only two degrees of freedom available for its motion and two independent coordinates are needed to specify completely its position on the sphere.

These coordinates could be taken as latitude and longitude, which denote positions on the spherical surface relative to an equator and a prime meridian as for the earth (see Figure 10.4), or we could choose the polar and azimuthal angles θ and ϕ as in the dumbbell example above.

In general, if there are N particles free to move in three-dimensional space but their $3N$ coordinates are connected by m conditions of constraint, then there exist $n = 3N - m$ independent generalized coordinates sufficient to describe uniquely the position of the N particles and n independent degrees of freedom available for the motion, provided the constraints are of the type described in the above examples. Such constraints are called *holonomic*. They must be expressible as equations of the form

$$f_j(x_i, y_i, z_i, t) = 0 \qquad i = 1, 2, \ldots, N \qquad j = 1, 2, \ldots, m \qquad (10.5)$$

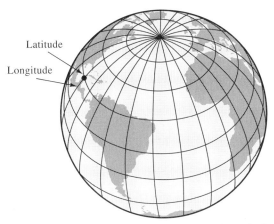

Figure 10.4 Coordinate of a point on earth marked
by latitude and longitude.

Note that these equations are equalities, they are integrable in form, and they may or may not be explicitly time-dependent.

Constraints that cannot be expressed as equations of equality or that are nonintegrable in form are called *nonholonomic,* and the equations representing such constraints cannot be used to eliminate from consideration any dependent coordinates describing the configuration of the system. As an example of such a constraint, consider a particle constrained to remain outside the surface of a sphere. (Humans on the earth capable of going to the moon but incapable of going more than a few miles underground represent a reasonable approximation to this situation.) This condition of constraint is given by the inequality

$$(x^2 + y^2 + z^2) - R^2 \geq 0$$

Clearly, this equation cannot be used to reduce below three the required number of independent coordinates of the particle when it lies outside the sphere. Inside the sphere is another matter. In this case the single constraint reduces the degrees of freedom to zero. It is difficult to handle such a situation using Lagrangian mechanics. We will ignore situations of this type.

Perhaps the classic example of a nonholonomic constraint in which the representative equation is nonintegrable appears in the case of a ball rolling along a rough, level surface without slipping. The "rolling" condition connects the coordinates. A change in the orientation of the ball cannot occur without an accompanying change in its position on the plane. However, the equation of constraint represents a condition on velocities, not coordinates. The ball's point of contact with the surface is instantaneously at rest. The desired constraint on coordinates can be generated only by integrating the equation representing the velocity constraint. This cannot be done unless the ball's trajectory is known. Unfortunately, this is the very problem we wish to solve. Hence, the constraint equation is nonintegrable and, as above, cannot be used to eliminate dependent coordinates from the problem. However, in contrast to the nonholonomic constraint represented by inequality conditions on the coordinates, this type of nonholonomic constraint can be handled tractably by the Lagrangian technique via the use of the method of Lagrange multipliers.[2] Again, we shall ignore such situations here.

10.3   CALCULATING KINETIC AND POTENTIAL ENERGIES IN TERMS OF GENERALIZED COORDINATES. AN EXAMPLE

The Lagrangian $L = T - V$ must be expressed as a function of the generalized coordinates and time derivatives (generalized velocities) appropriate for a given physical situation. (Sometimes, the Lagrangian may also be an explicit function of time, although we will not be concerned with such cases here.) We need to find out how to generate such an expression before deriving Lagrange's equations of motion from Hamilton's variational principle. It is not obvious a priori just how to do this. Almost always, the kinetic energy of an ensemble of particles can be written as a quadratic form in the velocities of the particles related to a Cartesian coordinate system.

Since Cartesian coordinates are orthogonal, there are no cross-coupled terms in such an expression. However, this is usually not true when the kinetic energy is expressed in terms of the generalized coordinates; that is, the coordinates chosen may lead to cross-coupled velocity terms of the form $\alpha \dot{q}_i \dot{q}_j$. There is no equivalent generalization regarding the expression for the potential energy of a system. In some cases, the expression contains cross terms even when expressed in Cartesian coordinates. Usually, however, the potential energy is expressible as some function of just one of the generalized coordinates, and it is easy to see just how it depends on that coordinate. One usually exploits this apparency in choosing the generalized coordinates for any particular situation. Unfortunately, such choices almost invariably lead to cross terms in the expression for the kinetic energy of the system.

As a specific example, let us take the fairly nasty case, depicted in Figure 10.5, of a pendulum of mass m attached to a support of mass M that is free to move in a single

---

[2]For example, see pages 38–44 of H. Goldstein, *Classical Mechanics,* Addison-Wesley, Reading, MA (1965).

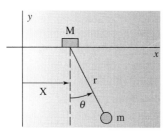

Figure 10.5   Simple pendulum attached to movable support.

dimension along a frictionless, horizontal surface. First, let us see just how many generalized coordinates are needed to specify the system configuration uniquely. Each mass needs three Cartesian coordinates, but there are four holonomic constraints

$$Z = 0 \qquad Y = 0$$
$$Z = 0 \qquad ((x - X)^2 + y^2) - r^2 = 0$$

The first two constraints ensure that the motion of the mass M lies along the x-axis. The second two constraints ensure that the pendulum swings in the xy plane along an arc of radius r, relative to its movable support. There are two degrees of freedom for the motion and two generalized coordinates necessary to describe the configuration of this system. We have chosen those coordinates to be X, which denotes the horizontal position mass of M, and θ, the angular displacement of the pendulum away from the vertical.

The potential and kinetic energies of this system can be expressed in terms of Cartesian coordinates and velocities as

$$T = \frac{1}{2} M\dot{X}^2 + \frac{1}{2} m(\dot{x}^2 + \dot{y}^2)$$

$$V = mgy \tag{10.6}$$

Obtaining the potential energy in terms of generalized coordinates requires a transformation of coordinates

$$x = X + r \sin \theta \qquad y = -r \cos \theta \qquad X = X \tag{10.7}$$

while obtaining the kinetic energy as a function of generalized velocities can be effected by differentiating the above

$$\dot{x} = \dot{X} + r\dot{\theta} \cos \theta \qquad \dot{y} = r\dot{\theta} \sin \theta \qquad \dot{X} = \dot{X} \tag{10.8}$$

Substituting these transformations into Equation 10.6 yields

$$T = \frac{M}{2} \dot{X}^2 + \frac{m}{2} [(\dot{X} + r\dot{\theta} \cos \theta)^2 + (r\dot{\theta} \sin \theta)^2]$$

$$= \frac{M}{2} \dot{X}^2 + \frac{m}{2} [\dot{X}^2 + (r\dot{\theta})^2 + 2\dot{X}r\dot{\theta} \cos \theta] \tag{10.9}$$

$$V = -mgr \cos \theta$$

There are several features of Equation 10.9 that illustrate our comments above:
1. The kinetic energy is expressible as a quadratic form in the generalized velocities, and there is a resulting cross term.
2. The potential energy is dependent on a single generalized coordinate, in this case, the cosine of an angle.

Although it might have been easy to see how to write the potential energy directly in terms of a single generalized coordinate, it might take some thought to write the kinetic energy directly in terms of the generalized velocities. It is worth doing. Let us give it a try.

The velocity of mass M relative to our fixed inertial frame of reference is

$$\mathbf{V}_M = \mathbf{i}\dot{X}$$

The velocity of mass m can be expressed as the velocity of mass M plus the velocity of mass m relative to that of mass M, that is,

$$\mathbf{v}_m = \mathbf{V}_M + \mathbf{v}_{m(rel)}$$

where

$$\mathbf{v}_{m(rel)} = \mathbf{e}_\theta r \dot{\theta}$$

and $\mathbf{e}_\theta$ is a unit vector tangent to the arc along which the pendulum swings. Hence, the velocity of m can now be written in component form directly as a function of the generalized coordinates X and θ

$$\dot{x} = \dot{X} + r\dot{\theta} \cos \theta \qquad \dot{y} = r\dot{\theta} \sin \theta$$

Plugging these expressions for the Cartesian velocities of the two masses into Equation 10.6 yields the kinetic energy in terms of the generalized coordinates as expressed in Equation 10.9.

An even more direct way to generate the correct expression for the kinetic energy can be obtained by noting that

$$T = \frac{1}{2} M\mathbf{V}_M \cdot \mathbf{V}_M + \frac{1}{2} m\mathbf{v}_m \cdot \mathbf{v}_m \tag{10.10}$$

where

$$\mathbf{V}_m = \mathbf{i}\dot{X} \qquad \mathbf{v}_m = \mathbf{i}\dot{X} + \mathbf{e}_\theta r \dot{\theta} \tag{10.11}$$

$$\mathbf{V}_M \cdot \mathbf{V}_M = \dot{X}^2 \qquad \mathbf{v}_m \cdot \mathbf{v}_m = \dot{X}^2 + r^2\dot{\theta}^2 + 2\dot{X}r\dot{\theta} \cos \theta$$

To obtain a correct expression for the kinetic energy of any given system, one will rarely go wrong by following the first procedure outlined above (Equations 10.6–10.9), namely, write the kinetic energy in terms of velocities relative to Cartesian coordinates, find the transformation relating Cartesian to generalized coordinates, and then differentiate. In many cases, however, it is easier to use the last procedure demonstrated above (Equations 10.10–10.11), particularly if one is able to visualize just what form the generalized velocities take in terms of the unit vectors corresponding to the selected generalized coordinates.

When problems involve only holonomic constraints, there always exist transformation equations that relate the Cartesian coordinates of an ensemble of particles to their generalized coordinates, and thus the required generalized velocities can be obtained by differentiation. For example, in the case of a single particle, we have:

Three degrees of freedom—unconstrained motion in space

$$x = x(q_1, q_2, q_3)$$
$$y = y(q_1, q_2, q_3) \tag{10.12}$$
$$z = z(q_1, q_2, q_3)$$

Two degrees of freedom—motion constrained to a surface

$$x = x(q_1, q_2)$$
$$y = y(q_1, q_2) \qquad (10.13)$$
$$z = z(q_1, q_2)$$

One degree of freedom—motion constrained to a line

$$x = x(q)$$
$$y = y(q) \qquad (10.14)$$
$$z = z(q)$$

And, as in the above example, we can obtain the velocity transformations by simply differentiating the coordinate transformations:

$$\dot{x} = \sum_{i=1}^{n} \frac{\partial x}{\partial q_i} \dot{q}_i$$

$$\dot{y} = \sum_{i=1}^{n} \frac{\partial y}{\partial q_i} \dot{q}_i \qquad (10.15)$$

$$\dot{z} = \sum_{i=1}^{n} \frac{\partial z}{\partial q_i} \dot{q}_i$$

where n is the number of degrees of freedom.

10.4 LAGRANGE'S EQUATIONS OF MOTION FOR CONSERVATIVE SYSTEMS

We are now ready to derive Lagrange's equations from Hamilton's variational principle. First, we should point out that all our examples thus far have consisted only of conservative systems whose motion is either unconstrained or, at worst, subject only to time-independent holonomic constraints. We will continue to confine our analysis to such systems. The interested reader who wishes to endure the agony of dealing with either nonconservative systems or systems suffering nonholonomic constraints is urged to seek out the many excellent presentations contained in other, more advanced, texts on this subject.[3]

Hamilton's principle is expressed by Equation 10.1. We will proceed from that point by carrying out the same variational procedure as we did in Section 10.1 for the specific case of an object freely falling in a gravitational field. Only this time we will carry out the process for any general conservative system. We begin by assuming that the Lagrangian is a known function of the generalized coordinates q_i and velocities $\dot{q}_i$. The variation in its time integral is

$$\delta \int_{t_2}^{t_2} L\, dt = \int_{t_1}^{t_2} \delta L\, dt = \int_{t_1}^{t_2} \sum_k \left(\frac{\partial L}{\partial q_k} \delta q_k + \frac{\partial L}{\partial \dot{q}_k} \delta \dot{q}_k \right) dt = 0 \qquad (10.16)$$

---

[3] See Footnote 1.

The q_k are functions of time; they change as the system "evolves" from t_1 to t_2. Therefore, the variation δq_k is equal to the difference between two slightly differing functions of time t. The $\delta \dot{q}$ can then be expressed as

$$\delta \dot{q}_k = \frac{d}{dt} \delta q_k$$

This result can be substituted into the last term of Equation 10.16, which can then be integrated by parts to obtain

$$\int_{t_1}^{t_2} \sum_k \frac{\partial L}{\partial \dot{q}_k} \frac{d}{dt} (\delta q_k) \ dt = \left[\sum_k \frac{\partial L}{\partial \dot{q}_k} \delta q_k \right]_{t_1}^{t_2} - \int_{t_1}^{t_2} \sum_k \frac{d}{dt} \left(\frac{\partial L}{\partial \dot{q}_k} \right) \delta q_k dt \quad (10.17)$$

The integrated term in brackets vanishes since the variation $\delta q_k = 0$ at the endpoints t_1 and t_2. Thus, we obtain

$$\delta \int_{t_1}^{t_2} L \ dt = \int_{t_1}^{t_2} \sum_k \left[\frac{\partial L}{\partial q_k} - \frac{d}{dt} \left(\frac{\partial L}{\partial \dot{q}_k} \right) \right] \delta q_k \ dt = 0 \quad (10.18)$$

Each generalized coordinate q_k is independent of the others, as is each variation δq_k. Moreover, the actual value of each variation δq_k is completely arbitrary. In other words we can vary each coordinate in any way we so choose as long as we make sure that its variation vanishes at the endpoints of the path. Consequently, the only way that we can ensure that the above integral vanishes, given all the infinite varieties of possible values for the δq_k's, is to demand that each bracketed term in the integrand of Equation 10.18 vanish separately; that is,

$$\frac{\partial L}{\partial q_k} - \frac{d}{dt} \left(\frac{\partial L}{\partial \dot{q}_k} \right) = 0 \qquad (k = 1, 2, \ldots, n) \quad (10.19)$$

These are the desired Lagrangian equations of motion for a conservative system subject to, at worst, only holonomic constraints.

10.5   SOME APPLICATIONS OF LAGRANGE'S EQUATIONS

Here we shall illustrate the great utility of Lagrange's equations of motion by using them to obtain the differential equations of motion for several different systems. The general problem-solving strategy proceeds as follows:

1. Select a suitable set of generalized coordinates that uniquely specifies the system configuration.

2. Find the equations of transformation relating the dependent Cartesian coordinates to the independent generalized coordinates.

3. Find the kinetic energy as a function of the generalized coordinates and velocities. If possible, use the prescription $T = m\mathbf{v} \cdot \mathbf{v}/2$ with $\mathbf{v}$ expressed in terms of unit vectors appropriate to the selected generalized coordinates. If necessary, express the kinetic energy in terms of Cartesian coordinates, then differentiate the coordinate transfor-

mations and plug the resulting velocity transformations into the kinetic energy expression.

4. Find the potential energy as a function of the generalized coordinates using, if necessary, the coordinate transformations.

EXAMPLE 10.1 *The Harmonic Oscillator*

Consider the case of a one-dimensional harmonic oscillator. Let x be the displacement coordinate. Step 2 above is not explicitly necessary. The single Cartesian coordinate x is obviously the single generalized coordinate. The Lagrangian is

$$L(x, \dot{x}) = T - V = \frac{1}{2} m\dot{x}^2 - \frac{1}{2} kx^2$$

We have ignored the other two coordinates y and z, since they are both constrained to be zero. Now carry out the Lagrange operations of Equation 10.19

$$\frac{\partial L}{\partial \dot{x}} = m\dot{x} \qquad \frac{\partial L}{\partial x} = -kx$$

$$\frac{d}{dt}\left(\frac{\partial L}{\partial \dot{x}}\right) - \frac{\partial L}{\partial x} = m\ddot{x} + kx = 0$$

This is the equation of motion of the undamped harmonic oscillator discussed in Chapter 3. ■

EXAMPLE 10.2 *Single Particle in a Central Force Field*

The problem here is to use the Lagrange equations to generate the differential equations of motion for a particle constrained to move in a plane subject to a central force. The single constraint is given by $z = 0$, so we need two generalized coordinates. We shall choose plane polar coordinates: $q_1 = r$, $q_2 = \theta$. The transformation equations and resultant velocities are

$$x = r \cos \theta \qquad\qquad y = r \sin \theta$$
$$\dot{x} = \dot{r} \cos \theta - r\dot{\theta} \sin \theta \qquad \dot{y} = \dot{r} \sin \theta + r\dot{\theta} \cos \theta$$

Thus, the Lagrangian is

$$T = \frac{1}{2} m(\dot{x}^2 + \dot{y}^2) = \frac{1}{2} m(\dot{r}^2 + r^2\dot{\theta}^2) \qquad V = V(r)$$

$$L = \frac{1}{2} m(\dot{r}^2 + r^2\dot{\theta}^2) - V(r)$$

We could have obtained the kinetic energy term more directly by expressing the velocity vector in terms of radial and tangential unit vectors

$$\mathbf{v} = \mathbf{e}_r \dot{r} + \mathbf{e}_\theta r\dot{\theta}$$

and, thus, the square of the particle's velocity is

$$\mathbf{v} \cdot \mathbf{v} = \dot{r}^2 + r^2\dot{\theta}^2$$

which is just what we obtained using the coordinate transformations.

The relevant partial derivatives needed to implement the Lagrangian equations are

$$\frac{\partial L}{\partial \dot{r}} = m\dot{r} \qquad \frac{\partial L}{\partial r} = mr\dot{\theta}^2 - \frac{\partial V}{\partial r} = mr\dot{\theta}^2 + f(r)$$

$$\frac{\partial L}{\partial \theta} = 0 \qquad \frac{\partial L}{\partial \dot{\theta}} = mr^2\dot{\theta}$$

The equations of motion are, thus,

$$\frac{d}{dt}\frac{\partial L}{\partial \dot{r}} = \frac{\partial L}{\partial r} \qquad \frac{d}{dt}\frac{\partial L}{\partial \dot{\theta}} = \frac{\partial L}{\partial \theta} = 0$$

$$m\ddot{r} = mr\dot{\theta}^2 + f(r) \qquad \frac{d}{dt}(mr^2\dot{\theta}) = 0$$

For future reference we might take note of the fact that since the time derivative of the quantity $mr^2\dot{\theta}$ is zero, it is a constant of the motion. This quantity is the angular momentum of the particle. We can easily see that its constancy arises naturally in the Lagrangian formalism as a consequence of the fact that the θ coordinate is missing from the Lagrangian function.   ∎

EXAMPLE 10.3   *Atwood's Machine*

Atwood's machine consists of two weights of mass m_1 and m_2 connected by an ideal massless, inextensible string of length l that passes over a frictionless pulley of radius a and moment of inertia I (Figure 10.6). The system has only one degree of freedom—one mass moves either up or down while the other is constrained to move in the opposite sense, always separated from the first by the length of string. The pulley rotates appropriately. There exist five holonomic constraints. Four pre-

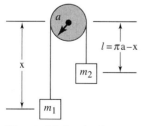

Figure 10.6   An Atwood's machine.

vent motion in either the y or z direction, whereas the fifth expresses the above mentioned constraint

$$(x_1 + \pi a + x_2) - l = 0$$

where x_1 and x_2 are the vertical positions of each mass relative to the center of the pulley. The Lagrangian is

$$T = \frac{1}{2} m_1 \dot{x}^2 + \frac{1}{2} m_2 \dot{x}^2 + \frac{1}{2} I \frac{\dot{x}^2}{a^2}$$

$$V = -m_1 g x - m_2 g(l - \pi a - x)$$

$$L = \frac{1}{2} \left(m_1 + m_2 + \frac{I}{a^2} \right) \dot{x}^2 + (m_1 - m_2) g x + m_2(l - \pi a)$$

where x is the single generalized coordinate of the system. The Lagrange equations of motion are

$$\frac{d}{dt} \frac{\partial L}{\partial \dot{x}} = \frac{\partial L}{\partial x}$$

$$\left(m_1 + m_2 + \frac{I}{a^2} \right) \ddot{x} = (m_1 - m_2) g$$

$$\ddot{x} = \frac{(m_1 - m_2)}{\left(m_1 + m_2 + \frac{I}{a^2} \right)} g$$

giving the final acceleration of the system. If $m_1 > m_2$, then m_1 falls with constant acceleration while m_2 rises with the very same acceleration. The converse is true if $m_2 > m_1$. If $m_1 = m_2$, each mass remains at rest (or moves at constant velocity). The effect of the moment of inertia of the pulley is to reduce the acceleration of the system. The reader probably recalls this result from analysis presented in more elementary physics classes. ∎

EXAMPLE 10.4   *The Double Atwood's Machine*

Consider the system shown in Figure 10.7. We have replaced one of the weights in the simple Atwood's machine by a second, simple Atwood's machine. There are two degrees of freedom for the motion of this system. Loosely speaking, they are the freedom of mass 1 (and the attached movable pulley) to move up and down about the fixed pulley and the freedom of mass 2 (and the attached mass 3) to move up and down about the movable pulley. No other motion is permissible. Thus, there must be ten holonomic constraints. Eight of those constraints limit the motion of all three masses plus the movable pulley to only a single dimension. There are two holonomic constraints connecting the x coordinates

$$(x_p + x_1) - l = 0 \qquad (2x_1 + x_2 + x_3) - (2l + l') = 0$$

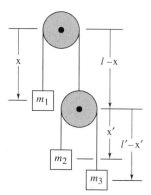

Figure 10.7   Compound
Atwood machine.

where x_i and x_p are the vertical positions of the masses and movable pulley relative
to the center of the fixed pulley. (The student should verify these equations of constraint.) The implication of these constraints regarding the reduction to the two
selected generalized coordinates x and x' is indicated in Figure 10.7. In this case
we have assumed that each pulley is massless. Hence, we can neglect the effects of
moments of inertia. We have also assumed that the radii of the pulleys are essentially zero (or small compared with l and l', the lengths of the constraining string).
This assumption allowed us to simplify the above equation of constraint by neglecting the length of string that goes around each pulley (contrary to the analysis presented in Example 10.3 for the simple Atwood's machine). We can now write down
the kinetic and potential energies for this system as well as its resultant Lagrangian

$$T = \frac{1}{2} m_1 \dot{x}^2 + \frac{1}{2} m_2 (-\dot{x} + \dot{x}')^2 + \frac{1}{2} m_3 (-\dot{x} - \dot{x}')^2$$

$$V = -m_1 g x - m_2 g(l - x + x') - m_3 g(l - x + l' - x')$$

$$L = \frac{1}{2} m_1 \dot{x}^2 + \frac{1}{2} m_2 (-\dot{x} + \dot{x}')^2 + \frac{1}{2} m_3 (\dot{x} + \dot{x}')^2$$

$$+ (m_1 - m_2 - m_3) g x + (m_2 - m_3) g x' + \text{constant}$$

The equations of motion are, thus,

$$\frac{d}{dt} \frac{\partial L}{\partial \dot{x}} = \frac{\partial L}{\partial x} \qquad \frac{d}{dt} \frac{\partial L}{\partial \dot{x}'} = \frac{\partial L}{\partial x'}$$

$$m_1 \ddot{x} + m_2 (\ddot{x} - \ddot{x}') + m_3 (\ddot{x} + \ddot{x}') = (m_1 - m_2 - m_3) g$$

$$m_2 (-\ddot{x} + \ddot{x}') + m_3 (\ddot{x} + \ddot{x}') = (m_2 - m_3) g$$

The accelerations can be obtained from an algebraic solution of the above
equations.   ∎

EXAMPLE 10.5 *Euler's Equations for the Free Rotation of a Rigid Body*

In this example, we shall use Lagrange's method to derive Euler's equations for the motion of a rigid body. We shall consider the case of torque-free rotation. No potential energy is involved, so the Lagrangian is equal to the kinetic energy

$$L = T = \frac{1}{2}(I_1\omega_1^2 + I_2\omega_2^2 + I_3\omega_3^2)$$

where the ω's are referred to principal axes of the body. In Section 9.5 we showed that the ω's can be expressed in terms of the Eulerian angles θ, ϕ, and ψ as follows:

$$\omega_1 = \dot{\theta} \cos \psi + \dot{\phi} \sin \theta \sin \psi$$

$$\omega_2 = -\dot{\theta} \sin \psi + \dot{\phi} \sin \theta \cos \psi$$

$$\omega_3 = \dot{\psi} + \dot{\phi} \cos \theta$$

Regarding the Eulerian angles as the generalized coordinates, the equations of motion are

$$\frac{d}{dt}\frac{\partial L}{\partial \dot{\theta}} = \frac{\partial L}{\partial \theta} \tag{10.20}$$

$$\frac{d}{dt}\frac{\partial L}{\partial \dot{\phi}} = \frac{\partial L}{\partial \phi} \tag{10.21}$$

$$\frac{d}{dt}\frac{\partial L}{\partial \dot{\psi}} = \frac{\partial L}{\partial \psi} \tag{10.22}$$

Now, by the chain rule,

$$\frac{\partial L}{\partial \dot{\psi}} = \frac{\partial L}{\partial \omega_3}\frac{\partial \omega_3}{\partial \dot{\psi}} = I_3\omega_3$$

so

$$\frac{d}{dt}\frac{\partial L}{\partial \dot{\psi}} = I_3\dot{\omega}_3$$

Again, using the chain rule, we have

$$\frac{\partial L}{\partial \psi} = I_1\omega_1\frac{\partial \omega_1}{\partial \psi} + I_2\omega_2\frac{\partial \omega_2}{\partial \psi}$$

$$= I_1\omega_1(-\dot{\theta}\sin\psi + \dot{\phi}\sin\theta\cos\psi) + I_2\omega_2(-\dot{\theta}\cos\psi - \dot{\phi}\sin\theta\sin\psi)$$

$$= I_1\omega_1\omega_2 - I_2\omega_2\omega_1$$

Consequently, Equation 10.22 becomes

$$I_3\dot{\omega}_3 = \omega_1\omega_2(I_1 - I_2)$$

which, as we showed in Section 9.3, is the third of Euler's equations for the rotation of a rigid body under zero torque. The other two of Euler's equations can be obtained by cyclic permutation of the subscripts: $1 \to 2$, $2 \to 3$, $3 \to 1$. This is valid because we have not designated any particular principal axis as being preferred.   ■

EXAMPLE 10.6   *Particle Sliding on a Movable Inclined Plane*

Consider the case of a particle of mass m free to slide along a smooth inclined plane of mass M. The inclined plane is not fixed but is free to slide along a smooth horizontal surface, as shown in Figure 10.8. There are only two degrees of freedom, since each object is constrained to move along a single dimension. We can most easily specify the position of the inclined plane by choosing its generalized coordinate to be x, the displacement of the plane relative to some fixed reference point. We can then complete the specification of the system's configuration by choosing x', the displacement down the plane relative to the top of the plane, to be the generalized coordinate of the particle.

We can calculate the kinetic energy of each mass in terms of the dot product of its respective velocity vector, with each velocity vector specified in terms of unit vectors directed along the relevant generalized coordinate.

$$\mathbf{V} = \mathbf{i}\dot{x} \qquad \mathbf{v} = \mathbf{i}\dot{x} + \mathbf{e}_\theta \dot{x}'$$

The unit vector $\hat{\mathbf{e}}_\theta$ is directed down the plane at an elevation angle θ relative to the horizontal (see Figure 10.8).

Thus, the kinetic energy is

$$T_M = \frac{1}{2} M \mathbf{V} \cdot \mathbf{V} = \frac{1}{2} M \dot{x}^2$$

$$T_m = \frac{1}{2} m \mathbf{v} \cdot \mathbf{v} = \frac{1}{2} m (\mathbf{i}\dot{x} + \mathbf{e}_\theta \dot{x}') \cdot (\mathbf{i}\dot{x} + \mathbf{e}_\theta \dot{x}')$$

$$= \frac{1}{2} m (\dot{x}^2 + \dot{x}'^2 + 2\dot{x}\dot{x}' \cos \theta)$$

$$T = T_M + T_m$$

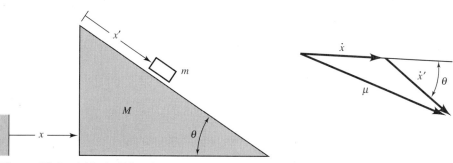

Figure 10.8   A block sliding down a movable wedge or inclined plane.

The potential energy of the system depends only upon the vertical position of the particle of mass m. We can choose it to be zero when the particle is at the top of the plane.

$$V = -mgx' \sin \theta$$

Therefore, the Lagrangian of the system is

$$L = \frac{1}{2} M\dot{x}^2 + \frac{1}{2} m(\dot{x}^2 + \dot{x}'^2 \, 2\dot{x}\dot{x}' \cos \theta) + mgx' \sin \theta$$

and the equations of motion are

$$\frac{d}{dt}\frac{\partial L}{\partial \dot{x}} = \frac{\partial L}{\partial x} = 0 \qquad \frac{d}{dt}\frac{\partial L}{\partial \dot{x}'} = \frac{\partial L}{\partial x'}$$

$$\frac{d}{dt}[m(\dot{x} + \dot{x}' \cos \theta) + M\dot{x}] = 0 \qquad \frac{d}{dt}(\dot{x}' + \dot{x} \cos \theta) = g \sin \theta$$

Note: The time derivative of the first term is zero. Hence, the first term in brackets is a constant of the motion. This situation occurs because the Lagrangian is independent of the coordinate x, similar to the situation that occurred in Example 10.2. Close examination of this term reveals that it is the total linear momentum of the system in the x direction.

The fact that it is a constant of the motion is also a reflection of the fact, from the Newtonian viewpoint, that there is no net force on the system in the x direction. Notice how naturally this result seemingly falls out of the Lagrangian formalism. We will have more to say about this in the following section.

We carry out the above time derivatives, obtaining

$$m(\ddot{x} + \ddot{x}' \cos \theta) + M\ddot{x} = 0 \qquad \ddot{x}' + \ddot{x} \cos \theta = g \sin \theta$$

Solving for $\ddot{x}$ and $\ddot{x}'$ we find

$$\ddot{x} = \frac{-g \sin \theta \cos \theta}{\dfrac{m + M}{m} - \cos^2 \theta} \qquad \ddot{x}' = \frac{g \sin \theta}{1 - \dfrac{m \cos^2 \theta}{m + M}}$$

This particular example illustrates nicely the ease with which complicated problems fall apart when attacked with the Lagrangian formalism. One could certainly solve the problem using Newtonian methods, but such an attempt would require a great deal more thought and physical insight than demanded in the Lagrangian "turn the crank" method shown above. Given that caveat, let us go ahead and crack this problem using our trusty Newtonian sledgehammer. ∎

EXAMPLE 10.7 *Particle Sliding on a Movable Inclined Plane*
Using Newtonian Methods

We have already indicated just how the first equation of motion for this problem might be obtained using Newtonian methods; that is, the total x momentum is conserved since there are no net forces in the x direction. Thus,

$$m(\dot{x} + \dot{x}' \cos \theta) + M\dot{x} = \text{constant}$$

$$\frac{d}{dt}[m(\dot{x} + \dot{x}' \cos \theta) + M\dot{x}] = 0$$

$$m(\ddot{x} + \ddot{x}' \cos \theta) + M\ddot{x} = 0$$

We can derive the second equation using $\mathbf{F} = m\mathbf{a}$, albeit with a little care. Let us examine the forces acting on the particle of mass m. We resolve them into components parallel and perpendicular to the inclined surface of the movable plane. Regardless of the state of motion of the movable plane, the only force parallel to this surface is

$$F_{x'} = mg \sin \theta$$

The acceleration of the particle down the incline includes a component of the horizontal acceleration of the plane directed along the incline as well as the acceleration of the particle down the incline relative to a point of reference fixed to the movable incline (see the vector diagram of the relevant velocities in Figure 10.8). The net acceleration is

$$a_{x'} = \ddot{x}' + \ddot{x} \cos \theta$$

The movability of the plane might have been confusing and made it difficult to obtain this latter result if we had not already seen the answer residing in the Lagrangian analysis above. Thus,

$$mg \sin \theta = m(\ddot{x}' + \ddot{x} \cos \theta)$$

which, indeed, is the same differential equation of motion obtained in the above Lagrangian analysis of the problem.   ■

10.6   GENERALIZED MOMENTA. IGNORABLE COORDINATES

A key feature of Examples 10.2 and 10.6 was the emergence of a momentum, conserved along the direction of a generalized coordinate not explicitly contained in the Lagrangian of the system. We would like to explore this situation in a little more detail. Perhaps the simplest example that illustrates such a condition is a free particle moving in a straight line, say, along the x-axis. Its kinetic energy is

$$T = \frac{1}{2}m\dot{x}^2$$

where m is the mass of the particle and $\dot{x}$ is its velocity. The Lagrangian for this system assumes the particularly simple form $L = T$. The Lagrangian equation of motion is, thus,

$$\frac{d}{dt}\frac{\partial L}{\partial \dot{x}} = \frac{d}{dt}\frac{\partial T}{\partial \dot{x}} = 0 \qquad \frac{d}{dt}(m\dot{x}) = 0$$

$$\therefore \ m\dot{x} = \text{constant}$$

$$(10.23)$$

As occurred in Examples 10.2 and 10.6, when the Lagrangian is independent of a co-ordinate, a solution to the equation of motion leads to the constancy of a quantity that can be identified with the momentum of the system referred to that missing coordinate. In this case we see that the constant quantity is exactly equal to the product $m\dot{x}$, the "Newtonian" linear momentum p_x of the free particle. Hence, we make the formal definition that

$$p_x = \frac{\partial L}{\partial \dot{x}} = m\dot{x}$$

is the momentum of the particle. In the case of a system described by the generalized coordinates $q_1, q_2, \ldots, q_k, \ldots, q_n$, the quantities p_k defined by

$$p_k = \frac{\partial L}{\partial \dot{q}_k} \tag{10.24}$$

are called the *generalized momenta conjugate to the generalized coordinate* q_k. Lagrange's equations for a conservative system can then be written as

$$\dot{p}_k = \frac{\partial L}{q_k} \tag{10.25}$$

It is now readily apparent that if the Lagrangian does not explicitly contain the coordinate q_k, then

$$\dot{p}_k = \frac{\partial L}{\partial q_k} = 0 \tag{10.26}$$

$$\therefore p_k = \text{constant} \tag{10.27}$$

The missing coordinate is ignorable, and its conjugate momentum is a constant of the motion.

EXAMPLE 10.8 *Pendulum Attached to a Movable Support*

Let us now continue the analysis of a pendulum attached to a movable support as outlined in Section 10.3 (See Figure 10.5). We have already calculated the kinetic and potential energies for this system in terms of the generalized coordinates X, the position of the movable support, and the angle θ that the pendulum makes with the vertical. They are given by Equation 10.9. The Lagrangian for this system is

$$L = \frac{1}{2}(M + m)\dot{X}^2 + \frac{1}{2}m(r^2\dot{\theta}^2 + 2\dot{X}r\dot{\theta}\cos\theta) + mgr\cos\theta$$

The equations of motion are

$$\dot{p}_x = \frac{d}{dt}\frac{\partial L}{\partial \dot{X}} = \frac{\partial L}{\partial X} = 0 \qquad \dot{p}_\theta = \frac{d}{dt}\frac{\partial L}{\partial \dot{\theta}} = \frac{\partial L}{\partial \theta}$$

$$\frac{d}{dt}[(M + m)\dot{X} + mr\dot{\theta}\cos\theta] = 0$$

$$\frac{d}{dt}[m(r^2\dot{\theta} + \dot{X}r\cos\theta)] = -mgr\sin\theta$$

$$\ddot{\theta} + \frac{\ddot{X}}{r}\cos\theta + \frac{(g - \dot{X}\dot{\theta})}{r}\sin\theta = 0$$

Note: The Lagrangian is independent of the generalized coordinate X. It is an *ignorable* coordinate. Its conjugate momentum is the term in brackets in the first equation above: the total linear momentum of the system in the X direction. This momentum is a constant of the motion. It must be a conserved quantity, since the potential energy of the system is independent of this coordinate (that is why the Lagrangian is missing that coordinate) and, therefore, there are no net external forces acting in this direction.

Sometimes the differential equations of motion look complicated. One might wonder whether or not the differential equations are infested with errors. Such could be the situation upon consideration of the second equation of motion derived above. As a check of the validity of such equations demand that the system adhere to certain limiting conditions, and then look closely at just what the derived equation of motion implies, given the imposed conditions. For instance, suppose that in the above problem, we "nail down" the moveable support; that is, we fix it firmly to the track along which it was previously allowed to move without friction. We, thus, reduce the above example to that of a simple pendulum, and the equation of motion had better reflect this fact. We will see that it does. The central term containing the acceleration $\ddot{X}$ and the mixed term containing the velocity $\dot{X}$ go to zero, since we have eliminated any X-direction motion of the mass M. The resultant equation of motion becomes

$$\ddot{\theta} + \frac{g}{r}\sin\theta = 0$$

which is that of a simple pendulum.

So far, so good. Still, we might continue to be bothered about the $\ddot{X}$ term. The previous condition led to its elimination from the equation of motion. That is not a good way to see whether its presence makes any sense. Let us play another trick similar to the one above, but this time with the angular acceleration and velocity terms $\ddot{\theta}$ and $\dot{\theta}$. Can we imagine a scenario in which they might be zero? What kind of motion would the system exhibit, given such a restriction? If those two terms are zero, the equation of motion reduces to a solution for θ in terms of the horizontal acceleration $\ddot{X}$ and g

$$\tan\theta = \frac{-\ddot{X}}{g}$$

In other words if we uniformly accelerate the support toward the right, the above solution describes a pendulum that hangs "motionless" at an angle θ to the "left" of the vertical relative to its support; that is, the system becomes a simple "linear accelerometer." Indeed, this is a possible scenario for the motion of this system. Such an analysis was presented in Chapter 5 and is also presented in several beginning primers on basic physics.[4] You see, it all makes sense.

We're still stuck with trying to make sense out of the mixed term $\ddot{X}\dot{\theta}$. It seems to have the effect of "reducing" g in the general motion of the system. It is left to the reader to ponder the meaning of this term.   ■

EXAMPLE 10.9   *The Spherical Pendulum, or Bar of Soap in a Bowl*

A classic problem in mechanics is that of a particle constrained to stay on a smooth spherical surface under gravity, such as a small mass sliding around inside a smooth spherical bowl. The case is also illustrated by a simple pendulum that is free to swing in any direction, Figure 10.9. This is the so-called spherical pendulum, mentioned in Section 5.5

There are two degrees of freedom, and we shall use generalized coordinates θ and ϕ, as shown. These are actually equivalent to spherical coordinates with $r = l =$ constant in which l is the length of the pendulum cord. The two components of the velocity are $v_\theta = l\dot{\theta}$ and $v_\phi = l \sin \theta \, \dot{\phi}$. The height of the bob, measured from the xy plane, is $l - l \cos \theta$, so the Lagrangian function is

$$L = \frac{1}{2} ml^2(\dot{\theta}^2 + \sin^2 \theta \, \dot{\phi}^2) - mgl(1 - \cos \theta)$$

---

[4]For example, see Example 6.8 in Raymond Serway, *Physics for Scientists and Engineers,* Saunders College Publishing, Philadelphia, 1990.

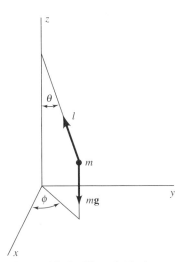

Figure 10.9   The spherical pendulum.

The coordinate ϕ is ignorable, so we have immediately

$$p_\phi = \frac{\partial L}{\partial \dot\phi} = ml^2 \sin^2 \theta \, \dot\phi = \text{constant}$$

This is the angular momentum about the vertical or z-axis. We are left with just the equation in θ:

$$\frac{d}{dt} \frac{\partial L}{\partial \dot\theta} = \frac{\partial L}{\partial \theta}$$

which reads

$$ml^2\ddot\theta = ml^2 \sin \theta \cos \theta \, \dot\phi^2 - mgl \sin \theta$$

Let us introduce the constant S, defined by

$$S = \sin^2 \theta \, \dot\phi = \frac{p_\phi}{ml^2} \tag{10.28a}$$

(This is the angular momentum divided by ml^2.) The differential equation of motion for θ then becomes

$$\ddot\theta + \frac{g}{l} \sin \theta - S^2 \frac{\cos \theta}{\sin^3 \theta} = 0 \tag{10.28b}$$

It is instructive to consider some special cases at this point. First, if the angle ϕ is constant, then $\dot\phi = 0$, and so $S = 0$. Consequently, Equation 10.28b reduces to

$$\ddot\theta + \frac{g}{l} \sin \theta = 0$$

which, of course, is just the differential equation of the simple pendulum. The motion takes place in the plane $\phi = \phi_0 = \text{constant}$.

The second special case is that of the *conical pendulum*. Here the bob describes a horizontal circle, so $\theta = \theta_0 = \text{constant}$. In this case $\dot\theta = 0$ and $\ddot\theta = 0$, so Equation 10.28b reduces to

$$\frac{g}{l} \sin \theta_0 - S^2 \frac{\cos \theta_0}{\sin^3 \theta_0} = 0$$

or

$$S^2 = \frac{g}{l} \sin^4 \theta_0 \sec \theta_0 \tag{10.29a}$$

From the value of l given by Equation 10.29b, we find from Equation 10.28a that

$$\dot\phi_0^2 = \frac{g}{l} \sec \theta_0 \tag{10.29b}$$

as the condition for conical motion of the pendulum.

Let us now consider the case in which the motion is almost conical; that is, the value of θ remains close to the value θ_0. If we insert the expression for L^2 given in Equation 10.28a into the separated differential equation of θ (Equation 10.28b), the result is

$$\ddot{\theta} + \frac{g}{l}\left(\sin\theta - \frac{\sin^4\theta_0}{\cos\theta_0}\frac{\cos\theta}{\sin^3\theta}\right) = 0$$

It is convenient at this point to introduce the new variable ξ defined as

$$\xi = \theta - \theta_0$$

The expression in parentheses, which we shall call $f(\xi)$, may be expanded as a power series in ξ according to the standard formula

$$f(\xi) = f(0) + f'(0)\xi + f''(0)\frac{\xi^2}{2!} + \cdots$$

We find, after performing the indicated operations, that $f(0) = 0$ and $f'(0) = 3\cos\theta_0 + \sec\theta_0$. Since we are concerned with the case of small values of ξ, we shall neglect higher powers of ξ than the first, and so we can write

$$\ddot{\xi} + \frac{g}{l}(3\cos\theta_0 + \sec\theta_0)\xi = 0$$

Thus, ξ oscillates harmonically about $\xi = 0$, or equivalently, θ oscillates harmonically about the value θ_0 with a period

$$T_1 = 2\pi\sqrt{\frac{l}{g(3\cos\theta_0 + \sec\theta_0)}}$$

Now the value of $\dot{\phi}$ does not vary greatly from the value given by the purely conical motion $\dot{\phi}_0$, so ϕ increases steadily during the oscillation of θ about θ_0. During one complete oscillation of θ, the value of the azimuth angle ϕ increases by the amount

$$\phi_1 \cong \dot{\phi}_0 T_1$$

From the values of $\dot{\phi}_0$ and T_1 given above, we readily find

$$\phi_1 = 2\pi(3\cos^2\theta_0 + 1)^{-1/2}$$

Now the quantity in parentheses is less than 4, for nonzero θ_0, so ϕ_1 is greater than π (180°). The excess $\Delta\phi$ is shown in Figure 10.10, which is a plot of the projection of the path of the pendulum bob on the xy plane. As the pendulum swings, it precesses in the direction of increasing ϕ, as indicated.

Finally, for the general case we can go back to the differential equation of motion (Equation 10.28b) and integrate once with respect to θ by using the fact that $\ddot{\theta} = \dot{\theta}d\dot{\theta}/d\theta = \frac{1}{2}d\dot{\theta}^2/d\theta$. The result is

$$\frac{1}{2}\dot{\theta}^2 = \frac{g}{l}\cos\theta - \frac{S^2}{2\sin^2\theta} + C = -U(\theta) + C$$

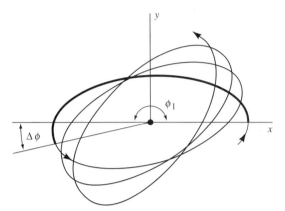

Figure 10.10   Projection on the *xy* plane of the path
of motion of the spherical pendulum.

in which C is the constant of integration and

$$U(\theta) = -\frac{g}{l} \cos \theta + \frac{S^2}{2 \sin^2 \theta}$$

is the effective potential. Actually, the integrated equation of motion is just the energy equation where the total energy $E = C \, ml^2$. For a given initial condition, the
motion of the pendulum is such that the bob oscillates between two horizontal
circles. These circles define the turning points of the θ-motion for which $\dot{\theta} = 0$ or
$U(\theta) = C$. This is illustrated in Figures 10.11 and 10.12.   ■

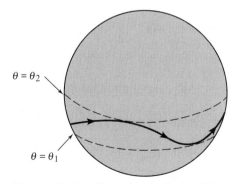

Figure 10.11   Illustrating the limits of the
motion of the spherical pendulum.

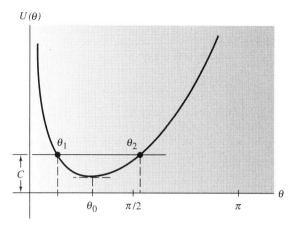

Figure 10.12 Graph of the effective potential for the spherical pendulum.

10.7 THE HAMILTONIAN FUNCTION. HAMILTON'S EQUATIONS

Consider the following function of the generalized coordinates:

$$H = \sum_k \dot{q}_k p_k - L$$

For simple dynamic systems the kinetic energy T is a homogeneous quadratic function of the $\dot{q}$'s, and the potential energy V is a function of the q's alone, so that

$$L = T(q_k, \dot{q}_k) - V(q_k)$$

Now, from Euler's theorem for homogeneous functions,[5] we have

$$\sum_k \dot{q}_k p_k = \sum_k \dot{q}_k \frac{\partial L}{\partial \dot{q}_k} = \sum_k \dot{q}_k \frac{\partial T}{\partial \dot{q}_k} = 2T$$

Therefore,

$$H = \sum_k \dot{q}_k p_k - L = 2T - (T - V) = T + V \qquad (10.30)$$

That is, the function H is equal to the total energy for the type of system we are considering.

---

[5] Euler's theorem states that for a homogeneous function f of degree n in the variables $x_1, x_2, \ldots, x_r$

$$x_1 \frac{\partial f}{\partial x_1} + x_2 \frac{\partial f}{\partial x_2} + \cdots + x_r \frac{\partial f}{\partial x_r} = nf$$

Suppose we regard the n equations

$$p_k = \frac{\partial L}{\partial \dot{q}_k} \qquad (k = 1, 2, \ldots, n)$$

as solved for the $\dot{q}$'s in terms of the p's and the q's:

$$\dot{q}_k = \dot{q}_k(p_k, q_k)$$

With these equations we can then express H as a function of the p's and the q's:

$$H(p_k, q_k) = \sum_k p_k \dot{q}_k(p_k, q_k) - L \qquad (10.31)$$

Let us calculate the variation of the function H corresponding to a variation δp_k, δq_k. We have

$$\delta H = \sum_k \left[p_k \delta \dot{q}_k + \dot{q}_k \delta p_k - \frac{\partial L}{\partial \dot{q}_k} \delta \dot{q}_k - \frac{\partial L}{\partial q_k} \delta q_k \right]$$

The first and third terms in the brackets cancel, because $p_k = \partial L / \partial \dot{q}_k$ by definition. Also, since Lagrange's equations can be written as $\dot{p}_k = \partial L / \partial q_k$, we can write

$$\delta H = \sum_k [\dot{q}_k \, \delta p_k - \dot{p}_k \, \delta q_k]$$

Now the variation of H must be given by the equation

$$\delta H = \sum_k \left[\frac{\partial H}{\partial p_k} \delta p_k + \frac{\partial H}{\partial q_k} \delta q_k \right]$$

It follows that

$$\frac{\partial H}{\partial p_k} = \dot{q}_k$$

$$\frac{\partial H}{\partial q_k} = -\dot{p}_k$$

(10.32)

These are known as *Hamilton's canonical equations of motion*. They consist of $2n$ first-order differential equations, whereas Lagrange's equations consist of n second-order equations. We have derived Hamilton's equations for simple conservative systems. It can be shown that Equations 10.32 also hold for more general systems, for example, nonconservative systems, systems in which the potential-energy function involves the $\dot{q}$'s, and systems in which L involves the time explicitly, but in these cases the total energy is no longer necessarily equal to H.

Hamilton's equations will be encountered by the student when studying quantum mechanics (the fundamental theory of atomic phenomena). Hamilton's equations also find application in celestial mechanics. For further reading the student is referred to the Selected References (under *Advanced Mechanics*) at the end of the book.

EXAMPLE 10.10

Obtain Hamilton's equations of motion for a one-dimensional harmonic oscillator.

Solution:
We have

$$T = \frac{1}{2} m\dot{x}^2 \qquad V = \frac{1}{2} Kx^2 \qquad L = T - V$$

$$p = \frac{\partial L}{\partial \dot{x}} = m\dot{x} \qquad \dot{x} = \frac{p}{m}$$

Hence,

$$H = T + V = \frac{1}{2m} p^2 + \frac{K}{2} x^2$$

The equations of motion

$$\frac{\partial H}{\partial p} = \dot{x} \qquad \frac{\partial H}{\partial x} = -\dot{p}$$

then read

$$\frac{p}{m} = \dot{x} \qquad Kx = -\dot{p}$$

The first equation merely amounts to a restatement of the momentum–velocity relationship in this case. Using the first equation, the second can be written

$$Kx = -\frac{d}{dt}(m\dot{x})$$

or, upon rearranging terms,

$$m\ddot{x} + Kx = 0$$

which is the familiar equation of the harmonic oscillator.   ■

EXAMPLE 10.11

Find the Hamiltonian equations of motion for a particle in a central field.

Solution:
Here we have

$$T = \frac{m}{2}(\dot{r}^2 + r^2\dot{\theta}^2)$$

$$V = V(r)$$

$$L = T - V$$

in polar coordinates. Hence,

$$p_r = \frac{\partial L}{\partial \dot{r}} = m\dot{r} \qquad \dot{r} = \frac{p_r}{m}$$

$$p_\theta = \frac{\partial L}{\partial \dot{\theta}} = mr^2\dot{\theta} \qquad \dot{\theta} = \frac{p_\theta}{mr^2}$$

Consequently,

$$H = \frac{1}{2m}\left(p_r^2 + \frac{p_\theta^2}{r^2}\right) + V(r)$$

The Hamiltonian equations

$$\frac{\partial H}{\partial p_r} = \dot{r} \qquad \frac{\partial H}{\partial r} = -\dot{p}_r \qquad \frac{\partial H}{\partial p_\theta} = \dot{\theta} \qquad \frac{\partial H}{\partial \theta} = -\dot{p}_\theta$$

then read

$$\frac{p_r}{m} = \dot{r}$$

$$\frac{\partial V(r)}{\partial r} - \frac{p_\theta^2}{mr^3} = -\dot{p}_r$$

$$\frac{p_\theta}{mr^2} = \dot{\theta}$$

$$0 = -\dot{p}_\theta$$

The last two equations yield the constancy of angular momentum:

$$p_\theta = \text{constant and } mr^2\dot{\theta} = ml$$

from which the first two give

$$m\ddot{r} = \dot{p}_r = \frac{ml^2}{r^3} - \frac{\partial V(r)}{\partial r}$$

for the radial equation of motion. This, of course, is equivalent to that found earlier in Example 10.2.   ■

PROBLEMS

Lagrange's method should be used in all of the following, unless stated otherwise.

10.1 Find the differential equations of motion of a projectile in a uniform gravitational field without air resistance.

10.2 Find the acceleration of a solid uniform sphere rolling down a perfectly rough fixed inclined plane. Compare with the result derived earlier in Section 8.6.

10.3 Two blocks of equal mass m are connected by a flexible cord. One block is placed on a smooth horizontal table, the other block hangs over the edge. Find the acceleration of the system assuming (a) the mass of the cord is negligible and (b) the cord is heavy, of mass m'.

10.4 Set up the equations of motion of a "double-double" Atwood's machine consisting of one Atwood's machine (with masses m_1 and m_2) connected by means of a light cord passing over a pulley to a second Atwood's machine with masses m_3 and m_4. Neglect the masses of all pulleys. Find the actual accelerations for the case $m_1 = m$, $m_2 = 4m$, $m_3 = 2m$, and $m_4 = m$.

10.5 A ball of mass m rolls down a movable wedge of mass M. The angle of the wedge is θ, and it is free to slide on a smooth horizontal surface. The contact between the ball and the wedge is perfectly rough. Find the acceleration of the wedge.

10.6 A particle slides on a smooth inclined plane whose inclination θ is increasing at a constant rate ω. If $\theta = 0$ at time $t = 0$, at which time the particle starts from rest, find the subsequent motion of the particle.

10.7 Show that Lagrange's method automatically yields the correct equations of motion for a particle moving in a plane in a rotating coordinate system Oxy. [*Hint:* $T = \frac{1}{2}m\mathbf{v} \cdot \mathbf{v}$, where $\mathbf{v} = \mathbf{i}(\dot{x} - \omega y) + \mathbf{j}(\dot{y} + \omega x)$, and $F_x = -\partial V/\partial x$, $F_y = -\partial V/\partial y$.]

10.8 Repeat the Problem 10.7 for motion in three dimensions.

10.9 Find the differential equations of motion for an "elastic pendulum": a particle of mass m attached to an elastic string of stiffness K and unstretched length l_0. Assume that the motion takes place in a vertical plane.

10.10 The point of support of a simple pendulum is being elevated at a constant acceleration a, so that the height of the support is $\frac{1}{2}at^2$, and its vertical velocity is at. Find the differential equation of motion for small oscillations of the pendulum by Lagrange's method. Show that the period of the pendulum is $2\pi[l/(g + a)]^{1/2}$ where l is the length of the pendulum.

10.11 Work Problem 8.10 by using Lagrange's equations. Show that the acceleration of the ball is $\frac{5}{7}g$.

10.12 A heavy elastic spring of uniform stiffness and density supports a block of mass m. If m' is the mass of the spring and k its stiffness, show that the period of vertical oscillations is

$$2\pi\sqrt{\frac{m + (m'/3)}{k}}$$

This problem shows the effect of the mass of the spring on the period of oscillation. (*Hint:* To set up the Lagrangian function for the system, assume that the velocity of any part of the spring is proportional to its distance from the point of suspension.)

10.13 **(a)** Find the general differential equations of motion for a particle in cylindrical coordinates: R, ϕ, z. Use the relation

$$v^2 = v_R^2 + v_\phi^2 + v_z^2$$
$$= \dot{R}^2 + R^2\dot{\phi}^2 + \dot{z}^2$$

(b) Find the general differential equations of motion for a particle in spherical coordinates: r, θ, ϕ. Use the relation

$$v^2 = v_r^2 + v_\theta^2 + v_\phi^2$$
$$= \dot{r}^2 + r^2\dot{\theta}^2 + r^2\sin^2\theta\,\dot{\phi}^2$$

(*Note:* Compare your results with the result derived in Chapter 1, Equations 1.52 and 1.60 by setting $\mathbf{F} = m\mathbf{a}$ and taking components.)

10.14 Find the differential equations of motion in *three* dimensions for a particle in a central field using spherical coordinates.

10.15 A bar of soap slides in a smooth bowl in the shape of an inverted right circular cone of apex angle 2α. The axis of the cone is vertical. Treating the bar of soap as a particle of mass m, find the differential equations of motion using spherical coordinates with $\theta = \alpha$ = constant. As is the case with the spherical pendulum, Example 10.9, show that the particle, given an initial motion with $\dot{\phi}_0 \neq 0$, must remain between two horizontal circles on the cone. (*Hint:* Show that $\dot{r}^2 = f(r)$ where $f(r) = 0$ has two roots that define the turning points of the motion in r.) What is the effective potential for this problem?

10.16 In Problem 10.15, find the value of $\dot{\phi}_0$ such that the particle remains on a *single* horizontal circle: $r = r_0$. Find also the period of small oscillations about this circle if $\dot{\phi}_0$ is not quite equal to the required value.

10.17 As stated in Chapter 4, the differential equation of motion of a particle of mass m and electric charge q moving with velocity $\mathbf{v}$ in a static magnetic field $\mathbf{B}$ is given by

$$m\ddot{\mathbf{r}} = q(\mathbf{v} \times \mathbf{B})$$

Show that the Lagrangian function

$$L = \frac{1}{2}mv^2 + q\mathbf{v} \cdot \mathbf{A}$$

yields the correct equation of motion where $\mathbf{B} = \nabla \times \mathbf{A}$. The quantity $\mathbf{A}$ is called the *vector potential*. (*Hint:* In this problem it will be necessary to employ the general formula $df(x, y, z)/dt = \dot{x}\partial f/\partial x + \dot{y}\partial f/\partial y + \dot{z}\partial f/\partial z$. Thus, for the part involving $\mathbf{v} \cdot \mathbf{A}$, we have

$$\frac{d}{dt}\left[\frac{\partial(\mathbf{v} \cdot \mathbf{A})}{\partial \dot{x}}\right] = \frac{d}{dt}\left[\frac{\partial}{\partial \dot{x}}(\dot{x}A_x + \dot{y}A_y + \dot{z}A_z)\right] = \frac{d}{dt}(A_x)$$

$$= \dot{x}\frac{\partial A_x}{\partial x} + \dot{y}\frac{\partial A_x}{\partial y} + \dot{z}\frac{\partial A_x}{\partial z}$$

and similarly for the other derivatives.)

10.18 Write the Hamiltonian function and find Hamilton's canonical equations for the three-dimensional motion of a projectile in a uniform gravitational field with no air resistance. Show that these equations lead to the same equations of motion as found in Chapter 4.

10.19 Find Hamilton's canonical equations for
(a) A simple pendulum
(b) A simple Atwood's machine
(c) A particle sliding down a smooth inclined plane

10.20 As we know, the kinetic energy of a particle in one-dimensional motion is $\frac{1}{2}m\dot{x}^2$. If the potential energy is proportional to x^2, say $\frac{1}{2}kx^2$, show by direct application of Hamilton's variational principle, $\delta\int L\, dt = 0$, that the equation of the simple harmonic oscillator is obtained.

COMPUTER/CALCULATOR APPLICATIONS

10.1 Assume that the spherical pendulum discussed in Section 10.6 is set into motion with the following initial conditions: $\phi_0 = 0$ rad, $\dot{\phi}_0 = 10.57$ rad/s, $\theta_0 = \pi/4$ rad, and $\dot{\theta}_0 = 0$ rad/s. Let the length of the pendulum be 0.284 m.

    (a) Calculate θ_1 and θ_2, the polar angular limits of the motion.

$$\frac{1}{2}\dot{\theta}^2 = -U(\theta) + C = 0$$

    (Hint: Solve the equation above numerically for the condition of $\dot{\theta} = 0$.)

    (b) Solve the equations of motion of the pendulum numerically and find the period of the θ-motion.

    (c) Plot θ as a function of the azimuthal angle ϕ over two azimuthal cycles.

    (d) Calculate the angle of precession $\Delta\phi$ that occurs during one complete cycle of θ.

10.2 A bead slides from rest down a smooth curve S in the xy plane from the point $(0, 2)$ to the point $(\pi, 0)$.

    (a) Show that the curve S for which the total time of travel is a minimum is a cycloid, described by the parametric equations $x = \theta - \sin\theta$ and $y = 1 + \cos\theta$ where θ ranges from 0 to π. *(Hint: The time of travel is given by $T = \int ds/v$ where ds is a differential element of displacement and v is the speed of the bead along the curve S. Express ds in terms of $y' = dy/dx$ and dx. Express the resultant integrand as an explicit function of y, y', and possibly x. The integral is a minimum when the integrand satisfies Lagrange's equation. Find the differential equation of the curve generated by the Lagrange equation and solve it.)*

    (b) Assume that the curve S can be approximated by a quadratic function, $y(x) = a_0 + a_1 x + a_2 x^2$. (This function must satisfy the boundary conditions given for the curve S.) Insert this function (and its derivative) into the above integral for the bead's time of travel. Find the constant coefficients a_i that minimize the integral.

    (c) Estimate the minimum time of transit.

    (d) Plot the function $y(x)$ obtained in part (b) along with the exact solution given by the above equations representing a cycloid (from $x = 0$ to $x = \pi$). How well do the two solutions agree?

11

DYNAMICS OF OSCILLATING SYSTEMS

The modern development of physics is continually enhancing Hamilton's name. His famous analogy between mechanics and optics virtually anticipated wave mechanics, which did not have to add much to his ideas, but only had to take them seriously—a little more seriously than he was able to take them, with the experimental knowledge of a century ago. The central conception of all modern theory in physics is "the Hamiltonian." If you wish to apply modern theory to any particular problem, you must start with putting the problem "in Hamiltonian form."

Thus Hamilton is one of the greatest men of science the world has produced."

(Irwin Schroedinger, A Collection of Papers in Memory of Sir William Rowan Hamilton, *ed. D. E. Smith,* Scripta Mathematica Studies, *no. 2, N.Y., 1945)*

In the preceding chapters we have studied some simple systems that can undergo oscillations about a configuration of equilibrium, including a simple pendulum, a particle suspended on an elastic spring, a physical pendulum, and so on, all being cases of one degree of freedom characterized by a single frequency of oscillation. When we consider more complicated systems—systems with several degrees of freedom—we shall find that not one but several different frequencies of oscillation are possible. In our analysis of oscillating systems, we shall find it very convenient to use generalized coordinates and to employ Lagrange's method for finding the equations of motion in terms of those coordinates.

11.1 POTENTIAL ENERGY AND EQUILIBRIUM. STABILITY

Before we take up the study of motion of a system with many degrees of freedom about an equilibrium configuration, let us first examine just what is meant by the term *equilibrium*. As a way of introduction, let us recall the oscillatory motion of a mass on a spring about its equilibrium position. It is a conservative system, and its restoring force is derivable from a potential energy function

$$V = V(x) = \frac{1}{2}kx^2$$

$$F(x) = -\frac{dV(x)}{dx} = -kx$$

The equilibrium position of the oscillator is at $x = 0$, the position where the restoring force vanishes or the derivative of the potential energy function is zero. If the oscillator was initially placed at rest at $x = 0$, it would remain there at rest.

Let us consider the motion of a simple pendulum of length r constrained to swing in a vertical plane (See Figure 10.2). As in our discussion of Section 10.2, let its position be described by the single generalized coordinate θ, the angle that it makes with the vertical. Taking the potential energy to be zero at $\theta = 0$, the potential energy function and the derived restoring "force" are given by

$$V = mgr(1 - \cos \theta)$$

$$N_\theta = -\frac{\partial V}{\partial \theta} = -mg(r \sin \theta) = -mgx$$

where x is the horizontal displacement of the pendulum bob from the vertical. The generalized coordinate of the pendulum is an angular variable and the restoring force is actually a restoring torque N_θ. The pendulum is in its equilibrium position when the restoring torque is equal to zero. In each of these two cases, regardless of whether the potential energy is a function of either a positional or an angular coordinate, equilibrium corresponds to the configuration where the derivative of the potential energy function vanishes.

Now let us generalize the above cases to a system with n degrees of freedom whose generalized coordinates $q_1, q_2, \ldots, q_n$ completely specify its configuration. The q's can be a mixture of both positional and angular variables. We shall assume that the system is conservative and that its potential energy function is a function of the q's alone:

$$V = V(q_1, q_2, \ldots, q_n) \tag{11.1}$$

All forces and torques acting on the system vanish when

$$\frac{\partial V}{\partial q_k} = 0 \quad (k = 1, 2, \ldots, n) \tag{11.2}$$

This more complicated system is said to be in equilibrium when Equations 11.2 hold true. These equations constitute a necessary condition for the system to remain at rest if it is initially at rest in such a configuration. However, if the system is given a small displacement from this configuration, it may or may not return to the equilibrium configuration. If it always tends to return to equilibrium, given a sufficiently small displacement, the equilibrium is said to be *stable; otherwise*, it is *unstable.* (If the system has no tendency to move either toward or away from equilibrium, the equilibrium is said to be *neutral.*)

A ball placed (1) at the bottom of a spherical bowl, (2) on top of a spherical cap, and (3) on a plane horizontal surface are examples of stable, unstable, and neutral equilibrium, respectively.

Intuition tells us that the potential energy must be a *minimum* in all cases for stable equilibrium. That this is so can be argued from energy considerations. If the system is conservative, the total energy $T + V$ is constant, so for a small change near equilibrium $\Delta T = -\Delta V$. Thus, T will decrease if V increases; that is, the motion tends to slow down and return to the equilibrium position, given a small displacement. The reverse is true if the potential energy is *maximum;* that is, any displacement causes V to decrease and T to increase, so the system tends to move away from the equilibrium position at an ever-increasing rate.

Extended Criteria for Stable Equilibrium

We consider first a system with one degree of freedom. Suppose we expand the potential energy function $V(q)$ as a Taylor series about the point $q = 0$, namely,

$$V(q) = V_0 + qV_0' + \frac{q^2}{2!}V_0'' + \frac{q^3}{3!}V_0''' + \cdots + \frac{q^n}{n!}V_0^{(n)} + \cdots \tag{11.3}$$

where we use the notation $V_0' = (dV/dq)_{q=0}$, and so on. Now if $q = 0$ is a position of equilibrium, then $V_0' = 0$. This eliminates the linear term in the expansion. Furthermore, the term V_0 is a constant whose value depends on the arbitrary choice of the zero of the potential energy, so without incurring any loss of generality we can set $V_0 = 0$. Consequently, the expression for $V(q)$ simplifies to

$$V(q) = \frac{q^2}{2}V_0'' + \cdots \tag{11.3a}$$

If V_0'' is not zero, then for a small displacement q from equilibrium the force is approximately linear in the displacement:

$$F(q) = -\frac{dV}{dq} = -qV_0''$$

This will be of a restorative or stabilizing type if V_0'' is positive, whereas, if V_0'' is negative, the force will be antirestoring and the equilibrium is unstable. If $V_0'' = 0$, then we must examine the first nonvanishing term in the expansion. If this term is of even order in n, then the equilibrium is again stable, or unstable, depending on whether the derivative $V_0^{(n)} = (d^n V/dq^n)_{q=0}$ is positive or negative, respectively. If the first nonvanishing derivative is of odd order in n, then the equilibrium is always unstable regardless of the sign of the derivative; this corresponds to the situation at point C in Figure 11.1. Clearly, if all derivatives vanish, then the potential energy function is a constant, and the equilibrium is neutral.

Similarly, for the case of a system with several degrees of freedom, we can effect a linear transformation so that $q_1 = q_2 = \cdots = q_n = 0$ is configuration of equilibrium, if an equilibrium configuration exists. The potential-energy function can then be expanded in the form

$$V(q_1, q_2, \ldots, q_n) = \frac{1}{2}(K_{11}q_1^2 + 2K_{12}q_1q_2 + K_{22}q_2^2 + \cdots) \qquad (11.4)$$

where

$$K_{11} = \left(\frac{\partial^2 V}{\partial q_1^2}\right)_{q_1=q_2=\cdots=q_n=0}$$

$$K_{12} = \left(\frac{\partial^2 V}{\partial q_1 \partial q_2}\right)_{q_1=q_2=\cdots=q_n=0}$$

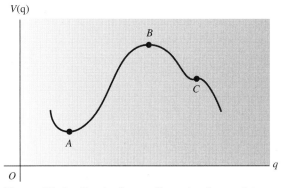

Figure 11.1 Graph of a one-dimensional potential energy function. The point A is one of stable equilibrium. Points B and C are unstable.

and so on. We have arbitrarily set $V(0, 0, \ldots, 0) = 0$. The linear terms in the expansion are absent because the expansion is about an equilibrium configuration.

The expression in parentheses in Equation 11.4 is known as a *quadratic form*. If this quadratic form is positive definite,[1] that is, either zero or positive for all values of the q's, then the equilibrium configuration $q_1 = q_2 = \cdots = q_n = 0$ is stable.

EXAMPLE 11.1 *Stability of Rocking Chairs, Pencils-On-End, and the Like*

Let us examine the equilibrium of a body having a rounded (spherical or cylindrical) base that is balanced on a plane horizontal surface. Let a be the radius of curvature of the base, and let the center of mass CM be a distance b from the initial point of contact, as shown in Figure 11.2(a). In Figure 11.2(b) the body is shown in a displaced position, where θ is the angle between the vertical and the line OCM (O being the center of curvature), as shown. Let h denote the distance from the plane to the center of mass. Then the potential energy is given by

$$V = mgh = mg[a - (a - b) \cos \theta]$$

where m is the mass of the body. We have

$$V' = \frac{dV}{d\theta} = mg(a - b) \sin \theta$$

which gives, for $\theta = 0$,

$$V'_0 = 0$$

Thus, $\theta = 0$ is a position of equilibrium. Furthermore, the second derivative is

$$V'' = mg(a - b) \cos \theta$$

---

[1] The necessary and sufficient conditions that the quadratic form in Equation 11.4 be positive definite are

$$K_{11} > 0 \qquad \begin{vmatrix} K_{11} & K_{12} \\ K_{21} & K_{22} \end{vmatrix} > 0 \qquad \begin{vmatrix} K_{11} & K_{12} & K_{13} \\ K_{21} & K_{22} & K_{23} \\ K_{31} & K_{32} & K_{33} \end{vmatrix} > 0 \qquad \text{and so on}$$

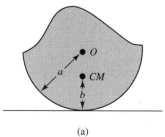 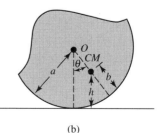

(a)	(b)

Figure 11.2 Coordinates for analyzing the stability of equilibrium of a round-bottomed object.

so, for $\theta = 0$

$$V_0'' = mg(a - b)$$

Hence, the equilibrium is stable if $a > b$, that is, if the center of mass lies below the center of curvature O. If $a < b$, the second derivative is negative and the equilibrium is unstable, such as with a pencil standing on end. If $a = b$, the potential energy function is constant, and the equilibrium is neutral. In this latter case, the center of mass coincides with the center of curvature. ∎

11.2 OSCILLATION OF A SYSTEM WITH ONE DEGREE OF FREEDOM ABOUT A POSITION OF STABLE EQUILIBRIUM

If a system has one degree of freedom, the kinetic energy may be expressed as

$$T = \frac{1}{2}M\dot{q}^2$$

where the coefficient M may be a constant or a function of the generalized coordinate q. In any case if $q = 0$ is a position of equilibrium, we shall consider q small enough so that $M = M(0) = $ constant is a valid approximation. From the expression for the potential energy (Equation 11.3a), we can write the Lagrangian function as

$$L = T - V = \frac{1}{2}M\dot{q}^2 - \frac{1}{2}V_0''q^2 \tag{11.5}$$

Lagrange's equation of motion

$$\frac{d}{dt}\frac{\partial L}{\partial \dot{q}} - \frac{\partial L}{\partial q} = 0$$

then becomes

$$M\ddot{q} + V_0''q = 0 \tag{11.6}$$

Thus, if $q = 0$ is a position of stable equilibrium, that is, if $V_0'' > 0$, then the system oscillates harmonically about the equilibrium position with angular frequency

$$\omega = \sqrt{\frac{V_0''}{M}} \tag{11.7}$$

EXAMPLE 11.2

Consider the motion of the round-bottomed object discussed in Example 11.1 (Figure 11.2). If the contact is perfectly rough, we have pure rolling, and the speed of the center of mass is approximately $b\dot{\theta}$ for small θ. The kinetic energy T is accordingly given by

$$T = \frac{1}{2}m(b\dot{\theta})^2 + \frac{1}{2}I_{cm}\dot{\theta}^2$$

where I_{cm} is the moment of inertia about the center of mass. Also, we can express the potential-energy function V as follows:

$$
\begin{aligned}
V(\theta) &= mg[a - (a - b)\cos\theta] \\
&= mg\left[a - (a - b)\left(1 - \frac{\theta^2}{2!} + \frac{\theta^4}{4!} - \cdots\right)\right] \\
&= \frac{1}{2}mg(a - b)\theta^2 + \text{constant} + \text{higher terms}
\end{aligned}
$$

We can then write

$$
L = \frac{1}{2}(mb^2 + I_{cm})\dot{\theta}^2 - \frac{1}{2}mg(a - b)\theta^2
$$

neglecting constants and higher terms. Comparing with Equation 11.5, we see that

$$
M = mb^2 + I_{cm}
$$
$$
V_0'' = mg(a - b)
$$

The motion about the equilibrium position $\theta = 0$ is, therefore, approximately simple harmonic with angular frequency

$$
\omega = \sqrt{\frac{mg(a - b)}{mb^2 + I_{cm}}} \qquad \blacksquare
$$

EXAMPLE 11.3 *Attitude Stability and Oscillation of an Orbiting Satellite*

In this example we shall analyze the oscillatory motion of a nonspherical satellite traveling in a circular orbit. For simplicity, we shall consider the satellite to be a dumbbell consisting of two small spheres, of mass $m/2$ each, connected by a light rod of length $2a$, Figure 11.3. Polar coordinates r, θ specify the center of mass CM (center of the rod), and the angle ϕ gives the "attitude" of the satellite axis relative to the radius vector OCM, O being the center of the earth. We shall treat the two end spheres as particles and assume that the motion is in a single plane, the plane of the orbit. For a circular orbit $r = r_0 = \text{constant}$, and $\dot{\theta} = \omega_0 = v_{cm}/r_0 = \text{constant}$.

The most important quantity to calculate in this example is the potential-energy function of the satellite. It is given by

$$
V = -\frac{GM_e m}{2}\left(\frac{1}{r_1} + \frac{1}{r_2}\right)
$$

in which M_e is the earth's mass and r_1 and r_2 are the distances from O to the respective end spheres, as shown. From the law of cosines we have

$$
r_{1,2} = (r_0^2 + a^2 \pm 2r_0 a \cos\phi)^{1/2} = (r_0^2 + a^2)^{1/2}(1 \pm \epsilon \cos\phi)^{1/2}
$$

where $\epsilon = 2r_0 a/(r_0^2 + a^2)$. Now $a \ll r_0$, so ϵ is a very small quantity. We

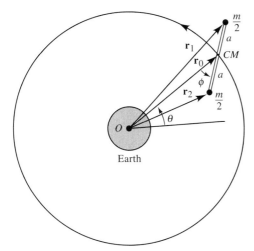

Figure 11.3 Dumbbell-shaped satellite in a circular orbit.

shall, therefore, express the potential-energy function by use of the binomial series $(1 + x)^{-1/2} = 1 - \frac{1}{2}x + \frac{3}{8}x^2 + \cdots$ where $x = \pm \epsilon \cos \phi$. The result, after collecting and canceling terms, is

$$V(\phi) = -\frac{GM_e m}{r_0}\left(1 + \frac{3a^2}{2r_0^2} \cos^2 \phi + \cdots\right)$$

where we have neglected a^2 compared to r_0^2 in all terms involving the quantity $r_0^2 + a^2$.

Taking the first and second derivatives with respect to ϕ, we find

$$V'(\phi) = \frac{GM_e m}{r_0^3}3a^2 \sin \phi \cos \phi$$

$$V''(\phi) = \frac{GM_e m}{r_0^3}3a^2 \cos(2\phi)$$

Thus, we have $\phi = 0$ and $\phi = \pi/2$ as two positions of equilibrium: $V'(0) = V'(\pi/2) = 0$. The first is stable, since $V''(0) > 0$. In this case the attitude of the satellite is such that the satellite's axis (line connecting the two masses) is along the radius vector OCM. The second position is an unstable equilibrium since $V''(\pi/2) < 0$; here the axis is at right angles to the radius vector.

The rocking motion of the satellite about the position of stable equilibrium is given by Equation 11.6 with $q = \phi$, $M = I_{cm} = ma^2$, and $V_0'' = 3a^2 GM_e m/r_0^3$. Thus, the angular frequency of the oscillation is

$$\omega = \sqrt{\frac{V_0''}{I_{cm}}} = \sqrt{\frac{3GM_e}{r_0^3}}$$

(Note that this is independent of m and a.) Now the angular frequency of the circular orbit around the earth is given by $\omega_0^2 = v_{cm}^2/r_0^2 = GM_e/r_0^3$. (See Example 6.3.) Thus, we can write

$$\omega = \omega_0 \sqrt{3}$$

For a synchronous earth satellite the orbital period $2\pi/\omega_0 = 24$ hr. Consequently, the rocking period of our dumbbell satellite in a synchronous orbit would be

$$\frac{2\pi}{\omega} = \frac{24}{\sqrt{3}} \text{ hr} = 13.86 \text{ hr} \qquad \blacksquare$$

11.3   COUPLED HARMONIC OSCILLATORS. NORMAL COORDINATES

Prior to developing the general theory of oscillating systems with any number of degrees of freedom, we shall study a simple specific example, namely, a system consisting of two harmonic oscillators that are coupled together.

For definiteness we use a model comprised of particles attached to elastic springs, although any type of oscillator could be used. For simplicity we assume that the oscillators are identical and are restricted to move in a straight line, Figure 11.4. The coupling is represented by a spring of stiffness K' as shown. The system has two degrees of freedom. We shall choose coordinates x_1 and x_2, the displacements of the particles from their respective equilibrium positions, to represent the configuration of the system.

Before plunging into the mathematics describing the motion of this system, we should like to consider just what sort of behavior we might expect. We would guess that the actual motion would depend critically upon the initial conditions of the system, whereas the vibrational frequency (or frequencies) would not. For example, suppose we held one mass at the position $x_1 = 0$, while we pulled the other mass a little to the right, say $x_2 = 1$, and then released them both from rest. Just after being released, m_2 is subject to a restoring force due to the compression of the right-hand spring and the stretching of the middle spring. m_1, even though at rest at $x_1 = 0$, is subject to a force due to the stretching of the middle spring. Hence, both masses will start to move, m_1 away from $x_1 = 0$ and m_2 toward $x_2 = 0$. The resulting motion looks to be fairly complex, but one thing is certain: Overall energy is conserved. Thus, as m_1, initially at rest, moves away from $x_1 = 0$, it gains energy at the expense of that of m_2. As time goes by we might anticipate that m_1 will eventually be displaced to the left at $x_1 = -1$ while m_2 will be at $x_2 = 0$, both instantaneously at rest. This configuration is completely symmetrical to

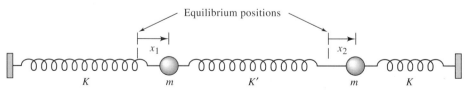

Figure 11.4   Model of two coupled harmonic oscillators.

the initial one, with m_1 and m_2 having exchanged energies. The system should continue to repeat this motion with m_1 and m_2 shuttling their energy back and forth through the coupling spring. The critical point here is that x_1 and x_2 are never simultaneously zero and that the coupling spring is never relaxed with m_1 and m_2 in that configuration. Hence, the two masses will continue to exchange energy.

A second important feature of this motion is that each mass vibrates in a *multifrequency* fashion. This can be most readily seen by analyzing the cause of single-frequency motion. Such motion occurs when the acceleration (or force per unit mass) of a mass is proportional to the negative of its displacement. In the situation here, each mass is subject to two forces, one from each connecting spring. The middle spring generates a force on each mass that is proportional to the difference in their displacements. Thus, we might anticipate that the general motion of each mass would be a composite of two different frequencies, and we will soon see that this is the case.

Figure 11.5 shows the motion of the two masses described above. The spring constants have values $K = 16$ and $K' = 1$ (arbitrary units), so this is a case of relatively weak coupling. The amplitude of oscillation of m_1 slowly builds up and then dies away in step with the dying away and buildup of the amplitude of oscillation of m_2. The motion has been plotted over a time interval equal to $\frac{1}{2}\tau$, where τ is the period of a full

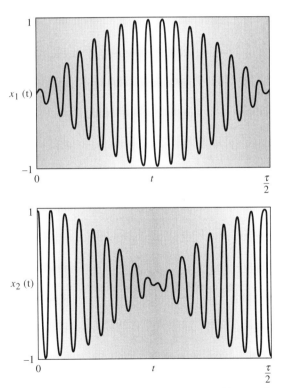

Figure 11.5 Displacement of two coupled harmonic oscillators.

energy transfer cycle. Each of these motions looks like a case of "beats" between two different single frequencies of the same amplitude. And that is exactly what they are.

The phenomenon of beats occurs when waves (or vibrations) of two different frequencies are added together. For example, let us assume that x_1 and x_2 can be represented by the sum (or difference) of two simple, equal-amplitude, harmonic motions whose frequencies are different. The resultant sum is equal to a product of sines (or cosines) of a sum and a difference of frequencies. For example, suppose we define Q_1 and Q_2 as

$$Q_1 = \frac{1}{\sqrt{2}}\cos \omega_a t \qquad Q_2 = \frac{1}{\sqrt{2}}\cos \omega_b t$$

and then add Q_2 to Q_1

$$x_2 = \frac{1}{\sqrt{2}}(Q_1 + Q_2) = \frac{1}{2}(\cos \omega_a t + \cos \omega_b t)$$

$$= \cos\left[\frac{(\omega_a + \omega_b)}{2}t\right]\cos\left[\frac{(\omega_a - \omega_b)}{2}t\right]$$

(11.7a)

The resulting sum is equal to x_2. (We inserted $1/\sqrt{2}$ for normalization purposes, a point to which we will later return.) Figure 11.5(b) is, in fact, a plot of this function for x_2. Note, at $t = 0$, it satisfies the condition that $x_2 = 1$.

Now suppose we subtract Q_2 from Q_1

$$x_1 = \frac{1}{\sqrt{2}}(Q_1 - Q_2) = \frac{1}{2}(\cos \omega_a t - \cos \omega_b t)$$

$$= \sin\left[\frac{(\omega_a + \omega_b)}{2}t\right]\sin\left[\frac{(\omega_a - \omega_b)}{2}t\right]$$

(11.7b)

In this case, the resulting difference is x_1, and it is plotted in Figure 11.5(a).

The fascinating thing here is that, although the coordinates x_1 and x_2 engage in this composite frequency dance of energy exchange, the values Q_1 and Q_2, whatever they mean, are functions only of the single frequencies ω_a and ω_b. Since x_1 and x_2 are expressible as sums and differences of Q_1 and Q_2, we can find the inverse solutions $Q_1 = Q_1(x_1, x_2)$ and $Q_2 = Q_2(x_1, x_2)$

$$Q_1 = \frac{1}{\sqrt{2}}(x_1 + x_2) \qquad Q_2 = \frac{1}{\sqrt{2}}(x_2 - x_1)$$

(11.7c)

We might think of these two, linearly related, coordinate pairs as equally valid sets that could be used to express the configuration of the system at any instant of time. Shown in Figure 11.6a is a plot of the motion of the two-mass system, with x_1 and x_2 picked as orthogonal coordinate axes. The duration of the plot is again one full energy exchange cycle. Each point on the plot represents the configuration (x_1, x_2) of the system at any point during its motion. Notice that the configuration point traces out trajectories that are confined to a region of space defined by a box whose boundaries make 45° lines

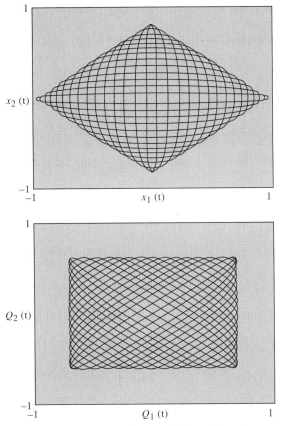

Figure 11.6 Motion of two coupled oscillators in configuration space. Coordinate axes are (a) the displacements x_1, x_2 of each mass from equilibrium (b) the linear sum $Q = 1/\sqrt{2}\,(x_1 + x_2)$ and $Q_2 = 1/\sqrt{2}\,(x_2 - x_1)$.

with the axes. This suggests that, if we were to rotate this coordinate system by 45°, the resultant configuration points would be confined to a box whose boundaries would be parallel to the axes of the rotated frame of reference. Let us perform such a rotation and see what happens.

$$\begin{bmatrix} x_1' \\ x_2' \end{bmatrix} = \begin{bmatrix} \cos 45° & \sin 45° \\ -\sin 45° & \cos 45° \end{bmatrix} \begin{bmatrix} x_1 \\ x_2 \end{bmatrix} = \begin{bmatrix} \dfrac{1}{\sqrt{2}} & \dfrac{1}{\sqrt{2}} \\ -\dfrac{1}{\sqrt{2}} & \dfrac{1}{\sqrt{2}} \end{bmatrix} \begin{bmatrix} x_1 \\ x_2 \end{bmatrix}$$

(11.7d)

$$\therefore \begin{bmatrix} Q_1 \\ Q_2 \end{bmatrix} = \begin{bmatrix} x_1' \\ x_2' \end{bmatrix}$$

Thus, Q_1 and Q_2 are the components of the double-mass vibrating system in a coordinate system rotated 45° with respect to the (x_1, x_2) coordinate system. Figure 11.6(b) is a plot of the motion of the system expressed in these rotated coordinates. The system configuration points are now confined to a box whose boundaries are parallel to the rotated axes.

Armed with this discussion regarding the behavior of the system, let us now solve its equations of motion. The kinetic energy of the system is

$$T = \frac{1}{2}m\dot{x}_1^2 + \frac{1}{2}m\dot{x}_2^2$$

and the potential energy is

$$V = \frac{1}{2}Kx_1^2 + \frac{1}{2}K'(x_2 - x_1)^2 + \frac{1}{2}Kx_2^2$$

Hence, the Lagrangian function L is given by

$$L = \frac{1}{2}m\dot{x}_1^2 + \frac{1}{2}m\dot{x}_2^2 - \frac{1}{2}Kx_1^2 - \frac{1}{2}K'(x_2 - x_1)^2 - \frac{1}{2}Kx_2^2 \tag{11.8}$$

Lagrange's equations

$$\frac{d}{dt}\frac{\partial L}{\partial \dot{x}_1} - \frac{\partial L}{\partial x_1} = 0 \qquad \frac{d}{dt}\frac{\partial L}{\partial \dot{x}_2} - \frac{\partial L}{\partial x_2} = 0$$

then read

$$m\ddot{x}_1 + Kx_1 - K'(x_2 - x_1) = 0 \tag{11.9}$$
$$m\ddot{x}_2 + Kx_2 + K'(x_2 - x_1) = 0$$

As anticipated, the resulting equations of motion are coupled.

We could now solve for x_1 and x_2, but the process would be moderately complicated. Instead, let us pursue the interpretation that x_1 and x_2 can be viewed as coordinates of a single vector whose endpoint in a two-dimensional space represents the instantaneous configuration of the system. We show such a vector in Figure 11.7. As time goes on, the endpoint of this vector traces out a path in configuration space whose coordinates are given by $x_1(t)$ and $x_2(t)$, the solutions to the above equations of motion. However, as per our discussion above, the motion in this space can be described equally well in a Q_1, Q_2 frame of reference rotated 45° with respect to that of the x_1, x_2 system.

Let us write the equations of motion in the rotated frame of reference. We will start by writing the original equations (Equations 11.9) in matrix form

$$\begin{bmatrix} \left(D^2 + \dfrac{K + K'}{m}\right) & -\dfrac{K'}{m} \\[2ex] -\dfrac{K'}{m} & \left(D^2 + \dfrac{K + K'}{m}\right) \end{bmatrix} \begin{bmatrix} x_1 \\ x_2 \end{bmatrix} = 0 \tag{11.10}$$

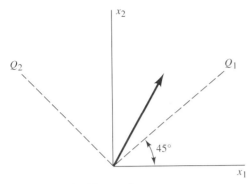

Figure 11.7 Vector whose components represent the displacements of two coupled oscillators. The Q_1, Q_2 coordinates are obtained by rotating the x_1, x_2 coordinates by 45°.

where we have used the operator notation $D^2 x = \ddot{x}$. We can express the column vector above, denoted by coordinates x_1 and x_2, in terms of coordinates Q_1 and Q_2 by inverting the rotation matrix in Equation 11.7d. The required inverse can be obtained by replacing the 45° rotation angle with $-45°$ and switching the x_1, x_2 and Q_1, Q_2 coordinates in Equation 11.7d. We can then substitute the vector, now expressed in terms of Q_1 and Q_2, into Equation 11.10 obtaining

$$\left[\begin{array}{cc} \left(D^2 + \dfrac{K + K'}{m}\right) & -\dfrac{K'}{m} \\ -\dfrac{K'}{m} & \left(D^2 + \dfrac{K + K'}{m}\right) \end{array}\right] \left(\frac{1}{\sqrt{2}}\begin{bmatrix} 1 & -1 \\ 1 & 1 \end{bmatrix}\begin{bmatrix} Q_1 \\ Q_2 \end{bmatrix}\right) = 0 \quad (11.10a)$$

Neglecting the multiplicative factor $1/\sqrt{2}$, we now multiply the two square matrices in Equation 11.10a to obtain

$$\left[\begin{array}{cc} \left(D^2 + \dfrac{K}{m}\right) & -\left(D^2 + \dfrac{K + 2K'}{m}\right) \\ \left(D^2 + \dfrac{K}{m}\right) & \left(D^2 + \dfrac{K + 2K'}{m}\right) \end{array}\right]\begin{bmatrix} Q_1 \\ Q_2 \end{bmatrix} = 0 \qquad (11.11)$$

We can carry out the above matrix operation to generate the equations of motion in terms of the coordinates Q_1, Q_2. The resulting equations are no longer coupled as they were in Equation 11.9. Each of the two equations of motion represented by Equation 11.11 are identically zero only when the two separate equations

$$(D^2 + \omega_a^2)Q_1 = 0 \qquad (D^2 + \omega_b^2)Q_2 = 0 \qquad (11.12)$$

are identically zero, where

$$\omega_a = \left(\frac{K}{m}\right)^{1/2} \qquad \omega_b = \left(\frac{K + 2K'}{m}\right)^{1/2} \tag{11.13}$$

We have seen these equations before in Chapter 3. They are the differential equations of motion for the simple harmonic oscillator whose solutions are

$$Q_1 = a \cos(\omega_a t) + a' \sin(\omega_a t) \tag{11.14}$$

$$Q_2 = b \cos(\omega_b t) + b' \sin(\omega_b t)$$

where a, a' and b, b' are constants of integration, that can be determined from the initial conditions of the problem. A solution in terms of x_1 and x_2 can be obtained by transforming back to those coordinates using Equation 11.7d

$$x_1 = \frac{1}{\sqrt{2}}(Q_1 - Q_2)$$

$$= A \cos(\omega_a t) + A' \sin(\omega_a t) - B \cos(\omega_b t) - B' \sin(\omega_b t) \tag{11.15}$$

$$x_2 = \frac{1}{\sqrt{2}}(Q_1 - Q_2)$$

$$= A \cos(\omega_a t) + A' \sin(\omega_a t) + B \cos(\omega_b t) + B' \sin(\omega_b t)$$

For the sake of simplicity, the factor $1/\sqrt{2}$ has been absorbed into the constants A, A' and B, B'.

The coordinates x_1 and x_2 represent the physical displacement of each mass from its respective equilibrium position. As such, they represent the "real" solution to the problem. The linear combinations Q_1 and Q_2 are called *normal coordinates*. Each oscillates only at a single frequency. The real displacements of any complex system of coupled oscillators can always be expressed as linear sums of normal coordinates. Any coupled oscillator problem is always much more tractable if the equations of motion are expressed in terms of their normal coordinates. There are many methods for finding normal coordinates. The reader is encouraged to pursue them in more advanced texts dealing with the subject of linear vector spaces.[2] Sometimes, with fairly simple problems, it is easy to guess the normal coordinates. The serendipity or futility of the guess becomes quickly apparent during problem solution.

Initial Conditions

Let us now proceed to solve the specific problem that initiated this discussion, namely, mass m_2 is initially displaced one unit to the right, mass m_1 is held at $x_1 = 0$, and they both are simultaneously released from rest. These are the specific *initial conditions* for this problem. We will first derive relations for the constants in Equation 11.15 in terms of any set of general initial conditions and then we will invoke the ones above in order

---

[2] For example, see J. Matthews and R. L. Walker, *Mathematical Methods in Physics*, W. A. Benjamin, New York, 1970.

to obtain the specific solution to our problem. Thus, at time $t = 0$, we have

$$x_1(0) = A - B \qquad x_2(0) = A + B \qquad (11.16)$$

Upon differentiating Equation 11.15 with respect to time and evaluating the result at $t = 0$, we obtain

$$\dot{x}_1(0) = \omega_a A' - \omega_b B' \qquad \dot{x}_2(0) = \omega_a A' + \omega_b B' \qquad (11.17)$$

We can now solve for the amplitudes in Equation 11.15

$$A = \frac{1}{2}[x_1(0) + x_2(0)] \qquad B = \frac{1}{2}[x_2(0) - x_1(0)] \qquad (11.18)$$

$$A' = \frac{1}{2\omega_a}[\dot{x}_1(0) + \dot{x}_2(0)] \qquad B' = \frac{1}{2\omega_b}[\dot{x}_2(0) - \dot{x}_1(0)] \qquad (11.19)$$

Now, for our specific case, we have the initial conditions

$$x_1(0) = 0 \qquad x_2(0) = 1 \qquad \dot{x}_1(0) = \dot{x}_2(0) = 0$$

Inserting these into Equations 11.18 and 11.19 gives

$$A = B = \frac{1}{2} \qquad A' = B' = 0$$

and inserting these into Equation 11.15 gives

$$x_1(t) = \frac{1}{2}(\cos \omega_a t - \cos \omega_b t)$$

$$x_2(t) = \frac{1}{2}(\cos \omega_a t + \cos \omega_b t)$$

for the specific solution to our problem. This solution should be compared to Equations 11.7a and 11.7b initially put forth during our more general discussion of the problem.

Normal Coordinates

At this point it is well worth asking the question, What happens if we start the system off differently? In particular, suppose we start it off in a single normal *mode* of vibration. Will the system behave any differently than it did before? Obviously, it will, but what we mean is will it behave in any fundamentally different way? The answer is yes. After a system is in a normal mode, it will stay that way. It will never get into another one. The two masses will not shuttle their energies back and forth. They each will keep what they have. We can arrive at these conclusions in several ways. Let us see what the initial condition equations tell us.

Suppose, for example, that initially the two particles are pulled from their equilibrium positions by equal amounts in the same direction and then released from rest. The initial conditions are

$$x_1(0) = x_2(0) = 1 \qquad \dot{x}_1(0) = \dot{x}_2(0) = 0$$

Inserting these conditions into Equations 11.18 and 11.19 leads to the result that all constants vanish except $A = 1$. The solution for the displacements as a function of time are

$$x_1(t) = x_2(t) = \cos \omega_a t \qquad (11.20)$$
$$Q_1(t) = \sqrt{2} \cos \omega_a t \qquad Q_2(t) = 0$$

These results are plotted in Figures 11.8 and 11.9. We also show, in Figure 11.10, a plot of the motion in configuration space. It is now obvious how the motion appears much simpler in a frame of reference rotated by 45° with respect to a frame described by the displacements x_1, x_2. This mode of oscillation is called the *symmetric mode.* Note that the central spring is never compressed and that the two masses vibrate as though they were completely independent simple harmonic oscillators whose frequencies are $\omega_a = \sqrt{K/m}$. Note also that during the course of the motion, each mass passes through its equilibrium position at a time when the central, connecting spring is neither stretched nor compressed. Thus, there can be no transfer of energy from one mass to the other. If initially put into the symmetric mode, the system stays in that mode.

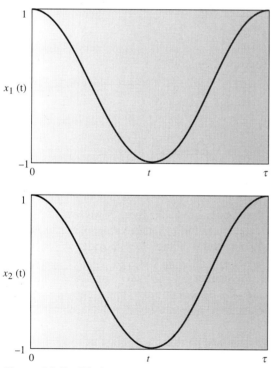

Figure 11.8 Displacement versus time for two coupled harmonic oscillators vibrating in the symmetric mode.

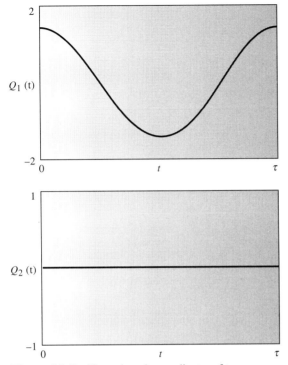

Figure 11.9 Normal mode coordinates of two coupled harmonic oscillators vibrating in the symmetric mode.

Now suppose that the two particles are pulled from their equilibrium positions by equal amounts in opposite directions and then released from rest. The initial conditions are

$$x_2(0) = -x_1(0) = 1 \qquad \dot{x}_1(0) = \dot{x}_2(0) = 0$$

These conditions imply that all constants in the solution for the displacements vanish except $B = 1$. Thus,

$$x_2(t) = -x_1(t) = \cos \omega_b t \tag{11.21}$$
$$Q_1(t) = 0 \qquad Q_2(t) = \sqrt{2} \cos \omega_b t$$

These results, as well as the motion in configuration space, are plotted in Figures 11.11, 11.12, and 11.13. This mode of oscillation is called the *antisymmetric mode* or, for obvious reasons, the "breathing" mode. The two masses again vibrate as though they were independent simple harmonic oscillators whose frequencies, in this case, are $\omega_b = \sqrt{(K + 2K')/m}$. How can this be? The central spring no longer remains flacid. Let us look at it very closely. Note that a point at the very center of the central spring never moves. It is a *nodal point* in the vibration. As far as either mass is concerned, we could

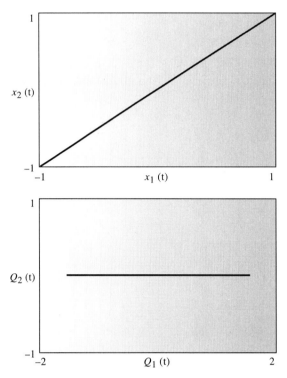

Figure 11.10 Motion of two coupled harmonic oscillators vibrating in symmetric mode, plotted in configuration space.

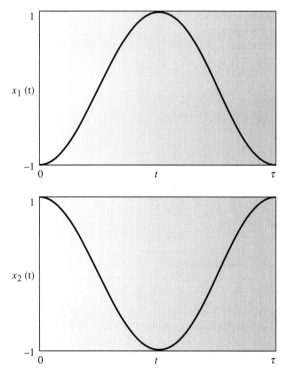

Figure 11.11 Displacement versus time for two coupled harmonic oscillators vibrating in the antisymmetric mode.

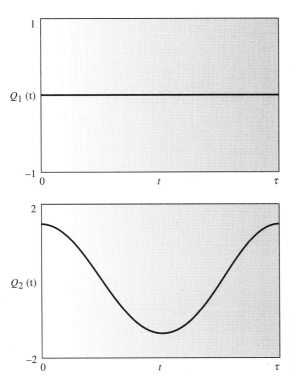

Figure 11.12 Normal mode co-ordinates of two coupled harmonic oscillators vibrating in the antisymmetric mode.

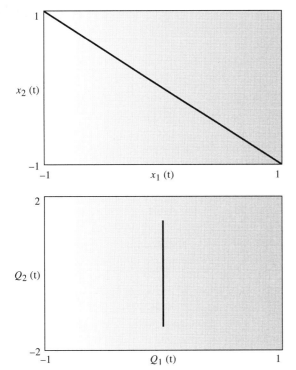

Figure 11.13 Motion of two coupled harmonic oscillators vibrating in the antisymmetric mode, plotted in configuration space.

cut the central spring in half and attach each of the resulting endpoints to a fixed, im-
mobile wall, exactly like the attachment points of the end springs. Each mass could then
be viewed as an independent harmonic oscillator attached by two springs between two
fixed walls. The resulting, effective spring constant would be $K + 2K'$ (it is left to the
student to determine that this is true). Note, again, that there is no energy transfer. Energy cannot be passed across a nodal point. Another way of looking at it is to note that
when the two masses pass through their equilibrium positions, the central spring is neither stretched nor compressed. No energy transfer is, therefore, possible. Once placed
in this antisymmetric, breathing mode, the system stays there.

It is instructive to express the Lagrangian function in terms of the normal coordinates. To do this we insert the expression for x_1, x_2 in terms of Q_1, Q_2 (Equations 11.15)
into Equation 11.8 to obtain

$$L = \frac{1}{2}m\dot{Q}_1^2 - \frac{1}{2}K_a Q_1^2 + \frac{1}{2}m\dot{Q}_2^2 - \frac{1}{2}K_b Q_2^2 \tag{11.22}$$

where K_a and K_b are the effective spring constants for the normal modes given by Equation 11.13. As a consequence, all possible cross terms of the sort $\alpha Q_1 Q_2$ and/or
$\beta \dot{Q}_1 \dot{Q}_2$ have vanished from the Lagrangian. The kinetic and potential energies have been
reduced to a sum of squares, precisely of the form that one would expect for two uncoupled, simple harmonic oscillators. Consequently, Lagrange's equations in normal
coordinates reduce to

$$\frac{d}{dt}\frac{\partial L}{\partial \dot{Q}_1} - \frac{\partial L}{\partial Q_1} = 0 \qquad \frac{d}{dt}\frac{\partial L}{\partial \dot{Q}_2} - \frac{\partial L}{\partial Q_2} = 0$$

and the resulting equations of motion are

$$m\ddot{Q}_1 + K_a Q_1 = 0 \qquad m\ddot{Q}_2 + K_b Q_2 = 0 \tag{11.23}$$

These are the differential equations of motion for two uncoupled, harmonic oscillators
identical to those previously derived as Equations 11.12 and 11.13 with solutions given
by Equation 11.14.

All this is well and good, but we have not yet presented any general, systematic way
of finding the normal coordinates. As mentioned before, sometimes one can guess the
coordinates, and if the guess is successful, the problem will be easy to solve. It is usually
worthwhile to make a guess. Even if one cannot get the exact values for the normal
coordinate components, one can usually get the form right, and the resulting mental
image should provide insight to the problem solution. As we will soon see, one can
always solve for the normal coordinate frequencies in a brute force way without knowledge of the normal coordinates. This information can be used to obtain the normal coordinates. They should agree fairly closely with those based on a guess. In the next
section, we will present an example that we will attempt to solve first by guessing the
normal coordinates. The problem will prove easy if the guess is correct. We will also
assume that our guess failed and will solve for the normal frequencies the brute force
way. Then we will find the corresponding normal coordinates and compare them with
our initial guess.

EXAMPLE 11.4 *The Double Pendulum, or Two Mountain Climbers*
Dangling on a Single Rope

Let us consider the motion of the so-called double pendulum consisting of a light inextensible cord of length $2l$ with one end fixed, the other supporting a bob (treated here as a particle) of mass m, and with a second bob, also of mass m, at the center as shown in Figure 11.14(a). Assuming that the system stays in a single plane, we can specify the configuration by two angles θ and ϕ, as shown.

For small oscillations about equilibrium, the kinetic energy of the system is given by

$$T = \frac{1}{2}m\mathbf{v}_1 \cdot \mathbf{v}_1 + \frac{1}{2}m\mathbf{v}_2 \cdot \mathbf{v}_2$$

We can express both $\mathbf{v}_1$ and $\mathbf{v}_2$ in terms of the angular velocities $\dot{\theta}$ and $\dot{\phi}$

$$\mathbf{v}_1 = \mathbf{e}_\theta l \dot{\theta} \qquad \mathbf{v}_2 = \mathbf{e}_\theta l \dot{\theta} + \mathbf{e}_\phi l \dot{\phi}$$

Check the validity of the last expression by noting that it gives the correct result for $\mathbf{v}_2$ when $\dot{\theta} = 0$ (pendulum 1 is fixed and pendulum 2 swings along an arc of radius l) or $\dot{\theta} = \dot{\phi}$ (pendula 1 and 2 swing together as though they were one long pendulum of length $2l$). The kinetic energy is, thus,

$$T = \frac{1}{2}m(l\dot{\theta})^2 + \frac{1}{2}m(\mathbf{e}_\theta(l\dot{\theta}) + \mathbf{e}_\phi(l\dot{\phi})) \cdot (\mathbf{e}_\theta(l\dot{\theta}) + \mathbf{e}_\phi(l\dot{\phi}))$$

$$\approx \frac{1}{2}m(l\dot{\theta})^2 + \frac{1}{2}m((l(\dot{\theta} + \dot{\phi}))^2$$

In carrying out the last step above, we have used the fact that the unit vectors $\mathbf{e}_\theta$ and

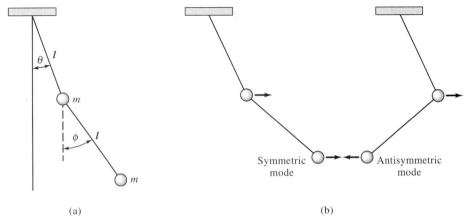

Symmetric mode   Antisymmetric mode

(a)   (b)

Figure 11.14   The double pendulum and its normal modes.

e_ϕ are essentially parallel if all angular displacements are assumed to be small. Thus, their dot product is one.

The sum of the potential energy of each mass relative to its equilibrium position is given by

$$V = mgl(1 - \cos \theta) + mgl(2 - (\cos \theta + \cos \phi))$$

We can use the small angle Taylor's expansion of the cosine function ($\cos \beta \approx 1 - \beta^2/2$) for both $\cos \theta$ and $\cos \phi$ to obtain the lowest order approximation of the potential energy

$$V \approx 2mgl\frac{\theta^2}{2} + mgl\frac{\phi^2}{2}$$

Lagrange's equations

$$\frac{d}{dt}\frac{\partial L}{\partial \dot\theta} - \frac{\partial L}{\partial \theta} = 0 \qquad \frac{d}{dt}\frac{\partial L}{\partial \dot\phi} - \frac{\partial L}{\partial \phi} = 0$$

generate the following equations of motion in θ and ϕ:

$$2\ddot\theta + 2\frac{g}{l}\theta + \ddot\phi = 0 \tag{11.24}$$

$$\ddot\theta + \ddot\phi + \frac{g}{l}\phi = 0$$

which is equivalent to the matrix equation

$$\begin{bmatrix} \left(2D^2 + 2\frac{g}{l}\right) & D^2 \\[2ex] D^2 & \left(D^2 + \frac{g}{l}\right) \end{bmatrix} \begin{bmatrix} \theta \\ \phi \end{bmatrix} = 0 \tag{11.24a}$$

At this point let us take a crack at guessing the normal coordinates. They should look something like the symmetric and antisymmetric modes pictured in Figure 11.14(b). The real question is what are the actual values of the θ and ϕ amplitudes for each mode; that is, if $\theta = 1$, what is ϕ? The answer to this question is not immediately obvious. One thing should be obvious, though. In each normal mode, the two masses ought to swing through equilibrium together. Recall our earlier discussion concerning the two coupled harmonic oscillators. Each passed through their respective equilibrium positions together. This is the only way the shuttling of energy back and forth between the two masses could be avoided. The same is true here. If we assume that we release this system from rest, initially in its correct normal mode configuration, both θ and ϕ ought to go to zero simultaneously. How do we ensure that this happens?

Let us use the previous example of the two coupled oscillators as a guide. We started them off in a normal mode by initially storing equal amounts of potential energy in each coordinate x_1 and x_2. This was true for each normal mode. We will assume that this is what we have to do here, too. In other words, we guess that we have to divvy out the initial potential energy in equal portions between the two coordinates θ and ϕ to ensure that they go to zero at the same time. The success of this guess (or lack thereof) will soon become apparent. Examining the potential energies tells us that the required initial displacement can be made by setting $\phi = \pm\sqrt{2}\theta$. This starts the system off with energies that are *equipartitioned*. We, therefore, guess that the two normal modes are given by

$$Q_1 = \frac{1}{\sqrt{3}}(\sqrt{2}\theta + \phi) \qquad Q_2 = \frac{1}{\sqrt{3}}(\sqrt{2}\theta - \phi) \qquad (11.25)$$

The $\sqrt{3}$ factor "normalizes" the normal mode vectors in θ, ϕ configuration space. (Its inclusion is not critical unless you want a description of the motion in terms of a set of *orthonormal coordinates* that span configuration space.) Q_1 and Q_2 form an orthonormal coordinate system that is rotated through an angle $\alpha = \cos^{-1}\sqrt{2/3} \approx 35°$ with respect to the θ, ϕ coordinate system. A picture of configuration space for this example is essentially the same as that of Figure 11.7 except for the difference in angles of rotation and the axis labels. The x_1, x_2 labels are to be replaced by θ, ϕ labels.

We can now solve the problem by inverting Equation 11.25 to obtain

$$\theta = \frac{\sqrt{3}}{2}\frac{1}{\sqrt{2}}(Q_1 + Q_2) \qquad \phi = \frac{\sqrt{3}}{2}(Q_1 - Q_2) \qquad (11.26)$$

and inserting the resulting solution of θ, ϕ in terms of Q_1, Q_2 into Equation 11.24a

$$\left[\begin{array}{cc} \left(2D^2 + 2\frac{g}{l}\right) & D^2 \\ \\ D^2 & \left(D^2 + \frac{g}{l}\right) \end{array} \right] \left[\begin{array}{cc} \frac{1}{\sqrt{2}} & \frac{1}{\sqrt{2}} \\ \\ 1 & -1 \end{array} \right] \left[\begin{array}{c} Q_1 \\ \\ Q_2 \end{array} \right] = 0$$

The factor $\sqrt{3}/2$ has been dropped in this last equation. Multiplying the two matrices together leads to the equations.

$$\left[\left(\frac{2}{\sqrt{2}} + 1\right)D^2 + \frac{2}{\sqrt{2}}\frac{g}{l}\right]Q_1 + \left[\left(\frac{2}{\sqrt{2}} - 1\right)D^2 + \frac{2}{\sqrt{2}}\frac{g}{l}\right]Q_2 = 0$$

$$\left[\left(\frac{1}{\sqrt{2}} + 1\right)D^2 + \frac{g}{l}\right]Q_1 - \left[\left(1 - \frac{1}{\sqrt{2}}\right)D^2 + \frac{g}{l}\right]Q_2 = 0$$

These two equations are identical, with the exception of a sign difference for the coefficients of Q_2. (This can be seen by multiplying the bottom equation by $2/\sqrt{2}$.) As before, the only way they can be satisfied is if the coefficients for Q_1 and Q_2 are identically zero. Thus, with a little algebra we get

$$\left[D^2 + (2 - \sqrt{2})\frac{g}{l}\right]Q_1 = 0 \qquad \left[D^2 + (2 + \sqrt{2})\frac{g}{l}\right]Q_2 = 0 \qquad (11.27)$$

The solution to this equation is the same as that given by Equation 11.14. In this case the normal mode frequencies are given by

$$
\omega_a = \left[(2 - \sqrt{2}) \frac{g}{l} \right]^{1/2} \quad symmetric \ mode
$$

$$
\omega_b = \left[(2 + \sqrt{2}) \frac{g}{l} \right]^{1/2} \quad antisymmetric \ mode
$$

(11.27a)

The normal modes are represented in Figure 11.14(b). The ratio of the two normal mode frequencies is independent of the length l and is given by

$$
\frac{\omega_a}{\omega_b} = \left[\frac{(2 + \sqrt{2})}{(2 - \sqrt{2})} \right]^{1/2} = 2.414
$$

so the oscillation in the fast, antisymmetric mode is about two and a half times the frequency of the slow, symmetric mode.

Now, suppose we could not guess the normal modes, or the guess that we did make was wrong, that is, it failed to eliminate the coupling between the resulting differential equations of motion. What to do? Go back to the fundamental equations of motion (Equation 11.24). Everything done up to that point was okay. Let us proceed from there. Let us make the assumption that a solution exists whose form is

$$
\theta = A_1 \cos \omega t \quad \phi = A_2 \cos \omega t
$$

We know from experience that a single frequency solution will not suffice. But that is okay. Our solution will actually yield two frequencies, the normal mode frequencies in fact, and the final solution will simply be a linear superposition of the two normal modes with their respective frequencies as arguments of the resulting cosine (and/or sine) functions.

Plugging these assumed solutions into the equations of motion yields

$$
\begin{bmatrix} \left(-2\omega^2 + 2\frac{g}{l} \right) & -\omega^2 \\ -\omega^2 & \left(-\omega^2 + \frac{g}{l} \right) \end{bmatrix} \begin{bmatrix} \theta \\ \phi \end{bmatrix} = 0
$$

(11.27b)

This equation is essentially the same one as the operator equation (Equation 11.24a) except that the factor $-\omega^2$ has replaced the double time derivative operator D^2, (it was generated by the operation of D^2 upon the assumed solution). The operator equation has, thus, been converted into an algebraic equation involving the frequency ω. A solution exists only if the determinant of the above matrix is zero. Thus, we have

$$
\omega^4 - 4\omega^2 \left(\frac{g}{l} \right) + 2 \left(\frac{g}{l} \right)^2 = 0
$$

The solution to this quadratic (in ω^2) equation yields the normal mode frequencies ω_a and ω_b, already obtained in Equation 11.27a. We can now get the normal modes by inserting these frequencies, one by one, into either of the equations generated by the matrix operation of Equation 11.27b. For example, the first equation reads

$$\left(-2\omega^2 + 2\frac{g}{l} \right)\theta = \omega^2\phi$$

Inserting the two normal mode frequencies ω_a and ω_b into this equation yields the following conditions on the normal modes

$$\phi = +\sqrt{2}\theta \qquad \omega = \omega_a \qquad symmetric\ mode$$
$$\phi = -\sqrt{2}\theta \qquad \omega = \omega_b \qquad antisymmetric\ mode$$

which are precisely the conditions originally guessed.

It is left as a problem to show that the kinetic and potential energies reduce to sums of squares when expressed in terms of the normal coordinates Q_1 and Q_2 instead of θ and ϕ. ∎

11.4 GENERAL THEORY OF VIBRATING SYSTEMS

Turning now to a general system with n degrees of freedom, we have shown in Section 10.3 that the kinetic energy T is a homogeneous quadratic function of the generalized velocities, namely,

$$T = \frac{1}{2}M_{11}\dot{q}_1^2 + M_{12}\dot{q}_1\dot{q}_2 + \frac{1}{2}M_{22}\dot{q}_2^2 + \cdots = \sum_j \sum_k \frac{1}{2}M_{jk}\dot{q}_j\dot{q}_k$$

provided there are no moving constraints. Since we are concerned with motion about an equilibrium configuration, we shall assume, as in Section 11.2, that the M's are constant and equal to their values at the equilibrium configuration. We shall further assume that the coordinate origins have been chosen such that the equilibrium configuration is given by

$$q_1 = q_2 = \cdots = q_n = 0$$

Accordingly, the potential energy V, from Equation 11.4, is given by

$$V = \frac{1}{2}K_{11}q_1^2 + K_{12}q_1q_2 + \frac{1}{2}K_{22}q_2^2 + \cdots = \sum_j \sum_k \frac{1}{2}K_{jk}q_jq_k$$

The Lagrangian function then assumes the form

$$L = \sum_j \sum_k \frac{1}{2}(M_{jk}\dot{q}_j\dot{q}_k - K_{jk}q_jq_k) \tag{11.28}$$

and the equations of motion

$$\frac{d}{dt}\frac{\partial L}{\partial \dot{q}_k} - \frac{\partial L}{\partial q_k} = 0$$

then read

$$\sum_j (M_{jk}\ddot{q}_j + K_{jk}q_j) = 0 \qquad (k = 1, 2, \ldots, n) \qquad (11.29)$$

If a solution of the form

$$q_k = A_k \cos \omega t \qquad (k = 1, 2, \ldots, n) \qquad (11.30)$$

exists, then by direct substitution the following equations must be satisfied:

$$\sum_j (-M_{jk}\omega^2 + K_{jk})A_j = 0 \qquad (k = 1, 2, \ldots, n) \qquad (11.31)$$

A nontrivial solution requires that the determinant of the coefficients of the A's vanish:

$$\begin{vmatrix} -M_{11}\omega^2 + K_{11} & -M_{12}\omega^2 + K_{12} & \cdots \\ -M_{21}\omega^2 + K_{21} & -M_{22}\omega^2 + K_{22} & \cdots \\ & \cdots & \end{vmatrix} = 0 \qquad (11.32)$$

The above secular equation is an equation of the nth degree in ω^2. The n roots are the squares of the normal frequencies of the system.

Thus, if a given system has n degrees of freedom there are, in general, n different possible frequencies of oscillation about the equilibrium configuration, each characterized by its own normal mode. The calculation of these normal frequencies often entails the tedious task of solving a high-order algebraic equation, cubic for $n = 3$, biquadratic for $n = 4$, and so on. There are some special situations for which the roots of the secular equation are either repeated or zero, or both. In such cases the mathematical problem of finding the normal frequencies may be simplified. An example will be given at the end of this section. In the next section we shall give a method of determining the normal frequencies for a linear array of coupled oscillators for which n may have any value.

As in the previous case of two coupled oscillators, the complete solution for n oscillators is given by a superposition of sinusoidal oscillations at the normal frequencies ω_N, namely,

$$q_k(t) = \sum_{N=1}^{n} A_{kN} \cos(\omega_N t - \phi_N) \qquad (k = 1, 2, \ldots, n)$$

The amplitudes A_{kN} are not independent but are related by the fact that for each normal frequency ω_N a set of equations of the type shown in Equation 11.31 must be satisfied. This allows us, in principle, to determine the amplitude ratios $A_{1N}:A_{2N}: \ldots :A_{nN}$ and, thus, to construct linear combinations

$$c_{1N}q_1 + c_{2N}q_2 + \cdots + c_{nN}q_n = Q_N = a_N \cos(\omega_N - \phi_N)$$
$$(N = 1, 2, \ldots, n)$$

defining the normal coordinates Q_N of the system as we did above. For $n > 2$ this is usually not the most efficient way to solve the problem. If you think that you can guess the normal coordinates Q_N, do so. The differential equations of motion will immediately "uncouple" when expressed in terms of the Q_N. Failing that, the coefficients c_{kN} that transform the problem's generalized coordinates to the normal coordinates are best found using the methods of matrix algebra. This involves diagonalizing the matrix of the determinental equation 11.32. We shall not pursue this point any further (See Appendix H). The interested reader is referred to texts on linear vector spaces.[3]

EXAMPLE 11.5 *Linear Motion of a Triatomic Molecule*

Let us consider the motion of a three particle system in which all the particles lie in a straight line. An example of such a collinear system is the carbon dioxide molecule CO_2, which has the structure O–C–O. We shall consider motion only in one dimension, along the x-axis (Figure 11.15). The two end particles, each of mass m, are bound to the central particle, mass M, via a potential function that is equivalent to that of two springs of stiffness K, as shown in Figure 11.15. The coordinates expressing the displacements of each mass are x_1, x_2, and x_3.

Solution:
In this problem we can easily guess the normal modes. They are pictured in Figures 11.15(a)–11.15(c). If you think about it a little while you should realize that what's going on here is that the center of mass of the molecule is not accelerating. In mode (c) the central mass is vibrating 180° out of phase with the two end masses. The ratio of the vibrational amplitudes is such that the center of mass remains at rest.

---

[3] See footnote 2.

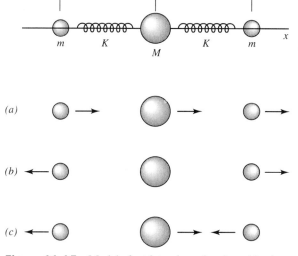

Figure 11.15 Model of a triatomic molecule and its three normal modes for motion in a single line.

Mode (b) obeys the same condition. The central mass remains at rest while the two equal end masses vibrate 180° out of phase with each other, with equal amplitudes, again fixing the center of mass. Mode (a) depicts overall translation of the center of mass at constant velocity.

We could go on and solve the problem using this guess. However, we will not. We will solve it using the general method introduced in the last example, in which we assume that the normal modes are not known. We will ultimately generate a secular equation that, in this example, will be of third order in ω^2. (There are three coordinates, hence three normal modes and frequencies in the solution.) It turns out that this particular third-order equation will be very easy to solve. Upon obtaining the frequencies of each normal mode, we will then insert them into any one of the equations relating the amplitudes of the displacement coordinates to each other (the matrix equivalent of the secular equation in ω^2), thus obtaining the normal modes.

The Lagrangian of the system is

$$L = T - V$$
$$= \left(\frac{m}{2}\dot{x}_1^2 + \frac{M}{2}\dot{x}_2^2 + \frac{m}{2}\dot{x}_3^2\right) - \left[\frac{K}{2}(x_2 - x_1)^2 + \frac{K}{2}(x_3 - x_2)^2\right]$$

and Lagrange's three equations of motion read

$$\begin{aligned}
m\ddot{x}_1 + Kx_1 \quad\quad - Kx_2 \quad\quad\quad\quad\quad &= 0 \\
- Kx_1 + M\ddot{x}_2 + 2Kx_2 \quad\quad - Kx_3 &= 0 \quad\quad (11.33)\\
- Kx_2 \quad\quad + m\ddot{x}_3 + Kx_3 &= 0
\end{aligned}$$

If a solution of the form $x_1 = A_1 \cos \omega t$, $x_2 = A_2 \cos \omega t$, $x_3 = A_3 \cos \omega t$ exists, then

$$\begin{aligned}
(- m\omega^2 + K)A_1 \quad\quad - KA_2 \quad\quad\quad\quad\quad &= 0 \\
- KA_1 + (-M\omega^2 + 2K)A_2 \quad\quad - KA_3 &= 0 \quad (11.34)\\
- KA_2 \quad\quad + (-m\omega^2 + K)A_3 &= 0
\end{aligned}$$

The secular equation is, thus,

$$\begin{vmatrix} - m\omega^2 + K & - K & 0 \\ - K & -M\omega^2 + 2K & - K \\ 0 & - K & -m\omega^2 + K \end{vmatrix} = 0 \quad\quad (11.35)$$

which, upon expanding the determinant and collecting terms, fortuitously becomes the product of three factors

$$\omega^2(-m\omega^2 + K)(-mM\omega^2 + KM + 2Km) = 0$$

Equating each of the three factors to zero gives the three normal frequencies of the system:

$$\omega_a = 0 \quad\quad \omega_b = \left(\frac{K}{m}\right)^{1/2} \quad\quad \omega_c = \left(\frac{K}{m} + 2\frac{K}{M}\right)^{1/2}$$

Let us discuss the modes corresponding to these three roots.

(a) The first mode is no oscillation at all but is *pure translation* of the system as a whole. If we set $\omega = 0$ in Equations 11.34, we find that $A_1 = A_2 = A_3$ for this mode.

(b) Setting $\omega = \omega_b$ in Equations 11.34 gives $A_2 = 0$ and $A_1 = -A_3$. In this mode the center particle is at rest while the two end particles vibrate in opposite directions (antisymmetrically) with the same amplitude.

(c) Finally, setting $\omega = \omega_c$ in Equations 11.34 we obtain the following relations: $A_1 = A_3$ and $A_2 = -2A_1(m/M) = -2A_3(m/M)$. Thus, in this mode the two end particles vibrate in unison while the center particle vibrates oppositely with a different amplitude. The three modes are illustrated in Figure 11.15.

It is interesting to note that the ratio ω_c/ω_b is independent of the constant K, namely,

$$\frac{\omega_c}{\omega_b} = \left(1 + 2\frac{m}{M}\right)^{1/2}$$

In the carbon dioxide molecule the mass ratio m/M is very nearly $16:12$ for ordinary CO_2 (C_{12} and O_{16} atoms). Thus, the frequency ratio

$$\frac{\omega_c}{\omega_b} = \left(1 + 2 \times \frac{16}{12}\right)^{1/2} = \left(\frac{11}{3}\right)^{1/2} = 1.915$$

∎

11.5 VIBRATION OF A LOADED STRING OR LINEAR ARRAY OF COUPLED HARMONIC OSCILLATORS

Any real, solid system contains many particles, each bound to a small region of space by atomic potentials, not just two or three particles coupled together by springs. However, the binding potential "felt" by each particle is well represented by a quadratic function of the difference between each particle's displacement from its equilibrium position and the corresponding displacement of its immediate neighbor. Thus, such a system is essentially one of many coupled oscillators. Its analysis can lead to a description of the oscillations of a continuous medium, the propagation of waves through a continuous medium, or the vibrations of a crystalline lattice. In this section, we will take the first step toward arming you with the theoretical weaponry necessary to attack such problems. We will consider the motion of a simple mechanical system consisting of a light elastic string, clamped at both ends and loaded with n particles, each of mass m, equally spaced along the length of the string. However, before proceeding with the analysis, we will make a brief historical digression on this subject.[4]

An analysis of the dynamics of a line of interconnected masses was first attempted by Newton, himself. Two of his successors, the remarkable Bernoullis (John and his son Daniel), were the ones who had ultimate success with the problem. They demonstrated

---

[4]See, for example, A. P. French, *Vibrations and Waves*, The MIT Introductory Physics Series, Norton, New York, 1971.

that a system of n masses has exactly n independent modes (for one-dimensional motion only). In 1753 Daniel Bernoulli demonstrated that the general motion of this vibrating system is describable as a superposition of its normal modes. According to Leon Brillouin, a major contributor to the theory of vibrations of a crystalline lattice[5]:

> *This investigation by the Bernoullis may be said to form the beginning of theoretical physics as distinct from mechanics, in the sense that it is the first attempt to formulate laws for the motion of a system of particles rather than for that of a single particle. The principle of superposition is important, as it is a special case of a Fourier series, and in time it was extended to become a statement of Fourier's theorem.*

Strong words, these! Let us now begin the analysis.

Let us label the displacements of the various particles from their equilibrium positions by the coordinates $q_1, q_2, \ldots, q_n$. Actually, there are two types of displacement that can occur, namely, a longitudinal displacement in which the particle moves along the direction of the string, and a transverse displacement in which the particle moves at right angles to the length of the string. These are illustrated in Figure 11.16. For simplicity we shall assume that the motion is either purely longitudinal or purely transverse,

---

[5]L. Brillouin, *Wave Propagation in Periodic Structures,* Dover, New York, 1953.

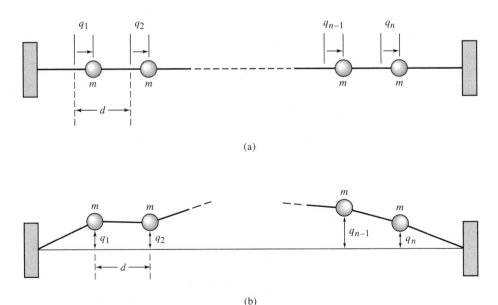

(a)

(b)

Figure 11.16   Linear array of vibrating particles or the loaded string. (a) Longitudinal motion. (b) Transverse motion.

although in the actual physical situation a combination of the two could occur. The kinetic energy of the system is then given by

$$T = \frac{m}{2}(\dot{q}_1^2 + \dot{q}_2^2 + \cdots + \dot{q}_n^2)$$

If we use the letter k to denote any given particle, then, in the case of longitudinal motion, the stretch of the section of string between particle k and particle $k + 1$ is

$$q_{k+1} - q_k$$

Hence, the potential energy of this section of the string is

$$\frac{1}{2}K(q_{k+1} - q_k)^2$$

in which K is the elastic stiffness coefficient of the section of string connecting the two adjacent particles.

For the case of transverse motion, the distance between particle k and $k + 1$ is

$$[d^2 + (q_{k+1} - q_k)^2]^{1/2} = d + \frac{1}{2d}(q_{k+1} - q_k)^2 + \cdots$$

in which d is the equilibrium distance between two adjacent particles. The stretch of the section of string connecting the two particles is then approximately

$$\Delta l = \frac{1}{2d}(q_{k+1} - q_k)^2$$

Thus, if T is the tension in the string, the potential energy of the section under consideration is given by

$$T\Delta l = \frac{T}{2d}(q_{k+1} - q_k)^2$$

It follows that the total potential energy of the system in either the longitudinal or the transverse type of motion is expressible as a quadratic function of the form

$$V = \frac{K}{2}[q_1^2 + (q_2 - q_1)^2 + \cdots + (q_n - q_{n-1})^2 + q_n^2]$$

in which

$$K = \frac{T}{d} = \frac{\text{tension}}{\text{separation}} \qquad \textit{transverse vibration}$$

or

$$K = \text{elastic constant} \qquad \textit{longitudinal vibration}$$

The Lagrangian function for either case is, thus, given by

$$L = \frac{1}{2} \sum_k [m\dot{q}_k^2 - K(q_{k+1} - q_k)^2] \qquad (11.36)$$

The Lagrangian equations of motion

$$\frac{d}{dt}\frac{\partial L}{\partial \dot{q}_k} = \frac{\partial L}{\partial q_k}$$

then become

$$m\ddot{q}_k = -K(q_k - q_{k-1}) + K(q_{k+1} - q_k) \tag{11.37}$$

where $k = 1, 2, \ldots, n$.

To solve the above system of n equations, we use a trial solution in which the q's are assumed to vary harmonically with time:

$$q_k = a_k \cos(\omega t)$$

where a_k is the amplitude of vibration of the kth particle. Substitution of the above trial solution into the differential equations (Equations 11.37) yields the following recursion formula for the amplitudes:

$$-m\omega^2 a_k = K(a_{k-1} - 2a_k + a_{k+1}) \tag{11.38}$$

This formula will include the endpoints of the string if we set

$$a_0 = a_{n+1} = 0$$

The secular determinant is, thus,

$$\begin{vmatrix} -m\omega^2 + 2K & -K & 0 & \cdots & 0 \\ -K & -m\omega^2 + 2K & -K & \cdots & 0 \\ 0 & -K & -m\omega^2 + 2K & \cdots & 0 \\ \cdots & \cdots & \cdots & \cdots & \cdots \\ 0 & 0 & 0 & \cdots & -m\omega^2 + 2K \end{vmatrix} = 0$$

The determinant is of the nth order, and there are thus n values of ω that satisfy the equation. However, rather than find these n roots by algebra, it turns out that we can find them by working directly with the recursion relation (Equation 11.38).

To this end we define a quantity ϕ related to the amplitudes a_k by the following equation:

$$a_k = A \sin(k\phi) \tag{11.39}$$

Direct substitution into the recursion formula then yields

$$-m\omega^2 A \sin(k\phi) = KA[\sin(k\phi - \phi) - 2\sin(k\phi) + \sin(k\phi + \phi)]$$

which easily reduces to

$$m\omega^2 = K(2 - 2\cos\phi) = 4K\sin^2\frac{\phi}{2}$$

or

$$\omega = 2\omega_0 \sin\frac{\phi}{2} \tag{11.40}$$

in which

$$\omega_0 = \left(\frac{K}{m}\right)^{1/2} \tag{11.41}$$

Equation 11.40 gives the normal frequencies in terms of the quantity ϕ, which we have not, as yet, determined. Now, as a matter of fact, the same relation would have been obtained by any of the following substitutions for the amplitude a_k: $A \cos (k\phi)$, $A e^{ik\phi}$, $A e^{-ik\phi}$, or any linear combination of these. However, only the substitution $a_k = A \sin (k\phi)$ satisfies the end condition $a_0 = 0$. In order to determine the actual value of the parameter ϕ, and thus find the normal frequencies of the vibrating string, we use the other end condition, namely, $a_{n+1} = 0$. This condition will be met if we set

$$(n + 1)\phi = N\pi \tag{11.42}$$

in which N is an integer, because we then have

$$a_{n+1} = A \sin(N\pi) = 0$$

Having found ϕ, we can now calculate the normal frequencies. They are given by

$$\omega_N = 2\omega_0 \sin \left(\frac{N\pi}{2n + 2}\right) \tag{11.43}$$

Furthermore, from Equations 11.39 and 11.42 we see that the amplitudes for the normal modes are given by

$$a_k = A \sin \left(\frac{N\pi k}{n + 1}\right) \tag{11.44}$$

Here the value of $k = 1, 2, \ldots, n$ denotes a particular particle in the linear array, and the value of $N = 1, 2, \ldots, n$ refers to the normal mode in which the system is oscillating.

The different normal modes are illustrated graphically by plotting the amplitudes as given by Equation 11.44. These fall on a sine curve as shown in Figure 11.17, which shows the case of three particles $n = 3$. The actual motion of the system, when it is vibrating in a single pure mode, is given by the equation

$$q_k = a_k \cos (\omega_N t) = A \sin \left(\frac{\pi N k}{n + 1}\right) \cos (\omega_N t) \tag{11.45}$$

The general type of motion is a linear combination of all the normal modes. This can be expressed as

$$q_k = \sum_{N=1}^{n} A_N \sin \left(\frac{N\pi k}{n + 1}\right) \cos (\omega_N t - \epsilon_N) \tag{11.46}$$

in which the values of A_N and ϵ_N are determined from the initial conditions.

Suppose we look at the case where the number of masses on the string is very large. A real string is an aggregate of a very large number of very closely spaced atoms. Let n

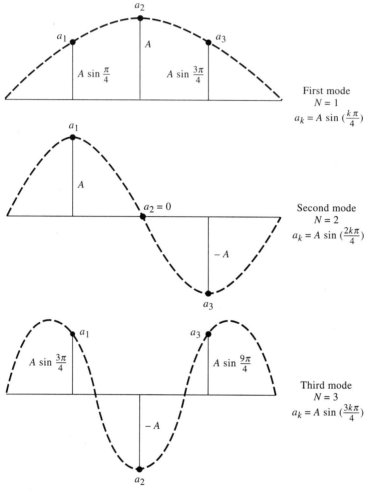

First mode
$N = 1$
$a_k = A \sin\left(\frac{k\pi}{4}\right)$

Second mode
$N = 2$
$a_k = A \sin\left(\frac{2k\pi}{4}\right)$

Third mode
$N = 3$
$a_k = A \sin\left(\frac{3k\pi}{4}\right)$

Figure 11.17 The three normal modes of a three-particle array.

increase, but at the same time, let the spacing d between neighboring particles decrease, such that the length of the string $L = (n + 1)d$ is held constant. Thus, for $N \ll n$, the argument $N\pi/(2(n + 1))$ of the sine term in Equation 11.43 is small. So we have approximately

$$\omega_N = \omega_0 \left(\frac{N\pi}{n + 1}\right) \tag{11.47}$$

but for the transverse oscillations we have

$$\omega_0 = \left(\frac{K}{m}\right)^{1/2} = \left(\frac{T}{md}\right)^{1/2}$$

and substituting this into Equation 11.47 gives

$$\omega_N \approx \left(\frac{T}{m/d}\right)^{1/2} \frac{N\pi}{(n+1)d} \tag{11.47a}$$

But $(n + 1)d = L$, the total length of the string, and m/d is its mass per unit length μ (the linear mass density). Thus, we have approximately

$$\omega_N = N\frac{\pi}{L}\left(\frac{T}{\mu}\right)^{1/2} \qquad (N = 1, 2, \ldots) \tag{11.47b}$$

In particular,

$$\omega_1 = \frac{\pi}{L}\left(\frac{T}{\mu}\right)^{1/2}$$

and $\omega_N = N\omega_1$. The normal frequencies are integral multiples of the lowest, or fundamental, frequency ω_1. Remember, this is only an approximation, but for $N \ll n$ it is an exceedingly good one.

Let us now examine the displacements of the particles under these conditions. What might we guess? They ought to very closely approximate the actual vibration of a real string. For the Nth mode the displacement of the kth particle is given by Equation 11.45. However, instead of denoting the particle by its k value, let us denote it by its distance down the string from the fixed end $x = kd$. Hence,

$$\frac{kN\pi}{(n+1)} = \frac{N\pi(kd)}{(n+1)d} = \frac{N\pi x}{L}$$

Replacing the argument of the sine term in Equation 11.45 by this factor permits us to rewrite that Equation as

$$q_N(x, t) = A \sin\left(\frac{N\pi x}{L}\right)\cos \omega_N t \qquad (N = 1, 2, \ldots)$$

$$= A \sin\left(\frac{2\pi x}{\lambda_N}\right)\cos (2\pi f_N t) \tag{11.47c}$$

where we have defined the wavelength λ_N and the frequency f_N by

$$\lambda_N = \frac{2L}{N} \qquad f_N = \frac{\omega_N}{2\pi}$$

The meaning of these terms applied to wave motion along a continuous medium will be made more precise in the next section. The above equation expresses the displacement of any point along a continuous string when it is vibrating in its Nth mode. It represents a standing wave of wavelength λ_N. Each vibrational mode consists of an integral number of half-wavelength units constrained to fit within the length L such that the endpoints are nodes; that is, they do not vibrate. The meaning of the term's fundamental frequency—first, second, third, and so on harmonics of something like a vibrating violin string—should also be clear. No wonder the early Pythagoreans had such a high regard for integers.

In the next section, we will treat this above situation directly as a continuous medium instead of one made up of a large number of discrete masses. We will develop a differential "wave" equation governing the motion of the continuous medium, and we will obtain standing wave solutions identical to the one shown above.

11.6 VIBRATION OF A CONTINUOUS SYSTEM. THE WAVE EQUATION

Let us consider the motion of a linear array of connected particles in which the number of particles is made indefinitely large and the distance between adjacent particles indefinitely small. In other words, we have a continuous heavy cord or rod. To analyze this type of system it is convenient to rewrite the differential equations of motion of a finite system (Equation 11.37) in the following form:

$$m\ddot{q} = Kd\left[\left(\frac{q_{k+1} - q_k}{d}\right) - \left(\frac{q_k - q_{k-1}}{d}\right)\right]$$

in which d is the distance between the equilibrium positions of any two adjacent particles. Now if the variable x represents general distances in the longitudinal direction, and if the number n of particles is very large so that d is small compared to the total length, then we can write

$$\frac{q_{k+1} - q_k}{d} \approx \left(\frac{\partial q}{\partial x}\right)_{x=kd+d/2}$$

$$\frac{q_k - q_{k-1}}{d} \approx \left(\frac{\partial q}{\partial x}\right)_{x=kd-d/2}$$

Consequently, the difference between the above two expressions is equal to the second derivative multiplied by d, namely,

$$\frac{q_{k+1} - q_k}{d} - \frac{q_k - q_{k-1}}{d} \approx d\left(\frac{\partial^2 q}{\partial x^2}\right)_{x=kd}$$

The equation of motion can, therefore, be written

$$\frac{\partial^2 q}{\partial t^2} = \frac{Kd^2}{m}\frac{\partial^2 q}{\partial x^2}$$

or

$$\frac{\partial^2 q}{\partial t^2} = v^2\frac{\partial^2 q}{\partial x^2} \tag{11.48}$$

in which we have introduced the abbreviation

$$v^2 = \frac{Kd^2}{m} \tag{11.49}$$

Figure 11.18 A running wave.

Equation 11.48 is a well-known differential equation of mathematical physics. It is called the *one-dimensional wave equation*. It is encountered in many different places. Solutions of the wave equation represent traveling disturbances of some sort. It is easy to verify that a very general type of solution of the wave equation is given by

$$q = f(x + vt)$$

or

$$q = f(x - vt)$$

where f is any differentiable function of the argument $x \pm vt$. The first solution represents a disturbance that is propagating in the negative x direction with speed v, and the second equation represents a disturbance moving with speed v in the positive x direction. In our particular problem, the disturbance q is a *displacement* of a small portion of the system from its equilibrium configuration, Figure 11.18. For the cord this displacement could be a kink that travels along the cord, and for a solid rod it could be a region of compression or of rarefaction moving along the length of the rod.

Evaluation of the Wave Speed

In the preceding section we found that the constant K, for transverse motion of a loaded string, is equal to the ratio T/d where T is the tension in the string. For the continuous string this ratio would, of course, become infinite as d approaches zero. However, if we introduce the linear density or mass per unit length μ, we have

$$\mu = \frac{m}{d}$$

Consequently, the expression for v^2 (Equation 11.49) can be written

$$v^2 = \frac{(T/d)d^2}{\mu d} = \frac{T}{\mu}$$

so that d cancels out. The speed of propagation for transverse waves in a continuous string is then

$$v = \left(\frac{T}{\mu}\right)^{1/2} \tag{11.50}$$

For the case of longitudinal vibrations, we introduce the elastic modulus Y, which is defined as the ratio of the force to the elongation per unit length. Thus, K, the stiffness of a small section of length d, is given by

$$K - \frac{Y}{d}$$

Consequently, Equation 11.49 can be written as

$$v^2 = \frac{(Y/d)d^2}{\mu d} = \frac{Y}{\mu}$$

and again we see that d cancels out. Hence, the speed of propagation of longitudinal waves in an elastic rod is

$$v = \left(\frac{Y}{\mu}\right)^{1/2} \qquad (11.51)$$

Sinusoidal Waves

In the study of wave motion, those particular solutions of the wave equation

$$\frac{\partial^2 q}{\partial t^2} = v^2 \frac{\partial^2 q}{\partial x^2}$$

in which q is a sinusoidal function of x and t, namely,

$$q = A \frac{\sin}{\cos} \left[\frac{2\pi}{\lambda}(x + vt) \right] \qquad (11.52)$$

$$q = A \frac{\sin}{\cos} \left[\frac{2\pi}{\lambda}(x - vt) \right] \qquad (11.53)$$

are of fundamental importance. These solutions represent traveling disturbances in which the displacement, at a given point x, varies harmonically in time. The amplitude of this motion is the constant A, and the frequency f is given by

$$f = \frac{\omega}{2\pi} = \frac{v}{\lambda}$$

Furthermore, at a given value of the time t, say $t = 0$, the displacement varies sinusoidally with the distance x. The distance between two successive maxima, or minima, of the displacement is the constant λ, called the *wavelength*. The waves represented by Equation 11.52 propagate in the negative x direction, and those represented by Equation 11.53 propagate in the positive x direction, as shown in Figure 11.19. They are special cases of the general type of solution mentioned earlier.

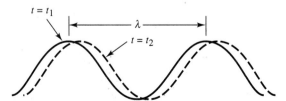

Figure 11.19   A sinusoidal wave.

Standing Waves

Since the wave equation (Equation 11.48) is linear, we can build up any number of solutions by making linear combinations of known solutions. One possible linear combination that is of particular significance is obtained by adding two waves of equal amplitude that are traveling in opposite directions. In our notation such a solution is given by

$$q = \frac{1}{2}A \sin\left[\frac{2\pi}{\lambda}(x + vt)\right] + \frac{1}{2}A \sin\left[\frac{2\pi}{\lambda}(x - vt)\right] \qquad (11.54)$$

By using the appropriate trigonometric identity and collecting terms, we find that the equation reduces to

$$q = A \sin\left(\frac{2\pi}{\lambda}x\right) \cos(\omega t) \qquad (11.54a)$$

in which $\omega = 2\pi v/\lambda$. Note that this equation is precisely identical to Equation 11.47c, which represents the motion of the loaded string in the limiting situation that the number of discrete masses approaches ∞ while their distance of separation approaches 0 in such a way that the total length of string remains constant. Again, Equation 11.54a represents a *standing wave*. The amplitude of the displacement varies continuously with x. At $x = 0, \lambda/2, \lambda, 3\lambda/2, \ldots$, the displacement of the string is always zero, since the sine term vanishes at those points. These points of zero amplitude are the *nodes*. At $x = \lambda/4, 3\lambda/4, 5\lambda/4, \ldots$, the amplitude of the vibrating string is a maximum. These points are the *antinodes*. The distance between two successive nodes (or antinodes) is $\lambda/2$. These facts are illustrated in Figure 11.20. Note again that there is a well-defined constraint on the values of allowable wavelengths λ. Since the endpoints of the string are fixed, we have as boundary conditions

$$q = 0 \qquad (x = 0, L)$$

that our solution (Equation 11.54a) must obey. The first condition at $x = 0$ is met automatically. The second boundary condition at $x = L$ is met if

$$L = N\left(\frac{\lambda}{2}\right) \qquad \lambda = \frac{2L}{N}$$

An integral number of half wavelengths must fit within the length L if the endpoints are

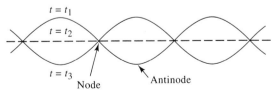

Figure 11.20 Standing waves.

to be nodes. This is precisely the condition obtained previously for the normal modes of the loaded string.

PROBLEMS

11.1 A particle of mass m moves in one-dimensional motion with the following potential energy functions:

(a) $V(x) = \dfrac{k}{2}x^2 + \dfrac{k^2}{x}$

(b) $V(x) = kxe^{-bx}$

(c) $V(x) = k(x^4 - b^2x^2)$

where all constants are real and positive. Find the equilibrium positions for each case and determine their stability.

(d) Find the angular frequency ω for small oscillations about the respective positions of stable equilibrium for parts (a), (b), and (c), and find the period in seconds for each case if $m = 1$ g, and k and b are each of unit value in cgs units.

11.2 A particle moves in two dimensions under the potential energy function

$$V(x,\ y) = k(x^2 + y^2 - 2bx - 4by)$$

where k is a positive constant. Show that there is one position of equilibrium. Is it stable or unstable?

11.3 The potential energy function of a particle of mass m in one-dimensional motion is given by

$$V(x) = -\frac{k}{2}x^2$$

and so the force is of the antirestoring type

$$F(x) = kx$$

with $x = 0$ as a position of unstable equilibrium when k is a positive constant. If the initial conditions are $t = 0$, $x = x_0$, and $\dot{x} = 0$, show that the ensuing motion is given by an exponential "runaway"

$$x(t) = x_0(e^{\alpha t} + e^{-\alpha t})/2$$

where the constant $\alpha = \sqrt{k/m}$.

11.4 A light elastic cord of length $2l$ and stiffness k is held with the ends fixed a distance $2l$ apart in a horizontal position. A block of mass m is then suspended from the midpoint of the cord. Show that the potential energy of the system is given by the expression

$$V(y) = 2k[y^2 - 2l(y^2 + l^2)^{1/2}] - mgy$$

where y is the vertical sag of the center of the cord. From this show that the equilibrium position is given by a root of the equation

$$u^4 - 2au^3 + a^2u^2 - 2au + a^2 = 0$$

where $u = y/l$ and $a = mg/4kl$.

11.5 A uniform cubical block of mass m and sides $2a$ is balanced on top of a rough sphere of radius b. Show that the potential energy function can be expressed as

$$V(\theta) = mg[(a + b) \cos \theta + b\theta \sin \theta]$$

where θ is the angle of tilt. From this, show that the equilibrium at $\theta = 0$ is stable, or unstable, depending on whether a is less than or greater than b, respectively.

11.6 Expand the potential-energy function of Problem 11.5 as a power series in θ. From this determine the stability for the case $a = b$.

11.7 A solid homogeneous hemisphere of radius a rests on top of a rough hemispherical cap of radius b, the curved faces being in contact. Show that the equilibrium is stable if a is less than $3b/5$.

11.8 Determine the frequency of vertical oscillations about the equilibrium position in Problem 11.4.

11.9 Determine the period of oscillation of the block in Problem 11.5.

11.10 Determine the period of oscillation of the hemisphere in Problem 11.7.

11.11 A small steel ball rolls back and forth about its equilibrium position in a rough spherical bowl. Show that the period of oscillation is $2\pi[7(b - a)/5g]^{1/2}$ where a is the radius of the ball and b is the radius of the bowl. Find the period in seconds if $b = 1$ m and $a = 1$ cm.

11.12 For an orbiting satellite in the form of a thin rod, show that the stable equilibrium attitude and period of oscillation are the same as those found in Example 11.3 for the dumbbell satellite.

11.13 In the system of two identical coupled oscillators shown in Figure 11.4, one oscillator is started with initial amplitude A_0 whereas the other is at rest at its equilibrium position, so that the initial conditions are

$$t = 0 \qquad x_1(0) = A_0 \qquad x_2(0) = 0 \qquad \dot{x}_1(0) = \dot{x}_2(0) = 0$$

Show that the amplitude of the symmetric component is equal to the amplitude of the antisymmetric component in this case and that the complete solution can be expressed as follows:

$$x_1(t) = \frac{1}{2}A_0(\cos \omega_a t + \cos \omega_b t) = A_0 \cos \bar{\omega}t \cos \Delta t$$

$$x_2(t) = \frac{1}{2}A_0(\cos \omega_a t - \cos \omega_b t) = A_0 \sin \bar{\omega}t \sin \Delta t$$

in which $\bar{\omega} = (\omega_a + \omega_b)/2$ and $\Delta = (\omega_b - \omega_a)/2$. Thus, if the coupling is very weak so that $K' \ll K$, then $\bar{\omega}$ will be very nearly equal to $\omega_a = (K/m)^{1/2}$, and Δ is very small. Consequently, under the stated initial conditions, the first oscillator will eventually come to rest while the second oscillator oscillates with amplitude A_0. Later, the system will return to the initial condition, and so on. Thus, the energy passes back and forth between the two oscillators indefinitely.

11.14 In Problem 11.13 show that, for weak coupling, the period at which the energy trades back and forth is approximately equal to $T_a(K/2K')$ where $T_a = 2\pi/\omega_a = 2\pi/(m/K)^{1/2}$ is the period of the symmetric oscillation.

11.15 Two identical simple pendulums are coupled together by a very weak force of attraction that varies as the inverse square of the distance between the two particles. (This force might be the gravitational attraction between the two particles, for instance.) Show that, for small departures from the equilibrium configuration, the Lagrangian can be reduced to the same mathematical form, with appropriate constants, as that of the two identical coupled oscillators treated in Section 11.3 and in Problems 11.13 and 11.15. (*Hint:* Consider Equation 11.4.)

11.16 Find the normal frequencies of the coupled harmonic oscillator system (Figure 11.4) for the general case in which the two particles have unequal mass and the springs have different stiffness. In particular, find the frequencies for the case $m_1 = m$, $m_2 = 2m$, $K_1 = K$, $K_2 = 2K$, $K' = 2K$. Express the result in terms of the quantity $\omega_0 = (K/m)^{1/2}$.

11.17 A light elastic spring of stiffness K is clamped at its upper end and supports a particle of mass m at its lower end. A second spring of stiffness K is fastened to the particle and, in turn, supports a particle of mass $2m$ at its lower end. Find the normal frequencies of the system for vertical oscillations about the equilibrium configuration. Find also the normal coordinates.

11.18 Consider the case of a double pendulum, Figure 11.14(a), in which the two sections are of different length, the upper one being of length l_1 and the lower of length l_2. Both particles are of equal mass m. Find the normal frequencies of the system and the normal coordinates.

11.19 Set up the secular equation for the case of three coupled particles in a linear array and show that the normal frequencies are the same as those given by Equation 11.43.

11.20 Illustrate the normal modes for the case of four particles in a linear array. Find the numerical values of the ratios of the second, third, and fourth normal frequencies to the lowest or first normal frequency.

11.21 A light elastic cord of natural length l and stiffness K is stretched out to a length $l + \Delta l$ and loaded with a number n of particles evenly spaced along its length. If m is the total mass of all n particles, find the speed of transverse and of longitudinal waves in the cord.

11.22 Work Problem 11.21 for the case in which, instead of being loaded, the cord is heavy with linear mass density μ.

COMPUTER/CALCULATOR APPLICATIONS

11.1 Consider a single pulse traveling down an infinitely long string. Assume that at $t = 0$, the shape of the pulse, or the vertical displacement of the string, is

$$y(x) = \frac{1}{1 + x^2} \tag{1}$$

Analogous to the discussion of Fourier series in Section 3.7, this pulse can be thought of as a superposition of harmonic waves of differing wave numbers k. However, the infinite sum of Section 3.7 that approximates a repetitive function needs to be replaced here by an integral over an infinite number of harmonic waves, each one weighted by an appropriate amplitude function, that is,

$$y(x) = \int_0^\infty a(k) \cos (kx) \, dk \tag{2}$$

We use cosine functions since $y(x)$ is an even function of x. The amplitude function $a(k)$ is given by

$$a(k) = \frac{2}{\pi} \int_{-\infty}^{\infty} y(x) \cos (kx) \, dx \tag{3}$$

(a) Calculate $a(k)$ using equation (3) above.
(b) Substitute $a(k)$ into Equation 2 and show that it yields $y(x)$.
(c) Integrate Equation 2 numerically for values of x ranging from 0 to 3 and show that the results agree with the exact values of $y(x)$.
    Assume that the speed of the pulse is given by $v = \omega/k = 1$. In such a case, the shape of the pulse is preserved as it travels down the string.
(d) Write down an exact expression for $y(x, t)$ assuming that at $t = 0$, $y(x, 0)$ is given by Equation 1.
(e) Write down the appropriate integral expression for $y(x, t)$ using equation 2.
(f) Now assume that the pulse is traveling down a "dispersive" string, for which the wave velocity is not a constant but depends on the wave number of the wave. Assume that $\omega/k = 1 + 0.25\ k^2$. The many waves of differing k that are superimposed to make the traveling pulse will change their phase relationship as each moves down the string. Thus, the shape of the pulse will change. To see this effect, modify the integral expression for $y(x, t)$ obtained in part (e) using the "dispersive" value of ω/k given above. Numerically integrate the resulting expression to obtain $y(x, t)$ for $t = 2.5$, 5.0, and 10.0 s. Pick a broad range of x about the location of the peak of the pulse at each of these times.
(g) Plot these resultant waveforms and compare them with $y(x, 0)$. Comment on the result.

APPENDIX A

UNITS

Basic SI (Système International) Units

Unit	Symbol	Physical Quantity
meter	m	length
kilogram	kg	mass
second	s	time
ampere	A	electric current
Kelvin	K	temperature
mole	mol	amount of substance
candela	cd	luminous intensity

Derived SI Units (not a complete list)

Unit	Symbol	Physical Quantity	Equivalent
newton	N	force	$kg \cdot m/s^2$
joule	J	work or energy	$N \cdot m$
watt	W	power	J/s
pascal	Pa	pressure	N/m^2
volt	V	electric potential difference	W/A
couloumb	C	electric charge	$A \cdot s$

Some Other Common Units

Unit	Metric Equivalent
inch	2.540 centimeters
foot (12 in)	0.3048 meter
yard (3 ft)	0.9144 meter
mile (5280 ft)	1.609 kilometers
feet/second	0.3048 meters/second
miles/hour	0.4470 meters/second
pound (mass)	0.4536 kilogram
gallon	3.785 liters (10^{-3} m^3)
cubic foot	0.02832 cubic meter
horsepower	746 watts

Prefixes for Multiplication by a Power of Ten

Name	Symbol	Factor	Name	Symbol	Factor
			centi	c	10^{-2}
kilo	k	10^3	milli	m	10^{-3}
mega	M	10^6	micro	μ	10^{-6}
giga	G	10^9	nano	n	10^{-9}
tera	T	10^{12}	pico	p	10^{-12}
peta	P	10^{15}	femto	f	10^{-15}
exa	E	10^{18}	atto	a	10^{-18}

COMPLEX NUMBERS

The quantity

$$z = x + iy$$

is said to be a *complex number* if x and y are real and $i = \sqrt{-1}$. The *complex conjugate* is defined as

$$z^* = x - iy$$

The *absolute value* $|z|$ is given by

$$|z|^2 = zz^* = x^2 + y^2$$

The following are true

$$z + z^* = 2x = 2\ \text{Re}\ z$$

$$z - z^* = 2y = 2\ \text{Im}\ z$$

Exponential Notation

$$z = x + iy = |z|e^{i\theta} = |z|(\cos\theta + i\sin\theta)$$

$$z^* = x - iy = |z|e^{-i\theta} = |z|(\cos\theta - i\sin\theta)$$

where

$$\tan\theta = \frac{y}{x}$$

(For a proof of the relation $e^{i\theta} = \cos\theta + i\sin\theta$ see under Series Expansions in Appendix D below.)

Circular and Hyperbolic Functions

The following relations are often useful

$$\cos\theta = \frac{e^{i\theta} + e^{-i\theta}}{2}$$

$$\sin\theta = \frac{e^{i\theta} - e^{-i\theta}}{2i}$$

$$\cosh\theta = \frac{e^{\theta} + e^{-\theta}}{2} \qquad \text{(hyperbolic cosine)}$$

$$\sinh\theta = \frac{e^{\theta} - e^{-\theta}}{2} \qquad \text{(hyperbolic sine)}$$

$$\tanh\theta = \frac{\sinh\theta}{\cosh\theta} = \frac{e^{\theta} - e^{-\theta}}{e^{\theta} + e^{-\theta}} \qquad \text{(hyperbolic tangent)}$$

Relations Between Circular and Hyperbolic Functions

$$\sin i\theta = i \sinh \theta$$
$$\cos i\theta = \cosh \theta$$
$$\sinh i\theta = i \sin \theta$$
$$\cosh i\theta = \cos \theta$$

Derivatives

$$\frac{d}{d\theta} \sin \theta = \cos \theta \qquad \frac{d}{d\theta} \sinh \theta = \cosh \theta$$

$$\frac{d}{d\theta} \cos \theta = -\sin \theta \qquad \frac{d}{d\theta} \cosh \theta = \sinh \theta$$

Identities

$$\cos^2 \theta + \sin^2 \theta = 1$$
$$\sin (\theta + \phi) = \sin \theta \cos \phi + \cos \theta \sin \phi$$
$$\cos (\theta + \phi) = \cos \theta \cos \phi - \sin \theta \sin \phi$$

$$\cosh^2 \theta - \sinh^2 \theta = 1$$
$$\sinh (\theta + \phi) = \sinh \theta \cosh \phi + \cosh \theta \sinh \phi$$
$$\cosh (\theta + \phi) = \cosh \theta \cosh \phi + \sinh \theta \sinh \phi$$

APPENDIX C

CONIC SECTIONS

Cartesian Coordinates
Ellipse:

$$\frac{x^2}{a^2} + \frac{y^2}{b^2} = 1 \qquad \text{area} = \pi ab$$

$$\text{eccentricity } e = \sqrt{1 - \frac{b^2}{a^2}}, \ a > b$$

$$\text{distance from origin to focus} = \sqrt{a^2 - b^2}$$

$$\text{major axis} = 2a$$

$$\text{minor axis} = 2b$$

Circle: Special case of ellipse with $a = b$ and $e = 0$.
Parabola:

$$y^2 = 4px \qquad \text{eccentricity } e = 1$$

$$\text{distance from origin to focus} = p$$

Hyperbola:

$$\frac{x^2}{a^2} - \frac{y^2}{b^2} = 1 \qquad \text{eccentricity } e = \sqrt{1 + \frac{b^2}{a^2}}$$

$$\text{distance from origin to focus} = \sqrt{a^2 + b^2}$$

Polar Coordinates

All cases:

$$r = \frac{r_0(1 + e)}{1 + e \cos \theta} \qquad \begin{array}{l} \text{focus is at the origin;} \\ r_0 \text{ is minimum value of } |r| \end{array}$$

Ellipse: $e < 1$, Parabola: $e = 1$, Hyperbola: $e > 1$, Circle: $e = 0$
For the ellipse the semimajor axis

$$a = \frac{r_0}{1 - e}$$

and the semiminor axis

$$b = a\sqrt{1 - e^2}$$

SERIES EXPANSIONS

Taylor's Series

$$f(x + a) = f(a) + xf'(a) + \frac{x^2}{2!} f''(a) + \cdots + \frac{x^n}{n!} f^n(a) + \cdots$$

$$f(x) = f(0) + xf'(0) + \frac{x^2}{2!} f''(0) + \cdots + \frac{x^n}{n!} f^n(0) + \cdots$$

where

$$f^n(a) = \frac{d^n}{dx^n} f(x) \bigg|_{x=a}$$

Often-Used Expansions

$$e^x = 1 + x + \frac{x^2}{2!} + \cdots + \frac{x^n}{n!} + \cdots$$

$$\sin x = x - \frac{x^3}{3!} + \frac{x^5}{5!} - \cdots$$

$$\cos x = 1 - \frac{x^2}{2!} + \frac{x^4}{4!} - \cdots$$

$$\sinh x = x + \frac{x^3}{3!} + \frac{x^5}{5!} + \cdots$$

$$\cosh x = 1 + \frac{x^2}{2!} + \frac{x^4}{4!} + \cdots$$

$$\ln(1 + x) = x - \frac{x^2}{2} + \frac{x^3}{3} - \cdots \qquad |x| < 1$$

$$\tan x = x + \frac{x^3}{3} + \frac{2}{15}x^5 + \cdots \qquad |x| < \frac{\pi}{2}$$

Complex Exponential

Setting $x = i\theta$ in the expansion for e^x gives

$$e^{i\theta} = 1 + i\theta + \frac{i^2\theta^2}{2!} + \frac{i^3\theta^3}{3!} + \cdots + \frac{i^n\theta^n}{n!} + \cdots$$

Since $i = \sqrt{-1}$

$$i^n = \begin{array}{l} +1\colon n = 0, 4, \ldots \\ -1\colon n = 2, 6, \ldots \\ +i\colon n = 1, 5, \ldots \\ -i\colon n = 3, 7, \ldots \end{array}$$

then

$$e^{i\theta} = \left(1 - \frac{\theta^2}{2!} + \frac{\theta^4}{4!} - \cdots\right) + i\left(\theta - \frac{\theta^3}{3!} + \frac{\theta^5}{5!} - \cdots\right)$$

$$= \cos \theta + i \sin \theta$$

from the series for the cosine and sine above.

Binomial Series

$$(a + x)^n = a^n + na^{n-1}x + \frac{n(n - 1)}{2!}a^{n-2}x^2 + \cdots + \binom{n}{m}a^{n-m}x^m + \cdots$$

where the binomial coefficient is

$$\binom{n}{m} = \frac{n!}{(n - m)!m!}$$

The series converges for $|x/a| < 1$.

Useful Approximations
For small x, the following approximations are often used

$$e^x \simeq 1 + x$$

$$\sin x \simeq x$$

$$\cos x \simeq 1 - \frac{1}{2}x^2$$

$$\sqrt{1 + x} \simeq 1 + \frac{1}{2}x$$

$$\frac{1}{1 + x} \simeq 1 - x$$

$$\frac{1}{1 - x} \simeq 1 + x$$

The last three are based on the binomial series, and the list can be extended for other values of the exponent:

$$(1 + x)^n = 1 + nx + \frac{1}{2}n(n - 1)x^2 + \cdots$$

APPENDIX E

SPECIAL FUNCTIONS

Elliptic Integrals

The elliptic integral of the *first kind* is given by the expressions

$$F(k,\ \phi) = \int_0^\phi \frac{d\phi}{(1\ -\ k^2 \sin^2 \phi)^{1/2}}$$

$$= \int_0^x \frac{dx}{(1\ -\ x^2)^{1/2}\ (1\ -\ k^2 x^2)^{1/2}}$$

and the elliptic integral of the *second kind* by

$$E(k,\ \phi) = \int_0^\phi (1\ -\ k^2 \sin^2 \phi)^{1/2}\ d\phi$$

$$= \int_0^x \frac{(1\ -\ k^2 x^2)^{1/2}}{(1\ -\ x^2)^{1/2}}\ dx$$

Both converge for $|k| < 1$. They are called *incomplete* if $x = \sin \phi < 1$, and *complete* if $x = \sin \phi = 1$. The complete integrals have the following series expansions

$$F(k) = F\left(k,\ \frac{\pi}{2}\right) = \frac{\pi}{2}\left(1\ +\ \frac{k^2}{4}\ +\ \frac{9}{64}k^4\ +\ \cdots\right)$$

$$E(k) = E\left(k,\ \frac{\pi}{2}\right) = \frac{\pi}{2}\left(1\ -\ \frac{k^2}{4}\ -\ \frac{9}{64}k^4\ -\ \cdots\right)$$

Gamma Function

The gamma function is defined as

$$\Gamma(n) = \int_0^\infty x^{n-1}\ e^{-x}\ dx$$

For any value of n

$$n\Gamma(n) = \Gamma(n + 1)$$

If n is a positive integer

$$\Gamma(n) = (n - 1)!$$

Special values

$$\Gamma\left(\frac{1}{2}\right) = \sqrt{\pi}$$

$$\Gamma(1) = 1$$

$$\Gamma\left(\frac{3}{2}\right) = \frac{1}{2}\sqrt{\pi}$$

$$\Gamma(2) = 1$$

Integrals expressible in terms of gamma functions

$$\int_0^1 \frac{dx}{\sqrt{1 - x^n}} = \frac{\sqrt{\pi}}{n} \frac{\Gamma(1/n)}{\Gamma[(1/2) + (1/n)]}$$

$$\int_0^1 (1 - x^2)^n x^m \, dx = \frac{\Gamma(n + 1)\Gamma[(m + 1)/2]}{2\Gamma[(2n + m + 3)/2]}$$

APPENDIX F

CURVILINEAR COORDINATES

We consider a general orthogonal system of coordinates u, v, and w with unit vectors $\mathbf{e}_1$, $\mathbf{e}_2$, and $\mathbf{e}_3$. The volume element is

$$dV = h_1 h_2 h_3 \, du \, dv \, dw$$

and the line element is

$$ds = \mathbf{e}_1 h_1 \, du + \mathbf{e}_2 h_2 \, dv + \mathbf{e}_3 h_3 \, dw$$

The gradient, divergence, and curl are as follows

$$\operatorname{grad} f = \frac{\mathbf{e}_1}{h_1} \frac{\partial f}{\partial u} + \frac{\mathbf{e}_2}{h_2} \frac{\partial f}{\partial v} + \frac{\mathbf{e}_3}{h_3} \frac{\partial f}{\partial w}$$

$$\operatorname{div} \mathbf{Q} = \frac{1}{h_1 h_2 h_3} \left[\frac{\partial}{\partial u} (h_2 h_3 Q_1) + \frac{\partial}{\partial v} (h_3 h_1 Q_2) + \frac{\partial}{\partial w} (h_1 h_2 Q_3) \right]$$

$$\operatorname{curl} \mathbf{Q} = \frac{1}{h_1 h_2 h_3} \begin{vmatrix} h_1 \mathbf{e}_1 & h_2 \mathbf{e}_2 & h_3 \mathbf{e}_3 \\ \dfrac{\partial}{\partial u} & \dfrac{\partial}{\partial v} & \dfrac{\partial}{\partial w} \\ h_1 Q_1 & h_2 Q_2 & h_3 Q_3 \end{vmatrix}$$

The h functions for some common coordinate systems are listed below.

Rectangular Coordinates: x, y, z

$$h_x = 1 \qquad h_y = 1 \qquad h_z = 1$$

Cylindrical Coordinates: R, ϕ, z

$$x = R \cos \phi \qquad y = R \sin \phi$$
$$h_R = 1 \qquad h_\phi = R \qquad h_z = 1$$

Spherical Coordinates: r, θ, ϕ

$$x = r \sin \theta \cos \phi \qquad y = r \sin \theta \sin \phi \qquad z = r \cos \theta$$
$$h_r = 1 \qquad h_\theta = r \qquad h_\phi = r \sin \theta$$

Parabolic Coordinates: u, v, θ

$$x = uv \cos \theta \qquad y = uv \sin \theta \qquad z = \frac{1}{2}(u^2 - v^2)$$

$$h_u = h_v = \sqrt{u^2 + v^2} \qquad h_\theta = uv$$

Example: The curl in spherical coordinates is

$$\text{curl } \mathbf{Q} = \begin{vmatrix} \mathbf{e}_r & r\mathbf{e}_\theta & r \sin \theta \, \mathbf{e}_\phi \\ \dfrac{\partial}{\partial r} & \dfrac{\partial}{\partial \theta} & \dfrac{\partial}{\partial \phi} \\ Q_r & rQ_\theta & r \sin \theta \, Q_\phi \end{vmatrix} \frac{1}{r^2 \sin \theta}$$

FOURIER SERIES

To find the coefficients of the terms in the trigonometric expansion

$$f(t) = \frac{a_0}{2} + \sum_{n=1}^{\infty} [a_n \cos(n\omega t) + b_n \sin(n\omega t)]$$

multiply both sides of the equation by $\cos(n'\omega t)$ and integrate over the interval $-\pi/\omega$ to $+\pi/\omega$:

$$\int_{-\pi/\omega}^{\pi/\omega} f(t) \cos(n'\omega t) \, dt = \frac{a_0}{2} \int_{-\pi/\omega}^{\pi/\omega} \cos(n'\omega t) \, dt$$

$$+ \sum_{n=1}^{\infty} [a_n \int_{-\pi/\omega}^{\pi/\omega} \cos(n'\omega t) \cos(n\omega t) \, dt + b_n \int_{-\pi/\omega}^{\pi/\omega} \cos(n'\omega t) \sin(n\omega t) \, dt]$$

Now if n' and n are integers, we have the general formulas

$$\int_{-\pi/\omega}^{\pi/\omega} \cos(n'\omega t) \, dt = 2\pi/\omega \qquad n' = 0$$

$$= 0 \qquad n' \neq 0$$

$$\int_{-\pi/\omega}^{\pi/\omega} \cos(n'\omega t) \cos(n\omega t) \, dt = \pi/\omega \qquad n' = n$$

$$= 0 \qquad n' \neq n$$

$$\int_{-\pi/\omega}^{\pi/\omega} \cos(n'\omega t) \sin(n\omega t) \, dt = 0 \qquad \text{for all } n' \text{ and } n$$

Thus, for a given n', all of the definite integrals in the summation vanish except the one for which $n' = n$. Consequently we can write

$$a_n = \frac{\omega}{\pi} \int_{-\pi/\omega}^{\pi/\omega} f(t) \cos(n\omega t) \, dt \qquad \text{for } n = 0, 1, 2, \ldots$$

Similarly, if the equation for $f(t)$ is multiplied by $\sin(n'\omega t)$ and integrated term by term, we use the general formula

$$\int_{-\pi/\omega}^{\pi/\omega} \sin(n'\omega t) \sin(n\omega t) \, dt = \pi/\omega \qquad n' = n$$

$$= 0 \qquad n' \neq n$$

in addition to the ones above. As before, all of the definite integrals vanish except for $n' = n$, and so we get

$$b_n = \frac{\omega}{\pi} \int_{-\pi/\omega}^{\pi/\omega} f(t) \sin (n\omega t) \, dt \qquad n = 1, 2, \ldots$$

Since the period $T = 2\pi/\omega$, the limits of integration can also be expressed as $-T/2$ to $T/2$. For more detailed information concerning continuity conditions, integrability, and so on, the reader should consult a text on Fourier series, such as R. V. Churchill, *Fourier Series and Boundary Value Problems,* McGraw-Hill, New York, 1963.

APPENDIX H

MATRICES

A *matrix* **A** is an array of elements a_{ij} arranged thus

$$\mathbf{A} = \begin{bmatrix} a_{11} & a_{12} & \cdots & a_{1j} & \cdots & a_{1m} \\ a_{21} & a_{22} & \cdots & a_{2j} & \cdots & a_{2m} \\ \vdots & \vdots & & \vdots & & \vdots \\ a_{i1} & a_{i2} & \cdots & a_{ij} & \cdots & a_{im} \\ \vdots & \vdots & & \vdots & & \vdots \\ a_{n1} & a_{n2} & \cdots & a_{nj} & \cdots & a_{nm} \end{bmatrix}$$

If $n = m$, it is called a *square* matrix. Unless stated otherwise, we shall consider only square matrices in this Appendix. A *symmetric* matrix is one such that $a_{ij} = a_{ji}$. If $a_{ij} = -a_{ji}$, it is *antisymmetric*.

The sum of two matrices is defined as

$$(\mathbf{A} + \mathbf{B})_{ij} = a_{ij} + b_{ij}$$

The product of two matrices is defined as

$$(\mathbf{AB})_{ij} = a_{i1}b_{1j} + a_{i2}b_{2j} + \cdots = \sum_k a_{ik}b_{kj}$$

The product **AB** is not, in general, equal to **BA**. If **AB** = **BA,** the two matrices are said to *commute*. A *diagonal matrix* is one whose nondiagonal elements are zero, $a_{ij} = 0$ for $i \neq j$. The *identity* matrix[1] is a diagonal matrix with all diagonal elements equal to unity,

$$\mathbf{1} = \begin{bmatrix} 1 & 0 & 0 & \cdots & 0 \\ 0 & 1 & 0 & \cdots & 0 \\ 0 & 0 & 1 & \cdots & 0 \\ & & \cdots & & \\ 0 & 0 & 0 & \cdots & 1 \end{bmatrix}$$

From the definition of the product, it is easily shown that

$$\mathbf{A1} = \mathbf{1A}$$

The *inverse* $\mathbf{A}^{-1}$ of a matrix **A** is defined by

$$\mathbf{AA}^{-1} = \mathbf{1} = \mathbf{A}^{-1}\mathbf{A}$$

---

[1] This should not be confused with the inertia tensor defined in Chapter 9.

The *transpose* $\mathbf{A}^T$ of a matrix $\mathbf{A}$ is defined as

$$(\mathbf{A}^T)_{ij} = (\mathbf{A})_{ji}$$

For two matrices $\mathbf{A}$ and $\mathbf{B}$, $(\mathbf{AB})^T = \mathbf{B}^T\mathbf{A}^T$.

The determinant of a matrix is the determinant of its elements,

$$\det \mathbf{A} = \begin{vmatrix} a_{11} & a_{12} & \cdots \\ a_{21} & a_{22} & \cdots \\ \cdot & \cdot & \cdots \end{vmatrix}$$

The determinant of the product of two matrices is equal to the product of the respective determinants,

$$\det \mathbf{AB} = \det \mathbf{A} \det \mathbf{B}$$

It can be shown that the inverse of a matrix $\mathbf{A}$ is given by the formula

$$\mathbf{A}^{-1} = \begin{bmatrix} \dfrac{\det \mathbf{A}_{11}}{\det \mathbf{A}} & \dfrac{\det \mathbf{A}_{21}}{\det \mathbf{A}} & \cdots \\ \dfrac{\det \mathbf{A}_{12}}{\det \mathbf{A}} & \dfrac{\det \mathbf{A}_{22}}{\det \mathbf{A}} & \cdots \\ \cdots & \cdots & \cdots \end{bmatrix}$$

where the matrix $\mathbf{A}_{ij}$ is the matrix left after the ith row and jth column have been removed from the matrix $\mathbf{A}$.

Matrix Representation of Vectors

A matrix with one row, or one column, defines a *row vector,* or *column vector,* respectively. If $\mathbf{a}$ is a column vector, then $\mathbf{a}^T$ is the corresponding row vector,

$$\mathbf{a} = \begin{bmatrix} a_1 \\ a_2 \\ \vdots \\ a_n \end{bmatrix} \qquad \mathbf{a}^T = [a_1, a_2, \ldots, a_n]$$

For two column vectors $\mathbf{a}$ and $\mathbf{b}$ with the same number of elements, the product $\mathbf{a}^T\mathbf{b}$ is a scalar, analogous to the dot product,

$$\mathbf{a}^T\mathbf{b} = [a_1, a_2, \ldots] \begin{bmatrix} b_1 \\ b_2 \\ \vdots \end{bmatrix} = a_1 b_1 + a_2 b_2 + \cdots$$

Two vectors $\mathbf{a}$ and $\mathbf{b}$ are *orthogonal* if $\mathbf{a}^T\mathbf{b} = 0$.

Matrix Transformations

A matrix $\mathbf{Q}$ is said to *transform* a vector $\mathbf{a}$ into another vector $\mathbf{a}'$ according to the rule

$$\mathbf{a}' = \mathbf{Q}\mathbf{a} = \begin{bmatrix} q_{11} & q_{12} & \cdots \\ q_{21} & q_{22} & \cdots \\ \cdot & \cdot & \cdots \\ \cdot & & \cdots \\ \cdot & \cdot & \end{bmatrix} \begin{bmatrix} a_1 \\ a_2 \\ \vdots \end{bmatrix} = \begin{bmatrix} q_{11}a_1 + q_{12}a_2 + \cdots \\ q_{21}a_1 + q_{22}a_2 + \cdots \\ \cdot & \cdot & \cdots \\ \cdot & \cdot & \cdots \\ \cdot & \cdot & \end{bmatrix}$$

The transpose of $\mathbf{a}'$ is then

$$\mathbf{a}'^T = \mathbf{a}^T\mathbf{Q}^T = [a_1, a_2, \ldots] \begin{bmatrix} q_{11} & q_{12} & \cdots \\ q_{21} & q_{22} & \cdots \\ \cdot & \cdot & \cdots \end{bmatrix}$$

$$= [q_{11}a_1 + q_{12}a_2 + \ldots, \, q_{21}a_1 + q_{22}a_2 + \ldots, \ldots]$$

A matrix $\mathbf{Q}$ is said to be *orthogonal* if $\mathbf{Q}^T = \mathbf{Q}^{-1}$. It defines an *orthogonal transformation*. It leaves $\mathbf{a}^T\mathbf{b}$ unchanged, since $\mathbf{a}'^T\mathbf{b}' = \mathbf{a}^T\mathbf{Q}^T\mathbf{Q}\mathbf{b} = \mathbf{a}^T\mathbf{Q}^{-1}\mathbf{Q}\mathbf{b} = \mathbf{a}^T\mathbf{b}.$

The transformation defined by the matrix product $\mathbf{Q}^{-1}\mathbf{A}\mathbf{Q}$ is called a *similarity transformation*. The transformation defined by the product $\mathbf{Q}^T\mathbf{A}\mathbf{Q}$ is called a *congruent transformation*.

If the elements of $\mathbf{Q}$ are complex, then $\mathbf{Q}$ is called *Hermitian* if $q_{ij}* = q_{ji}$, that is, $\mathbf{Q}^{T*} = \mathbf{Q}$. If $\mathbf{Q}^{T*} = \mathbf{Q}^{-1}$, then $\mathbf{Q}$ is called a *unitary* matrix, and the transformation $\mathbf{Q}^{-1}\mathbf{A}\mathbf{Q}$ is called a *unitary transformation*.

Eigenvectors of a Matrix

An *eigenvector* $\mathbf{a}$ of a matrix $\mathbf{Q}$ is a vector such that

$$\mathbf{Q}\mathbf{a} = \lambda\mathbf{a}$$

or

$$(\mathbf{Q} - \mathbf{1}\lambda)\mathbf{a} = 0$$

where λ is a scalar, called the *eigenvalue*. The eigenvalues are found by solving the *secular equation*

$$\det(\mathbf{Q} - \mathbf{1}\lambda) = \begin{vmatrix} q_{11} - \lambda & q_{12} & \cdots \\ q_{21} & q_{22} - \lambda & \cdots \\ \cdot & \cdot & \cdots \end{vmatrix} = 0$$

which is an algebraic equation of degree n (the number of rows or columns or order of the matrix).

If the matrix $\mathbf{Q}$ is diagonal, then the eigenvalues are its elements.

Consider two different eigenvectors $\mathbf{a}_\alpha$ and $\mathbf{a}_\beta$ of a symmetric matrix $\mathbf{Q}$. Then

$$\mathbf{Q}\mathbf{a}_\alpha = \lambda_\alpha\mathbf{a}_\alpha$$

$$\mathbf{Q}\mathbf{a}_\beta = \lambda_\beta\mathbf{a}_\beta$$

where λ_α and λ_β are the eigenvalues. Multiply the first by $\mathbf{a}_\beta^T$ and the second, transposed, by $\mathbf{a}_\alpha$. Then

$$\mathbf{a}_\beta^T \mathbf{Q} \mathbf{a}_\alpha = \lambda_\alpha \mathbf{a}_\beta^T \mathbf{a}_\alpha$$

$$\mathbf{a}_\beta^T \mathbf{Q}^T \mathbf{a}_\alpha = \lambda_\beta \mathbf{a}_\beta^T \mathbf{a}_\alpha$$

But if $\mathbf{Q}$ is symmetric, then $\mathbf{Q}^T = \mathbf{Q}$, so the two expressions on the left are equal. Hence

$$(\lambda_\beta - \lambda_\alpha)\mathbf{a}_\beta^T \mathbf{a}_\alpha = 0$$

If the eigenvalues are different, then the two eigenvectors must be orthogonal.

Reduction to Diagonal Form
Given a matrix $\mathbf{Q}$, we seek a matrix $\mathbf{A}$ such that

$$\mathbf{A}^{-1}\mathbf{Q}\mathbf{A} = \mathbf{D}$$

where $\mathbf{D}$ is diagonal. Now

$$\mathbf{D} - \lambda\mathbf{1} = \mathbf{A}^{-1}\mathbf{Q}\mathbf{A} - \lambda\mathbf{1} = \mathbf{A}^{-1}(\mathbf{Q} - \lambda\mathbf{1})\mathbf{A}$$

Hence the eigenvalues of $\mathbf{Q}$ are the same as those of $\mathbf{D}$, namely, the elements of $\mathbf{D}$. Let λ_k be a particular eigenvalue, found by solving the secular equation $\det (\mathbf{Q} - \lambda\mathbf{1}) = 0$. Then the corresponding eigenvector $\mathbf{a}_k$ satisfies the equation

$$\mathbf{Q}\mathbf{a}_k = \lambda_k \mathbf{a}_k$$

which is equivalent to n linear homogeneous algebraic equations

$$\sum_j q_{ij}a_{jk} = \lambda_k a_{ik} \qquad (i = 1, 2, \ldots, n)$$

These may be solved for the ratios of the a's to yield the components of the eigenvector $\mathbf{a}_k$. The same procedure is repeated for each eigenvalue in turn. We then form the matrix $\mathbf{A}$ whose columns are the eigenvectors $\mathbf{a}_k$, that is, $[\mathbf{A}]_{ik} = a_{ik}$. Thus, the matrix $\mathbf{A}$ must satisfy

$$\mathbf{Q}\mathbf{A} = \mathbf{A}\begin{bmatrix} \lambda_1 & 0 & \cdots \\ 0 & \lambda_2 & \cdots \\ \cdot & \cdot & \cdots \\ 0 & 0 & \cdots \lambda_n \end{bmatrix} = \mathbf{A}\mathbf{D}$$

so that $\mathbf{A}^{-1}\mathbf{Q}\mathbf{A} = \mathbf{D}$ as required. The above method can always be done if $\mathbf{Q}$ is symmetric and the eigenvalues are all different.

Application to Oscillating Systems

For a system with n degrees of freedom, the generalized displacement vector is

$$\mathbf{q} = \begin{bmatrix} q_1 \\ q_2 \\ \vdots \\ q_n \end{bmatrix}$$

In matrix notation the kinetic and potential energies (defined in Section 11.4 of the text) take the compact forms

$$T = \frac{1}{2}\mathbf{q}^T\mathbf{M}\mathbf{q} \qquad V = \frac{1}{2}\mathbf{q}^T\mathbf{K}\mathbf{q}$$

in which

$$\mathbf{M} = \begin{bmatrix} M_{11} & M_{12} & \cdots \\ M_{21} & M_{22} & \cdots \\ \cdot & \cdot & \cdots \end{bmatrix}$$

$$\mathbf{K} = \begin{bmatrix} \kappa_{11} & \kappa_{12} & \cdots \\ \kappa_{21} & \kappa_{22} & \cdots \\ \cdot & \cdot & \cdots \end{bmatrix}$$

We note that both $\mathbf{M}$ and $\mathbf{K}$ are symmetric matrices. The differential equations of motion of the system given by Equation 11.29 can then be written

$$\mathbf{M}\ddot{\mathbf{q}} + \mathbf{K}\mathbf{q} = 0$$

If a harmonic solution of the form

$$q_k = A_k \cos \omega t \qquad (k = 1, 2, \ldots, n)$$

exists, then $\ddot{q}_k = -\omega^2 q_k$ that is

$$\ddot{\mathbf{q}} = -\omega^2\mathbf{q}$$

Consequently

$$(-\mathbf{M}\omega^2 + \mathbf{K})\mathbf{q} = 0$$

A nontrivial solution requires the secular determinant to vanish

$$\det(-\mathbf{M}\omega^2 + \mathbf{K}) = 0$$

or

$$\left| -M_{ij}\omega^2 + \kappa_{ij} \right| = 0$$

The roots give the normal frequencies and the associated eigenvectors define the normal modes. For further reading, see any of the first seven titles under *Advanced Mechanics* in Selected References.

APPENDIX I

NUMERICAL SOLUTION OF COUPLED DIFFERENTIAL EQUATIONS

General Problem

Suppose a particle is subject to some complicated acceleration, and we want to figure out its trajectory. In general, the acceleration will be a known function of kinematic variables, say, $a = a(x, \dot{x}, t)$. The trick is to examine the behavior of the particle over a small enough time interval Δt that its acceleration doesn't change very much. We therefore divide the total elapsed time of the motion into N small intervals, each of duration Δt, and we take the acceleration of the particle during the ith time interval to be constant and equal to a_{i-1}, the acceleration it has upon "entry" to that time interval. We then use the equations of motion of a particle undergoing constant acceleration to generate the new values of velocity and position at the end of the ith time interval. These values can then be used to recompute the value of the acceleration, a_i, at the end of this time interval. That will be the acceleration of the particle for the next time interval Δt. Continuing to iterate this way, we can watch the motion of the particle unfold up to any arbitrary stopping point.

For example, if we are dealing with motion along the x-axis, the numerical equations of motion are

$$\dot{x}_{i+1} = \dot{x}_i + \ddot{x}_i \Delta t$$

$$x_{i+1} = x_i + \dot{x}_i \Delta t + \frac{1}{2} \ddot{x}_i \Delta t^2$$

$$= x_i + \frac{1}{2}(\dot{x}_i + \dot{x}_{i+1}) \Delta t \tag{I-1}$$

$$t_{i+1} = t_i + \Delta t$$

$$\ddot{x}_{i+1} = a(x_{i+1}, \dot{x}_{i+1}, t_{i+1})$$

It is a very straightforward problem to program this algorithm to numerically solve quite complicated equations of motion. Most of the numerical problems presented in this book are variations of this fundamental theme. In effect, you are being asked to numerically solve a differential equation. The method of solution indicated above is a slightly modified version of *Euler's method*. It is an easy one to remember since it is based on the simple model of constant acceleration for short time intervals. It is probably the simplest one to implement, but it is also the least accurate in terms of a given computation time. Other methods of solving differential equations numerically, such as the *Runge-Kutta method,*[1] are more accurate but involve a little more programming complexity. We will not discuss that technique here.

There are many ways to carry out numerical calculations of the type presented above. One could write a program in Basic, Fortran, C, or some other appropriate computer language. One could also use a reasonably powerful, programmable calculator, such as the HP48SX. Finally, one could use one of the many software packages available for Macintosh or IBM personal computers. Such packages include spreadsheet programs like Quattro Pro, Lotus 1-2-3, Excel, or many others too numerous to mention. One could also use one of the several available scientific software calculating tools, such as Mathcad, Maple, or Mathematica. We have used many of these tools in solving the Computer/Calculator problems given at the end of each chapter. We indicate below how we used some of them to solve the problem given as example 5.11 in the text. We urge the student to choose one or another of them and try to solve at least some of the Computer/Calculator problems.

Solving with Mathcad[2]

Mathcad is a program that can be run on an IBM PC. It is a scientific software calculating tool. It permits the user to easily evaluate formulae and generate plots. Your work can also be easily annotated with text. We will not attempt to describe everything that can be done with Mathcad. We will show you how it can be used to easily solve coupled differential equations of the type alluded to in Example 5.11. There, we generated the following differential equations of projectile motion as seen by a non-inertial observer inside a giant, rotating cylinder (See the Equations 5.26b of the Numerical Example of Section 5.4)

$$\ddot{r} = 2\omega r\dot{\phi} + (\omega^2 + \dot{\phi}^2)r$$

$$\ddot{\phi} = -\frac{2\dot{r}}{r}(\omega + \dot{\phi})$$

These equations are solved in Mathcad using the technique of seeded iteration on several variables described in their User's Guide. Both variables, r and ϕ, along with their first and second derivatives, are treated as components of a six-dimensional vector. Each step in the iteration generates new vector components from the old ones in the following way

---

[1] Numerically solving differential equations by Euler's method or the Runge-Kutta method are described in a number of texts. See, for example: P. Smith and R. C. Smith, *Mechanics,* Wiley, New York, 1990. G. L. Baker and J. P. Gollub, *Chaotic Dynamics,* Cambridge University Press, New York, 1992. S. I. Grossman and W. R. Derrick, *Advanced Engineering Mathematics,* Harper Collins, New York, 1988.

[2] Mathcad can be purchased from Mathsoft, Inc., One Kendall Square, Cambridge, MA 02139, or almost any computer software shop.

$$
\begin{bmatrix}
\ddot{r} \\
\dot{r} \\
r \\
\ddot{\phi} \\
\dot{\phi} \\
\phi
\end{bmatrix}_{i+1}
=
\begin{bmatrix}
2\omega r\dot{\phi} + (\omega^2 + \dot{\phi}^2)r \\
\dot{r} + \ddot{r}\Delta t \\
r + \dot{r}\Delta t + 0.5\ddot{r}\Delta t^2 \\
-2\dot{r}(\omega + \dot{\phi})/r \\
\dot{\phi} + \ddot{\phi}\Delta t \\
\phi + \dot{\phi}\Delta t + 0.5\ddot{\phi}\Delta t^2
\end{bmatrix}_{i}
$$

where the subscript i refers to the values of the vector components after the ith iteration respectively. Note that the accelerations are computed according to the equations of motion given above, while the velocities and positions are evolved using the modified Euler algorithm. The iteration is started off by assigning initial values to the vector components. For the problem outlined in Section 5.4, they are

$$
\begin{bmatrix}
\ddot{r} \\
\dot{r} \\
r \\
\ddot{\phi} \\
\dot{\phi} \\
\phi
\end{bmatrix}_{0}
=
\begin{bmatrix}
\omega^2 R \\
-\omega R \\
R \\
0 \\
0 \\
0
\end{bmatrix}
$$

Normally, one would terminate the iteration process after some test criterion had been met. In this case, that might be when the projectile "returns to the surface," in other words, when r_{i+1} exceeds R, the radius of the cylinder. Rather than set up such a test, in Mathcad, it is often simpler to just fix the number of iterations to some value N large enough to ensure the generation of a complete trajectory and let the calculation proceed. Initially, let N be small and Δt be large. One then obtains an estimate for how much total time, $T = N\Delta t$, is required to complete the trajectory. Then increase N and shrink Δt, preserving T, to increase the accuracy of your calculation.

Solving with Quattro Pro[3]

Most spreadsheets have the capability of carrying out simple mathematical operations upon arrays of data. One can think of the trajectory of the projectile illustrated in Example 5.11 as an $N \times 2$ subset (r, ϕ) of a large, $N \times 6$ array of the kinematical variables, $(\ddot{r}, \dot{r}, r, \ddot{\phi}, \dot{\phi}, \phi)$, where N is the number of iterations in time required to generate the trajectory. In the case of a spreadsheet, N is the number of cell rows and 6 is the number of columns, each column is devoted to one of the kinematic variables of the motion. Each row in the spreadsheet therefore represents the value of the entire set of kinematic variables at each new time interval. The values in each row are calculated from the values in the previous row according to the modified Euler method presented above.

Values in a given cell are referenced in a spreadsheet calculation by addresses, such

[3]Quattro Pro can be purchased from Borland International Inc., 1800 Green Hills Rd., P.O. Box 660001, Scotts Valley, CA 95066-0001 or from almost any computer software shop.

as G22. G refers to the seventh column and 22 refers to the twenty-second row. This is the location, or address, of a numerical value (or a formula that generates a numerical value) in that particular cell. Suppose, for example, that the value of the time increment, Δt, is 1 sec and that value has been written into cell A2. It is considered to be data that will not change during the calculation. Data, such as this, that remains invariant during a spreadsheet calculation, is usually loaded into upper row number areas. Now suppose, that column A, starting at row 10 and working downward, contains values for the radial acceleration, $\ddot{r}$, column B contains values for the radial velocity, $\dot{r}$, and so on, then the value of the radial velocity of the projectile in cell B21 is given by the formula +B20 +A20*A2. This means, take the value in cell B20 (the value of $\dot{r}$ at the end of the previous time interval) and add it to the estimated change in velocity (the value of the acceleration $\ddot{r}$ times Δt) to obtain the value of the velocity in the current cell. The cell in the next row, B22, contains the same formula but with cell addresses incremented by one row number. Therefore, when the spreadsheet finishes calculating, the complete radial velocity profile is contained in all the cells under column B from row 10 down to the point where the calculation is terminated. The same is true for all the other kinematical variables located under a given column.

A table of the first few values for a Quattro Pro spreadsheet calculation of the missile trajectory of Example 5.11 is given below.

The table continues on to row 650. At that point the calculation is terminated where the value of r returns to 1000 km (the missile strikes the "ground"). The total elapsed time of flight is about 320 s.

The key feature that makes such a calculation so simple to do in a spreadsheet is its "copy down" feature. One simply types in the formula that one wishes in an upper cell and "copies it down" to all the cells below, terminating on a row number deemed to be sufficient to see the desired endpoint of the calculation. In the spreadsheet above, all columns in row 10 contain the initial numerical values representing the starting point of

TABLE I-1 Quattro Pro Spreadsheet Kinematical Variables of Example 5.4: Missile Trajectory

2	dt(sec)		°/radian			
3	0.5		57.296			
4						
5	R(km)		ω(rad/s)		g(km/s^2)	
6	1000		0.00313		0.0098	
7						
8	A_r(km/s^2)	V_r(km/s)	r(km)	A_ϕ(rad/s^2)	V_ϕ(rad/s)	ϕ(deg)
9	—	—	—	—	—	—
10	.0098	−3.13	1000	0	0	0
11	.009782	−3.1251	998.435	$1.96e-05$	0	0
12	.009827	−3.1202	996.8725	$1.97e-05$	$9.8e-06$	0
13	.009874	−3.1153	995.3123	$1.97e-05$	$1.96e-05$	.000281
14	.009920	−3.1104	993.7547	$1.98e-05$	$2.95e-05$	.000843
15	.009967	−3.1054	992.1995	$1.98e-05$	$3.94e-05$	.001687

the trajectory. But all columns in row 11 actually contain the relevent modified Euler's method formulae (Equations I-1) that generate the values listed in row 11. These formulae are

A11: +2*C5*C11*E11 + (C5*C5 + E11*E11)*C11

B11: +B10 + A10*A2

C11: +C10 + B10*A2

D11: −2*B11*(C5 + E11)/C11

E11: +E10 + D10*A2

F11: +F10 + E10*A2*C2

The data are usually taken from rows immediately above along with the relevent permanent or absolute address data in the cells above row 8 (such as the time interval increment $dt = 0.5$s in the cell referenced by the absolute address A2). Each of the formulae were written into a cell in row 11 and then copied down to all following row numbers. The cell in each subsequent row under a given column contains the same formulae as in the cell directly above, but with all relative addresses updated by one. The spreadsheet automatically does this during the copy down operation. (The absolute addresses referenced by the $ sign are not changed.) Hence, the value calculated for each subsequent cell always uses data taken from cells located in the same relative way as for the cell from which the formula was originally copied.

Recalculation

Spreadsheets have one more nice feature that allows for a more accurate calculation to be performed with only a minor modification made to the above set of Equations I-1. However, the modifications necessitate the use of a technique called "recalculation." For example, instead of using the initial acceleration a_i of a particle upon "entry" to a given time interval Δt_i as indicated in Equations I-1, we use an estimated average value instead. Thus, Equations I-1 become

$$\dot{x}_{i+1} = \dot{x}_i + \frac{1}{2}(\ddot{x}_i + \ddot{x}_{i+1})\Delta t$$

$$x_{i+1} = x_i + \frac{1}{2}(\dot{x}_i + \dot{x}_{i+1})\Delta t$$

(I-2)

$$t_{i+1} = t_i + \Delta t$$

$$\ddot{x}_{i+1} = a(x_{i+1}, \dot{x}_{i+1}, t_{i+1})$$

These equations represent an *improved Euler method* and are equivalent to Equations I-1 in the case of constant acceleration. However, if the acceleration is not constant, they yield slightly more accurate results for a given number of iterations. In essence, these equations approximate the area under a curve according to the *trapezoidal* rule.

Note, though, that we are now faced with a circular calculation. For example, the position and velocity at the end of a given time interval depend on the acceleration at the end of that same time interval. And the acceleration depends on the velocity and position. The spreadsheet can handle this type of circularity if it is carried out in the "manual" as opposed to "automatic" mode of calculation. This feature can be invoked as one of the menu options in the spreadsheet. Set the calculation mode to manual before setting up the algorithm. The spreadsheet will not calculate anything from this point on, until you tell it to do so, by typing the key F9. When you tell it to "go" in manual mode, by typing F9, the spreadsheet will calculate the value in every cell, carrying out a fixed number of iterations or passes through the entire spreadsheet. Usually, one sets the number of iterations to some small value, say 10, scrolls to the end of the spreadsheet so that the values in the bottom row can be examined when the calculating grinds to a halt. Repeat the calculation process, until the values in those cells no longer change. Convergence has been achieved and you can now graph the results using Quattro Pro's graph menu.

Solving with a Programmable Calculator—HP48SX

One can solve problems such as this one with a programmable calculator, such as the HP48SX, but calculators are notoriously slow. The use of a different algorithm, requiring a smaller number of iterations to achieve a given accuracy might be required, unless the problem is not too involved. The fourth-order Runge-Kutta scheme usually represents the next method to try. The interested reader who wishes to pursue such an analysis can examine the texts recommended in footnote 1. Here, we will indicate how the benefit of the "recalculation" feature described above for spreadsheets can be closely approximated by a simple modification of Equations I-1.

Basically, we modify those equations by expanding the velocity around the middle of the time interval Δt, instead of the beginning. Thus, we write the equations as

$$\dot{x}\left(t_i + \frac{\Delta t}{2}\right) = \dot{x}\left(t_i - \frac{\Delta t}{2}\right) + \ddot{x}_i \Delta t$$

$$x_{i+1} = x_i + \dot{x}\left(t_i + \frac{\Delta t}{2}\right)\Delta t$$

$$\dot{x}_{i+1} = \dot{x}_i + \ddot{x}_i \Delta t \qquad\qquad (\text{I-3})$$

$$\ddot{x}_{i+1} = a(x_{i+1}, \dot{x}_{i+1}, t_{i+1})$$

Note, that the acceleration, $\ddot{x}_i$, is used above to calculate the change in velocity between midpoints of the time intervals. Thus, it is an estimate of the average and is analogous to the average acceleration used in Equations I-2 above. Moreover, the midpoint velocities that are generated by this acceleration are used to generate changes in position between the endpoints of the time intervals $t_{i+1} - t_i$. These velocities are therefore used in a way analogous to the average velocities in Equations I-2. The velocity at the end of the time interval, $\dot{x}_{i+1}$, is used to calculate the acceleration at the end of that interval and this is the weak point of the algorithm. It generates an acceleration which is not quite as good an estimate of the average acceleration generated by the process of recalculation above. However, these equations are essentially equal to those of I-2 and the algorithm is almost equivalent to the one implemented by the recalculation feature. In the case here, though, all newly calculated values depend entirely upon previously calculated ones. Hence, there is no circularity in the calculation. It proceeds in a completely linear fashion.

There are two minor problems with this algorithm that we need to address: First, how do we start things off? In other words, what is the velocity $\dot{x}(\Delta t/2)$? We are given $\dot{x}(0)$, not $\dot{x}(\Delta t/2)$. At the start, we estimate its value according to $\dot{x}(\Delta t/2) = \dot{x}(0) + \ddot{x}(0)\Delta t/2$. From that point on, the midpoint velocities are evolved using Equations I-3. Second, we now have to keep track of velocities calculated at the midpoint of each time interval, while keeping track of accelerations and positions at the endpoints. That's the cost of the increased accuracy of the algorithm represented by Equations I-3 as opposed to the one represented by Equations I-1—namely, increased complexity. Still, the method is a simple one and easy to implement.

Even a problem as complex as that of Example 5.11 can be easily programmed into a calculator such as the HP48SX. Its execution requires no more than several minutes for obtaining trajectory parameters (say, azimuthal drift angle and maximum altitude reached) accurate to 0.1%. Here we present our solution using this calculator.

Programming

When writing a program in any language, it is good practice to express the algorithm to be implemented in terms of a flowchart. A flowchart outlining the algorithm represented by Equations I-3 is shown in Figure I-1. Each box represents an action (or group of similar actions) to be taken. Each diamond represents a decision to be made. Each circle represents a control step, in this case, a program halt or stop. Basically, the algorithm is a loop which is repeated until some condition is met. In this case, the procedure is halted when the radial distance of the missile exceeds the radius of the rotating cylinder, in other words, when the missile impacts the "ground." Note, that the fourth execution block states "store minimum R." During each iteration of the loop, the value of R will change. The minimum value corresponds to the maximum altitude $h_{max} = R_0 - R_{min}$ obtained. This value, along with the final value of the azimuthal angle of drift, ϕ_N, and the total elapsed time of travel, T, will be displayed at the end of program execution.

The chart below is a list of the program for the HP48SX. We recommend that the interested reader examine the HP48 Owner's Manual[4] in order to understand the details of the HP48SX programming language.

Before executing the above program, four global variables (constants) were defined and stored in the calculator. They were: DT = 10 s, the time increment, G = .0098 km/s^2, the value of the initial radial acceleration of the rotating cylinder, RO = 1000 km, the radius of the cylinder, and $O = \sqrt{G/RO} = 3.1304951685 \cdot 10^{-3}$, the rotational angular velocity of the cylinder.

[4]HP 48 Owner's Manual, Hewlett Packard Co., Documentation Dept., Corvallis Div., 100 NE Circle Blvd., Corvallis, OR 97330-9973 (1991).

The chart below lists the results of program execution for different values of the time increment DT. Note that it takes about 16 minutes for the calculator to solve this problem to an accuracy of the order of about 0.1%. You could improve the accuracy by interpolation. Sometimes it is simpler to just use small increments so that interpolation is not necessary. Also, you could plot the trajectory on the HP48SX display by simply storing the $(r, \phi)_i$ values on the stack as a $2 \times N$ array and then using the HP48SX plot feature to generate a scatter plot of those values.

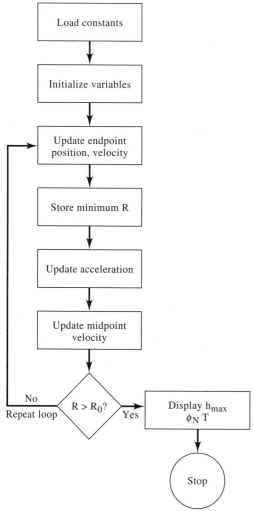

Figure I–1   Flowchart to carry out iterative calculation of Equation I–3.

TABLE I-2   HP48SX Program to calculate missile trajectory of Example 5.4.

Program Code	Comment
< < 0 G O RO 'RMIN' STO RO * NEG	Initialize T, $\ddot{r}$, $\dot{r}$; Store r_{min}
DUP G DT 2 / * + RO 0 0 0 0	Initialize $\bar{v}_r$, r, $\ddot{\phi}$, $\dot{\phi}$, $\bar{v}_f$, ϕ
→ T AR RDOT VR R AF FDOT VF F	Define local variables
< < DO	Start do loop
T DT + 'T' STO	Increment time, Store result
DT VR * R + 'R' STO	Update endpoint radial position
R RMIN MIN 'RMIN' STO	Calculate and store r_{min}
DT AR * RDOT + 'RDOT' STO	Update endpoint radial velocity
DT VF * F + 'F' STO	Update " azimuthal position
DT AF * FDOT + 'FDOT' STO	Update " azimuthal velocity
R FDOT RDDOT 'AR' STO	Update " radial acceleration
R RDOT FDOT FDDOT 'AF' STO	Update " azimuthal acceleration
DT AR * VR + 'VR' STO	Update midpoint radial velocity
DT AF * VF + 'VF' STO	Update " azimuthal velocity
UNTIL	Begin conditional test
R RO >	R > RO ? (No—back to DO)
END	End loop
T F 180 * π / RO RMIN -	Yes—display T, ϕ(deg), H_{max}
> >	End of local procedure
> >	End of program

TABLE 1-3   Results of Trajectory Calculation of Program in Table I-2.

Execution Time (s)	Time DT (s)	Max Altitude H_{max} (km)	Azimuthal Drift ϕ (deg)	Elapsed Time T (s)
13	10	315.35	34.76	330
26	5	304.03	33.70	325
61	2	297.32	33.10	322
122	1	295.10	32.90	321
241	0.5	294.00	32.80	320
480	0.25	293.45	32.75	319.75
960	0.125	293.17	32.73	319.625

SELECTED REFERENCES

Mechanics

Barger, V., and Olsson, M., *Classical Mechanics,* McGraw-Hill, New York, 1973.
Becker, R. A., *Introduction to Theoretical Mechanics,* McGraw-Hill, New York, 1954.
Lindsay, R. B., *Physical Mechanics,* Van Nostrand, Princeton, N.J., 1961.
Rossberg, K., *A First Course in Analytical Mechanics,* Wiley, New York, 1983.
Rutherford, D. E., *Classical Mechanics,* Interscience, New York, 1951.
Slater, J. C., and Frank, N. H., *Mechanics,* McGraw-Hill, New York, 1947.
Symon, K., *Mechanics,* 3rd ed., Addison-Wesley, Reading, Mass., 1971.
Synge, J. L., and Griffith, B. A., *Principles of Mechanics,* McGraw-Hill, New York, 1959.

Advanced Mechanics

Corbin, H. C., and Stehle, P., *Classical Mechanics,* Wiley, New York, 1950.
Desloge, E., *Classical Mechanics* (two volumes), Wiley-Interscience, New York, 1982.
Goldstein, H., *Classical Mechanics,* 2nd ed., Addison-Wesley, Reading, Mass., 1980.
Hauser, W., *Introduction to the Principles of Mechanics,* Addison-Wesley, Reading, Mass., 1965.
Landau, L. D., and Lifshitz, E. M., *Mechanics,* Addison-Wesley, Reading, Mass., 1960.
Marion, J. B., *Classical Dynamics,* Academic Press, New York, 1965.
Moore, E. N., *Theoretical Mechanics,* Wiley, New York, 1983.
Wells, D. A., *Lagrangian Dynamics,* Shaum, New York, 1967.
Whittaker, E. T., *Advanced Dynamics,* Cambridge University Press, London and New York, 1937.

Mathematical Methods

Churchill, R. V., *Fourier Series and Boundary Value Problems,* McGraw-Hill, New York, 1963.
Jeffreys, H., and Jeffreys, B. S., *Methods of Mathematical Physics,* Cambridge University Press, London and New York, 1946.
Kaplan, W., *Advanced Calculus,* Addison-Wesley, Reading, Mass., 1952.
Mathews, J., and Walker, R. L., *Methods of Mathematical Physics,* W. A. Benjamin, New York, 1964.
Margenau, J., and Murphy, G. M., *The Mathematics of Physics and Chemistry,* 2nd ed., Van Nostrand, New York, 1956.
Wylie, C. R., Jr., *Advanced Engineering Mathematics,* McGraw-Hill, New York, 1951.

Tables

Dwight, H. B., *Mathematical Tables,* Dover, New York, 1958.

Pierce, B. O., *A Short Table of Integrals,* Ginn, Boston, 1929.

Handbook of Chemistry and Physics, Mathematical Tables, Chemical Rubber Co., Cleveland, Ohio, 1962 or after.

ANSWERS TO SELECTED ODD-NUMBERED PROBLEMS

CHAPTER 1

1.1 (a) $\sqrt{6}$, (b) $3\mathbf{i} + \mathbf{j} - 2\mathbf{k}$, (c) 1, (d) $\mathbf{i} - \mathbf{j} + \mathbf{k}$

1.3 $\cos^{-1}\sqrt{5/14} = 53.3°$

1.5 $q = 1$ or 2

1.11 $3.232\mathbf{i}' + 1.598\mathbf{j}' - \mathbf{k}'$

1.13 $b\omega(\sin^2 \omega t + 4\cos^2 \omega t)^{1/2}, 2b\omega, b\omega$

1.17 $b\omega\left[\cos^2\left(\dfrac{\pi}{8}\cos 4\omega t\right) + \dfrac{\pi^2}{4}\sin^2(4\omega t)\right]^{1/2}$

1.21 For Problem 1.15 $a_n = bc(k^2 + c^2)^{1/2}\, e^{kt}, a_\tau = bk(k^2 + c^2)^{1/2}\, e^{kt}$

CHAPTER 2

2.1 (a) $\dot{x} = (F_0/m)t + (c/2m)t^2$, $x = (F_0/2m)t^2 + (c/6m)t^3$
   (b) $\dot{x} = (F_0/cm)(1 - \cos ct)$, $x = (F_0/c^2m)(ct - \sin ct)$
   (c) $\dot{x} = -(F_0/cm)(1 - e^{ct})$, $x = -(F_0/c^2m)(ct - e^{ct} + 1)$

2.3 (a) $V = -F_0 x - (c/2)x^2 + C$, (b) $V = (F_0/c)e^{-cx} + C$
   (c) $V = -(F_0/c)\sin cx + C$

2.5 (a) 541, (b) 87, (c) 454

CHAPTER 3

3.1 6.43 m/s, 2.07×10^4 m/s^2

3.3 $x(t) = 0.25\cos(20\pi t) + 0.00159\sin(20\pi t)$ in meters

3.5 $[(\dot{x}_2^2 - \dot{x}_1^2)/(x_1^2 - x_2^2)]^{1/2}, [(x_1^2\dot{x}_2^2 - x_2^2\dot{x}_1^2)/(\dot{x}_2^2 - \dot{x}_1^2)]^{1/2}$

3.17 (a) $T = 2\pi(l/g)^{1/2} \times 1.041$, (b) g will come out to be about 8 percent low,
   (c) $B/A = 0.0032$

CHAPTER 4

4.1 (a) $\mathbf{F} = -c(yz\mathbf{i} + xz\mathbf{j} + xy\mathbf{k})$, (b) $\mathbf{F} = -2(\alpha x\mathbf{i} + \beta y\mathbf{j} + \gamma z\mathbf{k})$
   (c) $\mathbf{F} = ce^{-(\alpha x + \beta y + \gamma z)}(\alpha\mathbf{i} + \beta\mathbf{j} + \gamma\mathbf{k})$, (d) $\mathbf{F} = -cnr^{n-1}\mathbf{e}_r$

4.3 (a) $c = 1/2$, (b) $c = -1$

4.9 $m\ddot{x} = -c_2\dot{x}\dot{s}, m\ddot{y} = -c_2\dot{y}\dot{s}, m\ddot{z} = -mg - c_2\dot{z}\dot{s}$

4.11 Long axis: $\psi = 80.8°$, Short axis: $\psi = -9.2°$

4.17 $v = (2gb)^{1/2}, R = 3mg$

CHAPTER 5

5.1 Up: 150 lb, Down: 90 lb

5.3 1.005 mg, about 5.7°

5.5 (a) $g/6$ forward, (b) $g/3$ toward rear

5.7 $(V_0^2/\rho)\mathbf{i}' + [(V_0^2/b) + (V_0^2 b/\rho^2)]\mathbf{j}$

CHAPTER 6

6.1 About 2×10^{-9}

6.3 About 1.4 h

6.13 $(120/61)^{1/2} = 1.4026$. Orbit is hyperbolic: $v_0^2/v_c^2 > 2$.

6.19 $\psi = \pi(1 + c/1 + 4c)^{1/2}$ where $c = \rho 4\pi a^3/3M_{sun}$

6.21 $a > (\epsilon/k)^{1/2}$

6.25 $\psi = 180.7°$ for orbits near the earth.

6.27 $\theta = -30°$

CHAPTER 7

7.1 $\mathbf{r}_{cm} = (\mathbf{i} + 2\mathbf{j} + 2\mathbf{k})/3$, $\mathbf{v}_{cm} = (3\mathbf{i} + 2\mathbf{j} + \mathbf{k})/3$, $\mathbf{p} = 3\mathbf{i} + 2\mathbf{j} + \mathbf{k}$

7.5 Direction: downward at an angle of 26.6° with the horizontal, Speed: 1.118 v_0

7.7 Car: $v_0/2$, Truck: $v_0/8$. Both final velocities are in the direction of the initial velocity of the truck.

7.13 Proton: $v_x' = v_y' = 0.657\, v_0$, Helium: $v_x' = 0.086\, v_0$, $v_y' = -0.164\, v_0$

7.15 Approximately 55.2°

7.19 $\ddot{z} = g - 3\dot{z}^2/(z_1 + z)$ where z_1 is a constant proportional to the initial radius of the drop. ($z_1 = 0$ for this problem.)

CHAPTER 8

8.1 (a) $b/3$ from center section, (b) $x_{cm} = y_{cm} = 4b/3\pi$ where lamina is in xy plane, (c) $x_{cm} = 0$, $y_{cm} = 3b/5$, (d) $x_{cm} = y_{cm} = 0$, $z_{cm} = 2b/3$, (e) $b/4$ from base

8.3 $a/14$ from center of large sphere

8.5 $(31/70)ma^2$

8.9 $2\pi\,(2a/g)^{1/2}$, $2\pi(3a/2g)^{1/2}$

8.13 $g(m_1 - m_2)/(m_1 + m_2 + I/a^2)$

8.17 $v_0 t - \dfrac{1}{2}\, gt^2\,(\sin\theta + \mu\,\cos\theta)$

$(2v_0^2/g)(\sin\theta + 6\mu\,\cos\theta)/(2\,\sin\theta + 7\mu\,\cos\theta)^2$

8.21 $\left[2g(1 - \cos\theta_0)\left(\dfrac{m}{3}l^2 + m'l'^2\right)\left(\dfrac{m}{2}l + m'l'\right)\right]^{1/2}/m'l'$

CHAPTER 9

9.1 (a) $I_{xx} = \dfrac{m}{3}a^2$, $I_{yy} = \dfrac{4m}{3}a^2$, $I_{zz} = \dfrac{5m}{3}a^2$

$$I_{xz} = I_{yz} = 0, \ I_{xy} = \frac{m}{2}a^2$$

(b) $\dfrac{2}{15}ma^2$, (c) $\mathbf{L} = (ma^2\omega/6\sqrt{5})(\mathbf{i} + 2\mathbf{j})$, (d) $T = \dfrac{1}{15}ma^2\omega^2$

9.3 (a) Inclination of the l-axis is $\dfrac{1}{2}\tan^{-1} I = 22.5°$

(b) Principal axes in the xy plane are parallel to the edges of the lamina.

9.9 (a) 1.414 s, 0.632 s; (b) 1.603 s, 0.663 s

9.13 $\alpha - \tan^{-1}[(I/I_s)\tan\alpha] \approx \alpha(I_s - I)/I_s = 0.00065$ arc sec

9.17 $S > \left[\dfrac{128 \ ga}{b^4}\left(\dfrac{a^2}{3} + \dfrac{b^2}{16}\right)\right]^{1/2} \approx 2910$ rps

CHAPTER 10

10.1 Use $L = \dfrac{m}{2}(\dot{x}^2 + \dot{y}^2 + \dot{z}^2) - mgz$

10.3 (a) $g/2$, (b) $g(m + m'z/b)/(2m + m')$ where b is the length of the cord, and z is the length hanging over the table at any instant.

10.5 $-mg \sin\theta \cos\theta/[(7/5)(m + M) - m\cos^2\theta]$

10.9 $d^2r/dt^2 = r\dot{\theta}^2 + g\cos\theta - (k/m)(r - l_0)$

$d(r^2\dot{\theta})/dt = -gr\sin\theta$

10.15 $U(r) = \dfrac{mh^2 \sin^2\alpha}{2r^2} + mgr\cos\alpha$ where $h = r^2\dot{\phi} = constant$

10.19 (a) $\dot{\theta} = p_\theta/ml^2, \ \dot{p}_\theta = -mgl\sin\theta$

(b) $\dot{x} = p_x/(m_1 + m_2), \ \dot{p}_x = g(m_1 - m_2)$

(c) $\dot{x} = p_x/m, \ \dot{p}_x = mg\sin\theta$

CHAPTER 11

11.1 (a) $x = k^{1/3}$ stable

(b) $x = 1/b$ unstable

(c) $x = 0$ unstable, $x = \pm b/\sqrt{2}$ stable

(d) $(3k/m)^{1/2}$, 3.628 sec; $b(k/m)^{1/2}$, 6.283 sec for parts (a) and (c), respectively.

11.9 $2\pi a[5/3g(b - a)]^{1/2}$

11.11 2.363 sec

11.17 $\omega = (k/m)^{1/2} \dfrac{(5 \pm \sqrt{17})^{1/2}}{2}$

11.21 $v_{long} = (K/m)^{1/2}(l + \Delta l), \ v_{trans} = (K/m)^{1/2}[(l + \Delta l)\Delta l]^{1/2}$

INDEX